NATURE
AND
VALUE

NATURE
AND
VALUE

EDITED BY
AKEEL BILGRAMI

Columbia University Press
New York

Columbia University Press
Publishers Since 1893
New York   Chichester, West Sussex
cup.columbia.edu

Columbia University Press wishes to express its appreciation for assistance
given by the NOMIS Foundation in the publication of this book.

Library of Congress Cataloging-in-Publication Data
Names: Bilgrami, Akeel, 1950– editor.
Title: Nature and value / edited by Akeel Bilgrami.
Description: New York : Columbia University Press, 2019. |
    Includes bibliographical references and index.
Identifiers: LCCN 2019026939 (print) | LCCN 2019026940 (ebook) |
    ISBN 9780231194624 (cloth) | ISBN 9780231194631 (paperback) |
    ISBN 9780231550901 (ebook)
Subjects: LCSH: Philosophy of nature. | Human ecology.
Classification: LCC BD581 .N36985 2019 (print) | LCC BD581 (ebook) |
    DDC 179/.1—dc23
LC record available at https://lccn.loc.gov/2019026939
LC ebook record available at https://lccn.loc.gov/2019026940

Cover design: Julia Kushnirsky
Cover photograph: American Smelting and Refining, Garfield, Utah, 1942,
by Andreas Feininger

This volume of essays is dedicated to the memory of Jonathan Schell ❧

CONTENTS

Preface   *ix*
AKEEL BILGRAMI

Acknowledgments   *xvii*

1. Nature and Value   1
   JONATHAN SCHELL

2. The Human Shadow   13
   JONATHAN SCHELL

3. The Anthropocene and Global Warming: A Brief Update   25
   JAN ZALASIEWICZ

4. The Extraordinary Strata of the Anthropocene   29
   JAN ZALASIEWICZ

5. The Anthropocene Dating Problem: Disciplinary
   Misalignments, Paradigm Shifts, and the Possibility
   for New Foundations in Science   46
   KYLE NICHOLS AND BINA GOGINENI

6. Disciplinary Variations on the Anthropocene: Temporality
   and Epistemic Authority. Response to Kyle Nichols
   and Bina Gogineni   63
   NIKOLAS KOMPRIDIS

7. Value and Alienation: A Revisionist Essay on Our Political Ideals   68
    AKEEL BILGRAMI

8. Equality and Liberty: Beyond a Boundary. Response
    to Akeel Bilgrami   89
    SANJAY G. REDDY

9. Experimenting with Other People   97
    JOANNA PICCIOTTO

10. The Green Growth Path to Climate Stabilization   117
    ROBERT POLLIN

11. All Too Human: Orienting Environmental Law in
    a Remade World   127
    JEDEDIAH BRITTON-PURDY

12. Life Sustains Life 1: Value, Social and Ecological   163
    JAMES TULLY

13. Life Sustains Life 2: The Ways of Reengagement with
    the Living Earth   181
    JAMES TULLY

14. The Value of Sustainability and the Sustainability of Value   205
    ANTHONY SIMON LADEN

15. Varieties of Agency: Comment on Anthony Laden   224
    CAROL ROVANE

16. Nonhuman Agency and Human Normativity   240
    NIKOLAS KOMPRIDIS

17. Natural Piety and Human Responsibility   261
    DAVID BROMWICH

    List of Contributors   277
    Index   279

PREFACE

AKEEL BILGRAMI

Some years ago, I had given a lecture in Istanbul, asking the question, how and when did the concept of nature get transformed into the concept of natural resources? The lecture was partly genealogical (tracing the transformation to historical changes, in political economy, in theology, and in the rise of institutions such as the Royal Society in England) and partly analytical (expounding the evacuation of *value* properties from our understanding of nature as a result of an outlook that had emerged around the new science of the seventeenth century, which equated nature exhaustively with the properties that the natural sciences study).

After the lecture, Heini Thyssen, who was in the audience and whom I had not met before, asked if I thought this question could be explored in further detail in some sort of research project. He himself had very interesting ideas of his own on the subject, which he put to me in a long boat ride we took that afternoon, in particular whether this transformation I was exploring generated a kind of unpaid, even unacknowledged, *debt* that we owed to nature; and he was curious about whether the whole question of debt in political economy might have a broader canvas—to include a debt of this kind not just in our relations with one another but in our relations with nature as well. This immediately struck me as much more than a promising analogy since part of the transformation I was keen to stress was that though we have always taken from nature, in all social worlds prior to the modern period there were rituals to show attitudes of respect and restorative return to nature before cycles of planting and even hunting. It's only in the last three hundred years or so that we have come to think that we might take from nature *with impunity*.

Mr. Thyssen encouraged me to propose an ongoing research project with scholars and public intellectuals meeting once a year to discuss these issues and see through to their significance for ecology, for politics and political economy, and in general for how our sensibilities may be shaped to be *ethically* responsive to growing concerns about our relations to nature. So about five years ago, I sent the NOMIS Foundation a three-year project proposal under the overarching title "Nature and Value." I invited people from a wide range of disciplines and interests to join us and to bring their distinct and diverse angles to this theme: geologists, climate scientists, philosophers, literary critics, sociologists, political scientists, economists, an art historian, a couple of activists, and one of the great public intellectuals of our time, the late Jonathan Schell.

We met three times over roughly three years in London—for two packed days of conversation on each occasion. The format differed from year to year, sometimes taking the form of freestanding papers presented by the participants, sometimes taking the form of a commentator presenting each paper by some author, with the author responding. Intense discussions followed each presentation. These were altogether worthwhile occasions—serious, instructive, learned, and throughout raising issues of the most fundamental importance and concern, and often reaching to depths well below the conventional thoughts available on the surface. The depth and originality of these writings would not by itself have warranted the publication of these essays, if there wasn't also an integrity to them, by which I mean a close integration not just of themes but of points of view and conclusions. It is rare to find scholars coming together on a systematic study of a topic, each from their diverse disciplinary location, and converging on a broadly unified standpoint on it. If I were asked to say in terms of the highest generality—abstracting from the wealth of detail in these essays—what that topic and what that standpoint is, I could do no better than to say that it is the wider significance of the concept of nature such that a proper understanding of that significance reaches all the way to an ethical and political position. It is not this topic and standpoint that are novel. It has surfaced recurrently in intellectual history for two thousand years (Romanticism in Germany and England, for example, is only a late occurrence of it). What is novel is how such widely situated thinkers can come to a position of overlapping consensus on it, each presenting quite distinct, though fascinatingly related and entirely coherent, considerations for embracing the consensus.

The volume begins with two short pieces by the late Jonathan Schell, author of that rightly celebrated classic on the destructive power of nuclear weapons, *The Fate of the Earth* (Stanford University Press, 1982). It is tempting to think that a major difference between the nuclear danger and the threat of climate change is one of public knowledge. Hiroshima and Nagasaki spectacularly awakened the world's public to the former but there has been no counterpart to make vivid the latter. We may feel hotter over summers, we may witness increasingly destructive storms and the rise of tides, but these remain in many sections of the public mind mere spikes in the natural course of things; they do not yet match the decisive knowledge of a devastation of the sort that was conveyed by what the United States visited upon Japan with the atom bomb. Corporations can continue to undermine the climate scientists' alarm bells by expending minor fortunes to propagandize against climate change "alarmists," and right-wing populist leaders can still successfully serve these corporate efforts. But after Hiroshima and Nagasaki no manipulative efforts of this sort can hide what the atom bomb (and its successor) can wreak. Climate science is still relatively esoteric knowledge and the experts who possess it have simply not yet reached the public mind with the same directness and display. Schell's short essays here are a good corrective to the temptation of believing in such a differential. He is particularly well situated to do so since, over a remarkable life of study, he has written on both crises with enormous concern and care. What is more, it would seem, from what he says, that the nuclear potential might in fact be more within our powers to constrain than the inexorable path that the earth and its atmosphere are now set on. As he says, if we survive the nuclear threat as a species it is not at all clear that we will be able to shrink our own power down to the point where we can reverse the effects of climate change. The reason for this is that we have the scientific knowledge whose application has made us so systematically destructive of nature and once we are possessed of such knowledge we are, just by its possession, *predisposed* to go on a path from which there is no turning back. In other words, we have all sorts of strengths that come with the knowledges we possess but what we lack, paradoxically, is precisely, in this respect of knowledge, the one strength we need, "the strength to make ourselves weak."

Schell's presence in our discussions was inspiring. His two brief contributions are from notes for a lecture and from an introduction to a long work on which he was embarked when he died. As editor, I have not indulged in the impertinence of revising them to fit the canonical format for an academic paper suited for a volume of this kind. Though the first is, just in one or two places, impressionistic as notes to oneself can be, and the second, again in just one or two places, promissory in tone as introductions to longer works can be, there is no obscurity in them that needs removing or improving by an editorial hand. They are highly readable and a real pleasure to read for their eloquence and their great insight. They are published here by the kind permission of his widow, Irena Gross.

In these writings, Schell mentions and comments on the excellent motivations for (as well as some of the shortcomings of) the demand for a shift in nomenclature recently sought by geologists for how to describe our present era—no longer the "Holocene," but the "Anthropocene"—to mark the scarcely credible fact that human beings have, by the application of the knowledges we possess, not only made myriad artifacts over the centuries but indeed *converted the earth we inhabit itself into an artifice of our making.* This grand and radical idea of the "Anthropocene" is expounded and discussed in detail by the two essays and one comment that follow immediately after Schell in the volume.

The eminent geologist Jan Zalasiewicz is a major figure in the forefront of the stratigraphic campaign for this name change. His contribution in two of our meetings was invaluable tutorials, not just regarding the motivations for the marking out of a new era for the earth, but to relate it, first, to global warming and, then, to ask the vital question of when to draw the boundary that separates this era from the previous one. This last question is scrupulously addressed by him in the second of his two essays here—as is the tantalizing question about how to relate the perspective of the geologist, which is concerned in general to place the blame (if that is the right word) on the causal efficacies of humanity as a collective actor, to the perspective of the social scientist whose instinct is to blame more specific "groups of humans (rich ones, in developed countries), who drive the process more acutely than others (impoverished communities in developing countries)."

Kyle Nichols, another geologist, and Bina Gogineni, a literary critic and theorist, take up both these questions in detail and draw important lessons for the necessity of an interdisciplinary approach to this last question, with some aptly formulated anxiety about how things have gone when the perspective of the humanities in particular was left out of the ongoing efforts to grasp the Anthropocene; and, in his comment, Nikolas Kompridis takes their argument further in a more explicitly philosophical vein raising questions about the divergent "temporalities" that different disciplinary analysts have focused on, as well as different conceptions of the "epistemic authority" that are reflected in each disciplinary perspective. In a separate contribution he adds a considered argument for how to understand the relations human beings bear to the nonhuman, an argument in which the notion of agency is central, repudiating a familiar picture owing to Bruno Latour and others on how human agency must be conceived in the first place.

Philosophy (as well as intellectual history) is the disciplinary focus of Akeel Bilgrami's, Sanjay Reddy's, and Joanna Picciotto's interest in the subject of nature in the wake of our current ecological concerns.

Bilgrami explores how it is that as a result of the desacralization of nature in the transformations that Weber described with his term *disenchantment,* a quite illicit extrapolation was made that evacuated nature of all *value* properties as well, even though these value properties were not conceived as getting their grounding in a divine or sacral source. He then proceeds to draw a politics out of what he describes as a "secular enchantment" of the world that still remains possible, one that if deeply explored could radically reconfigure our political ideals, fundamentally altering our understanding of the relations between liberty and equality so as to line them up with an ideal of the "unalienated life" that went missing with the transformation of the concept of nature into the concept of natural resources. These explorations are done via a critique of John Locke's contractualist arguments for privatization of the commons and more recent game-theoretic consolidations of Locke. After a superbly comprehending comment on these ideas, Sanjay Reddy, a capaciously minded economist, writes a sympathetic supplement to Bilgrami while at the same time raising a question for him that suggests new directions of work.

Joanna Picciotto focuses on roughly the same period of intellectual history but takes things in a different direction. Making a distinction between value and meaning, she looks with learning at the literature of "physico-theology" to uncover how, in this very time, sentient nonhuman creatures more generally came to be included in a more accommodating conception of the relations that hold between the inhabitants of nature, a measurable advance in the abilities of human beings to widen their ethical and political horizons. Picciotto's paper is a sharp revision of the intellectual history to which Bilgrami appeals and deepens our understanding of the early modern period's relevance to our current concerns about morals and nature.

For some (Naomi Klein, for instance) it has become an article of faith that there is no *sufficient* way to address the crisis of climate change without politically addressing the tendencies of capital, quite possibly to the point of bringing about a terminus of capitalism as we know it. While taking no stand on that point, Robert Pollin's essay is a careful and rigorous plea by a radical political economist that, at least in the case of the United States, a great deal can be done to control carbon emissions *within* the commitment to growth that is so central to capitalist economic formations. His blueprint for "green growth" as a path to climate stabilization is a document well worth studying for the prospects it holds out for a saner political economy.

The political economy angle in Pollin is linked to a legal perspective in Jedediah Purdy's essay. It incorporates economic questions regarding nature's resources not only with the law in general, but with environment law in particular, understood in the context of what he calls "infrastructure," a systemic notion that has consequences for the political ideals that were discussed in the earlier papers in the volume. Purdy reveals the extent to which the law, so understood, can lead to two things: forms of coordination that underlie the social cooperation needed to constrain our exploitation of nature, as well as the achievement of the distributive goals that we seek. It thus gives an institutional grounding for points made more abstractly in other papers up to this point about the relationship between our ideals of liberty and equality and the ideal of overcoming our alienation (from nature and from one another).

James Tully's two offerings continue the twin themes of this book (nature and value) by looking to how ecological issues are inseparable from the social issues of alienation, inequalities, and individualistic, competitive ideals of political economy and social life. The first of them explicitly combines the projects of Schell and Bilgrami in diagnosing modernity's undermining of ecosystems via the alienated disengagement from the participatory stance with others and with nature. He then develops extraordinarily ambitious ideas about what is entailed by the overcoming of alienation. There are a penetrating account of alienated responses to the environmental crisis and a sensitive presentation of an alternative form of response, deriving from a very wide range of scholarly sources, including indigenous traditions of thought and the anthropology of "gift"-giving traditions within which nature itself, in the form of the Earth, may be conceived of as an overarching gift in our grateful possession. Essential to this alternative form of responding to our crisis is a subject that recurs throughout this volume: how to integrate ethics with cognition. As Tully puts it, bridging the gap between "knowing the good" and "being good." If a stance of disengagement was a primary source of alienation in the modern period, what would reengagement even so much as look like after human relations (to nature, to one another) have been so deeply defined by such a disengagement? His second essay turns to answering this question in abundant detail. It is a vital contribution to our current crisis of "nature and value."

Anthony Laden's paper discusses the "gift of nature" conception and offers a notion of agency that is in keeping with what is needed to retain the gift with no other goal than sustainability of the *system* with which we are gifted. This amounts to an eschewal of productive goals not only in our political economy but in the very notion of agency and action. Agency, on this account, is open-ended, without a terminus in "ends" that would count as the achievement of a goal. "Keep things going" is the point of agency—as in a conversation—and it underlies the ideal of keeping nature going in the form of the preservation of a gift. In her comment, Carol Rovane gives a sympathetic exposition and yet critical response to Laden's essay, in which she first covers the rich conceptual terrain of his essay (relating the discussion of nature to social tradition, moral realism, and cultural relativism) and then suggests a different direction for the notion of agency, one that requires conceiving of groups as agents (or "persons") and which may provide yet another basis for the collective action problem of the environmental commons that was discussed in earlier papers.

The literary critic David Bromwich seeks inspiration in Wordsworth and Hans Jonas for a view of human nature as continuous with physical nature without the transcendent elevated status that we have come to give it in our familiar contrasts between nature and culture or natural tendency and moral commitment. The poet's term *natural piety*, intended to mark conscience as the exercise of a *second* nature rather than ethic's gratuitously metaphysical "noumenal" surpassing of nature's constraints, is expounded with a close reading of a range of poems. The essay's argument makes possible that we may see our moral natures as both natural and at the same time irreducible, unlike, for instance, as in Freud, where conscience is sometimes reductively presented as a second-order natural drive intended to curb our destructive first-order drives. This irreducibility of the normative to the natural, even as the human nature from which the normative emerges is continuous with physical nature, is a philosophical outcome that perhaps only a literary exploration of this kind could reveal.

There is a great deal being written on nature these days, as a result of the environmental crisis that confronts us, but it is really very unusual to have in one place such a wide range of approaches to the topic of nature and our failures with regard to it—approaches coming, as I said, from geology and climate science, economics, the law, political science, literature, philosophy, and intellectual history. It is more unusual yet to find writings prompted by the crisis refusing the tendency to see the ecological as a self-standing theme of concern, one that sequesters it from the larger questions of alienation, liberty, and equality. But, above all, I will say once again as I close this much too brief and breathless survey, it is almost unique to find such diverse approaches all converging on a more or less unified and coherent response to these failures with a mutual sympathy for one another's arguments, frequently building on the ideas of one another, even as they register disagreements and points of departure. Questions about nature have long needed this range of regard and this convergence of conviction and it has been immensely gratifying to summon the work of such fine and distinguished scholars to these ends.

ACKNOWLEDGMENTS

The essays in this volume are revised versions of papers presented in a series of workshops on the theme of "Nature and Value," which were generously funded by the Nomis Foundation. They have been very conscientiously assembled by Cosima Crawford and Sarah Stoeter from the Nomis team and ably copyedited by Robert Demke from Columbia University Press. The editor would like to thank them all.

NATURE
AND
VALUE

CHAPTER 1

NATURE AND VALUE

JONATHAN SCHELL

W hat I have to offer are definitely thoughts in progress, more questions than answers.

Now the inescapable backdrop of this discussion of value, social and ecological, is the overall ecological predicament that we now rather unexpectedly find ourselves in.

I propose to do two things. First, to offer some very general reflections on the character or nature of the ecological crisis that we face and that try to define a few of the questions it forces upon in the context of our discussion of value.

The Predicament

Two key things have become salient.

1. The obvious one: We are now technically prepared to destroy ourselves and an unknown proportion of other forms of life on Earth, amounting to what scientist are called the *sixth great extinction* of life on Earth, the other five, including the demise of the dinosaurs, being geological events that occurred tens of millions of years ago.

Not long ago, the word *nihilism* meant rejection of all principles, beliefs, morals, meanings. Now modern civilization has become nihilistic in a much more tangible and literal sense of the word: it is poised physically to annihilate itself, bring itself to nothing, as if, without anyone quite noticing what was happening, the old nihilism had supplied itself with a program of action.

Once, the destructive powers of human beings exhausted themselves within the human order, injuring and destroying human beings and their works. Now they have spilled out of those boundaries and have begun to mutilate, crowd out, devour, deplete, degrade, disfigure, dismantle, and destroy the natural order as well.

Since human civilization, already threatened directly with self-annihilation with the weapons of modern science, depends entirely for its continuance on this natural world, it is also mortally threatened indirectly by this route as well.

The Old Testament said, "One generation passeth away and another generation cometh, but the earth abideth forever." Now, it is no longer so.

2. *The second*: Few people seem to care very much about no. 1, at least not to the point at which they are ready to do anything about it.

The new state of affairs has proved bewildering to the human imagination. The particulars of the various menaces, actual and prospective—the spiraling temperatures, the melting ice caps and glaciers, the rising seas, the butchered landscapes, the polluted lakes and streams, the rivers drained of water before they reach the sea, the dissolving coral reefs, the dying oceans, the punishing hurricanes, the killing heat waves, the parched croplands, the falling water tables, the disappearing species, the blasted plains of Hiroshima and Nagasaki—all these are grossly and shockingly manifest, having been presented and re-presented for over more than a half-century in a thousand articles, books, films, and unimpeachable scientific reports.

But the responses on the scale that at first glance seemed warranted have not come. With notable exceptions, reactions have generally ranged from indifference to denial. The annals of preventive action are all but bare across the board.

Viz. *Doha and Copenhagen*. Decided to decide later, and implement in 2020.

This response or lack of it obviously raises the question of value in the largest possible way, since the question on the table is both the comparative and the combined value of both the human and the natural world.

I. The Predicament

1. Begin by asking what period we have entered.

In order to decide on the proper period we'd have to decide *first* in which of the traditional stories we human beings have told about ourselves or nature does the new crisis find its place?

Is it in history—the story of what human beings do and suffer? Not exactly. What we are experiencing cannot be understood merely as the passage from one historical period to another—say, from the post–Cold War era to something else, or even from the modern period to the post-modern period, or from the post-modern period to the next one.

This crisis clearly *transcends* human doings. Or perhaps it would be better to shun the word *transcend*, with its suggestion of rising above something, perhaps into a spiritual realm or a utopia; for history has not hereby soared into any empyrean, whether of faith or ideals, but rather has broken through downward, into the Earth underfoot, into life's primordial foundations, previously the province not of historians but of geologists and scholars of evolution. Not any transcendence but *biopolitics*—a new fusion of all that is artificial with all that is organic—is born.

This is a *radical* and unexpected reversal whose implications are not easy to absorb, either conceptually or practically. For example, in the political thought of classical Greece, politics was conceived as that which lifted humans above the organic level, regarded as bestial and low. As Aristotle commented, *"To have the sensation of the good and the bad and of the just and the unjust is what is proper to men as opposed to other living beings, and the community of these things makes dwelling and the city."* To live in a city—to be political—was considered the hallmark of the human as distinct from the bestial. But now politics has, evidently quite inadvertently, turned its steps backward and downward toward life's biological foundations. It is both burdened with menacing life and also charged with the responsibility of physically preserving it. We are no longer called to dream of splendid goals—to be reached, perhaps, via the long the path of "progress," which is modernity's version of transcendence; instead, politics is pushed to busy itself with safeguarding its all-too-physical origins, with holding fast to the bare minimum.

"To exist"—that is the new utopia. In this sense, a new existentialism has been born.

Seeking to capture our new reality in a word, the climate scientist Paul Crutzen has proposed that the current geological epoch, the Holocene, which began roughly 10,000 years [ago] with the end of the last ice age, has been succeeded by a new epoch, which should be called the *Anthropocene*. And yet if, according to traditional understandings, the new dispensation is not a new historical age, it seems to me that neither is it precisely a new geological age. His argument is that humankind has intervened so forcefully in nature that it has become the dominant influence even in geological terms. The new epoch therefore should be named after ourselves, who now are the equal of tectonic shifts, glaciation, volcanic eruptions, asteroid collisions, and the like. The term struck a chord, and the geophysical societies are now discussing whether to make the nomenclature official. Proponents argue that when geologists of the future look back at the record of our time, what they will find are the consequences of our activity, such as a deposit of radioactivity from the nuclear tests of the 1950s and early 1960s, a layer of plastic on the sea-bottom, the mass extinction of species caused by our activity.

This term has the great virtue of shining a clear light on the central feature of the crisis, namely humankind's emergence, unequalled in the annals of any other species in the whole 2.5 billion year history of life, as a decisive actor in the fortunes of life on Earth.

Crutzen is surely right. Earth and humanity are wedded in a new union. The walls of separation that once divided the human artifice from nature have come down. History has flooded into evolution. And evolution has returned the favor and flooded history. Ecology and economics are merging. Politics and natural selection are conjoined. Events from now on will transpire on a canvas that weaves together the very short term and the very long term—the "deep time" of geology and the daily news cycle. The doings of the ozone layer require headlines alongside the latest political gossip. We will have to respond—and quickly!—to news flashes from the Permian period, 250 million years in the past. Bulletins from the Carboniferous, when the organic carbon deposits that we now loft into the atmosphere were laid down, have especially urgent import.

So the stegosaurus is in the garden. The tyrannosaurus pokes his nose into the living room. Their fortunes foreshadow ours. The algae and the plankton are caucusing in the ocean stream, and their votes will matter as much for what happens in our world as the Iowa caucus in the next presidential election. (And yet—heaven help us!—the vote in Iowa will also determine the fate of the algae and plankton.) Let me borrow from W. H. Auden:

> The glacier knocks in the cupboard
> The desert sighs in the bed
> And the crack in the tea-cup opens
> A lane to the land of the dead.
> (W. H. Auden, "As I Walked Out One Evening")

And yet the new name, for all its aptness, creates risks of *new misunderstandings*.

For it seems to me that with the transformation that is now underway we in fact leave behind conventional geology just as decisively as we leave behind conventional history. I don't think the changes now afoot can be gauged merely by ascertaining that the human traces we are leaving in the stratigraphic record are on a scale with those left in the past by natural processes.

It will not suffice simply to give the new period a geological name instead of an historical one. If that were enough, we could expect the Anthropocene to be succeeded by the next geological epoch in the series, perhaps named after some other species—some Piscenecene, or Arachnocene epoch—and that one by another, and so on, as in ages past, time out of mind.

But now, unless we humans destroy ourselves, that can never happen. For the Anthropocene is irreversible in a way that no previous geological period, no matter how protracted, ever was. *The historical and the geological can no more be re-separated into two independent streams than can two rivers that have flowed together.* That will not occur even if we humans do finish ourselves off, for in that event it will not be that the streams have parted but that one of them will have dried up. In this world without us, traditional evolution would indeed revive, and the procession of geological ages would resume, though without anyone around to give them names (unless in the fullness of time a new creature, perhaps some gifted fish, evolves to the point at which it can assume the task).

But if on the other hand we save ourselves there is little reason to think the Anthropocene will ever end. For that to happen, humankind would have to shrink its power down to the point that it *could* no longer destroy itself and other forms of life. But if there is one thing that is beyond the capacity of humankind, new-laden with power though we may be, it is [to] do away with that power in any final sense. We know how to acquire knowledge, but we don't know how to get rid of it, especially the scientific knowledge that makes us powerful.

We lack the strength to make ourselves weak. The reason is not mere pig-headedness but something over which we have no control. It is the *extraordinary persistence of scientific knowledge* in a human order in which knowledge is power. The knowledge that underlay the advent of nuclear weapons provides a good example. Humanity can and should abolish nuclear weapons. What in all likelihood it can never abolish are the great advances in physics in the twentieth century—including Einstein's relativity theory and quantum mechanics—that made and will forever make the construction of nuclear weapons possible. The world can be nuclear-weapon-free, but it will always be nuclear-weapon-capable. The same paradox holds true for all the technology that provides the foundation of humanity's power to control, for better and worse, the natural world.

We are left with a curious suspicion. Evolution, with its characteristic time-periods of tens and hundreds of millions and even billions of years, seems to us the scene of all on Earth that is unimaginably long-lasting, while by comparison perishable human history, measured at the most in millennia, seems the scene of all that is short-lived, evanescent. But is it possible that if history, so far a mere flash in the geological pan, is now allowed to unfold into deep time, human achievement, in the form of scientific findings, will prove to be longer-lasting even than evolution's productions? The suspicion increases when we reflect that, as so many scholars of evolution have recently pointed out, science plays a role in human civilization that is highly analogous to the role that DNA plays in evolution. Both provide a foundation of evolving "information" that is passed along from generation to generation, making possible their respective artificial and natural productions. It probably is not a coincidence that DNA is the most durable component of natural evolution. As Richard Dawkins comments, in the evolutionary process, "Genes are the immortals." Might scientific findings be the immortals of human civilization? We don't know how far this analogy may extend. What we do know is that the intellectual foundations of our power are as deep-rooted as anything in our civilization.

That they are rooted so deep should not be surprising. It was arguably the human capacity to inherit knowledge that, at the very dawn of human civilization, put us on a collision course with the rest of nature. Civilization originated in natural selection but at the same time parted ways with it. The origins of civilization are notoriously murky, but it appears that our ability to pile up scientific findings is not so much an extension of natural selection as a process that runs parallel to it and rivals it. Human control has been won at the expense of natural control. The DNA-information that lies at the heart of every creature on Earth has been steadily pushed aside by the information in the books and computers and all the inherited lore of civilization. The decisions that were made by that mysterious force of evolution, the Selector of Darwinian natural selection, and registered in DNA, have been supplanted more and more by decisions that are made in human councils, or by the invisible hand of collective human action in default of councils. The orchestra of life, so to speak, now has two conductors, and to the extent that the new, human one is obeyed, the old, natural one is not.

In other words, the deepest of the changes dividing the Anthropocene from the Holocene—a line sharper than any since the first appearance of life on Earth—flows from nothing less than a revolution now in progress in the way the terrestrial biosphere works and is destined to continue to work from now on. This revolution can be seen as the consequence of a sort of *coup d'état*, a *biocoup* by humankind, or if you prefer, a biorevolution, initiating a kind of regime change for life [on] Earth. But as in all revolutions it cannot be taken for granted that success in overthrowing the old regime will be followed by success in establishing a new one. Whether this coup, dethroning natural selection and enthroning humanity, will be followed by a new regime viable for either ruler or ruled is anything but clear.

Thus to call the new dispensation the Anthropocene epoch risks misleading us in another fundamental respect. Even as it rightly emphasizes the growth in human powers, it could lead to an overestimation of those powers. The biocoup is an accomplished fact in the sense that, at least since the invention of nuclear weapons, humankind has stood poised, if it so chose, to administer unlimited destruction to nature. Humankind is in the position of a great empire that has conquered vast territories and now must decide whether to seek to govern them, and, if so, how. But our power of destruction of nature is by no means equaled by constructive power, which is rudimentary. Above all, sufficient knowledge is lacking. The achievements of human knowledge are awesome but the extent of the unknown and the uncontrolled, though by definition hard or even impossible to measure, remains more awesome still. Almost every human intervention into the global life-processes of Earth, whether unintentional or unintentional, is more or less a lunge in the dark.

We are not very good at nation-building in other people's countries after we have practiced regime change. Neither do I imagine we will be very good Earth-builders if we overthrow the regime of mother nature.

In conclusion, the boundary between the Holocene and the Anthropocene, then, is not a line separating one epoch from another, whether historical or geological. It is the point of origin of something more radical: we can find the proper term on the American dollar bill, which announces a *novus ordo seclorum*, a new order of the ages. Like the division between BC and AD in the Christian calendar, it sunders all time in two, dividing it into a before and after. A mere age or an epoch is destined to end. Like the generations in Ecclesiastes, the dinosaurs came but the dinosaurs passed away. The Middle Ages arose out of history but then sank back into it. By contrast, a new order of the ages has no necessary end. It is not itself one in a series; it is the start of a series.

The one we are now entering involves a change more radical still: it transforms the composition of what any "age" is—or, for that matter, what any order of the ages is.

History and evolution alike ended as independent processes and a hybrid process mingling both has begun. For this new process we as yet have no name.

It's obvious that our new predicament raises all kinds of new questions regarding value and the whys and wherefores of valuing things.

Let [me] try at least [to] crystallize some of the questions.

—Usually when we raise the question of the value of something we proceed in a conceptually simple way. We begin with certain standards of value, such as goodness, or beauty, or usefulness, and ask how high some object in question, such as a painting or a government policy, rises in the scale.

But in reference to our ecological crisis, the issue arises in at least two ways that are more complicated and elemental.

—Since Hiroshima, the issue of human extinction has been on the human agenda. This raises an entirely new question for the whole realm of value. The reason is that humanity is more than just an invaluable object, possessing a certain value that we might discuss.

It is, in addition, the principal seat of valuation, of the very activity of valuing. So in human extinction, there is, so to speak, a simultaneous double loss of value. We lose the invaluable object that we constitute, but also we lose the Earth's chief valuer, the one in whose eyes things can have value or not. For anything to have value, there must be a valuer. But that valuer is destroyed in human extinction. So we are left asking what is the value of valuing, and this sends us running in cognitive circles. But I think it's important to run in those circles.

—Second, all of the above refers to human "valuing." What valuing by others. The only candidates are animals. A stone does no valuing. I don't think a tree does either. And an animal might be a valuer, but only in a limited sense. For example, you might say it values food, or absence of pain, or certain animal pleasures. But it is not equipped to value goodness and beauty, for which it is not equipped.

—In other words, when it comes to valuing, there is an abyss of inequality that separates us from the rest of creation. This is very troubling in a time when we threaten that creation with mutilation. It raises the question of how we are to adequately value those other elements of the creation. Are they to be considered valuable only insofar as they serve us? Or do they have a value in themselves as people say? And if we say that, what do we mean? If it is true that all value implies a valuer, then in what eyes but human eyes might creatures and plants have value "in themselves"? And what, in any case, does this expression, "value in themselves" have?

—Briefly, a little thumbnail presentation of the role of valuing in the human and natural scheme of things.

Let us consider the qualities that things have. We can say they are of two kinds. One kind discloses their physical reality. These are tangible qualities, such as temperature, motion, extent, that are revealed to our senses or to scientific instruments, which are extensions and refinements of our senses.

The second kind of quality discloses the value of the things—their beauty, their goodness, their usefulness, even their tastiness.

—At first sight, this perceived value may seem "subjective" because it so clearly relates to our preferences, our needs, our human nature. This value thus may seem like something human added to the things, like a sauce poured over a dish, and not like something that is actually in the things.

And yet on second thought that conclusion seems wrong. Subjectivity is related to the individual. But we as humans share taste and judgment. Certainly, utility, maybe even beauty. We say, not just "it seems beautiful to me," but "it *is* beautiful." This argues that there is, after all, something in the things that we recognize as invaluable, not just some superadded element fixed on them by us.

—And yet it remains true that these valuable qualities disclose themselves mostly to humankind. Not true of taste, which we might share with a lion. But surely no lion sighed over the beauty of Jacob van Ruisdael's landscapes or admired anyone's virtue

—But there is one more difference between the tangible and physical qualities, and those denoting value.

The physical qualities, such as the heat of a stone, or the force of hurricane winds, directly effect other things in the physical world. They are indeed a part of the broad system of interacting physical things, in which they have their life and have to make their mark if we are to discern them at all. The warmth of the stone warms the pail of water it is in; the wind of the hurricane tears the roof off the house. Those things would occur whether we perceived them or not.

It is not so with beauty and goodness. These effect only us. In other words, except insofar as *we* are moved, they move nothing.

The beauty of Van Ruisdael's landscape doesn't disturb a molecule anywhere in the universe, near or far. It stares out without effect, until it meets a human eye and mind. It, so to speak, gives us creatures a private audience.

It is *this* fact, I want to suggest, that makes us think of these value qualities as subjective, only to reflect on second thought that this is not quite so, because we share our reactions in these matters.

But it's very hard to know how to weigh these values. The reason is a notable existential fact about human life: in a sense that is hard to define yet also impossible to deny, we stand alone in our experience as sentient beings. In this respect Earth does not contain our equal, insofar as we can determine, nor does the rest of the universe, although that point remains open.

This superiority of the human over all known things creates an abyss of inequality that is a great stumbling block in weighing the value of ourselves relative to nature. Our solitude in this respect places us at risk of an arrogance that would be as fatal to ourselves as other species.

But now let us imagine, as we can easily do, that one day we find ourselves face to face with some creature from another planet in another solar system that is our equal. In that case the private audience of value-qualities will, in a way, be at an end. Perhaps what is good will also seem to them good. Perhaps this will prove true with a hundred other species that float down among us in their flying source, at which point values would have an objective quality that now we can only guess at.

At that time, we will be able to declare a basis for the objectivity of our values that we can now pronounce only for scientific laws.

But for now, we cannot. The relevance for our subject today is that all valuing is strictly human based. Behavior is virtuous because we say it is virtuous.

In other words, we are left as judges in our own case. And that's a problem, because it could be asked of us—if only there were someone to do the asking—who are *you* to judge all creation?

Sometimes I think we have invented God, at least in part, in order to find an escape from this dilemma, because we wanted to escape from being judge in our case; we wanted a more objective judge who could separate out what was truly good in us, from, say, what was just disguised self-interest.

This conflict obviously becomes a major ecological problem if we judge and manipulate the entire creation to serve our notions of what is good, literally, what is good for us.

So one of the great quests has to be to find a broader criterion of valuing that is not anthropocentric in this pejorative sense.

But that is not so easy. It's a question that haunts our ecological crisis. Now I turn to James Tully to give us the answer.

II. What Sort of Response?

What sort of response is appropriate to such a new threat?

Offer only a *few fragments*. Turn to Arendt.

One thought that arises is the venerable idea that our notion of the Good is tied with Being per se while Evil is tied with nothingness.

This is to be contrasted with the more common notion that evil occurs when damage is done or suffering is brought to existing people.

Willing, 118. [Hannah Arendt, *The Life of the Mind*, vol. 2, *Willing* (New York: Harcourt Brace Jovanovich, 1978), 118.] Quotes and paraphrases Aquinas's elaboration of Augustine's idea that Goodness and Being are the same, and that evil is absence of this Being/Good: Arendt writes, "Because of its privative character, absolute or *radical evil cannot exist*. No evil exists in which one can detect 'the total absence of good.' For '*if the wholly evil could be, it would destroy itself*.'"

The relevance to our subject, if I am right that the defining character of the present danger is a kind of spreading nothingness of the future, whose generativity is foreclosed by extinctions, and if that is so the question doesn't have to do with suffering but with diminishment, not so much future hurricanes as the nullification at the root of future species, ecosystems, biosystems, and even of human beings.

Perhaps this carries us some of the distance toward understanding why we don't react to dangers of extinction? We are, in some literal sense, nonplussed. Plus = Being.

I'm reminded of the Philip Larkin poem on death as distinct from suffering in life.

> This is a *special* way of being afraid
> No trick dispels. Religion used to try,
> That vast moth-eaten musical brocade
> Created to pretend we never die,
> And specious stuff that says *No rational being*
> *Can fear a thing it will not feel*, not seeing
> That this is what we fear—no sight, no sound,
> No touch or taste or smell, nothing to think with,
> Nothing to love or link with,
> The anaesthetic from which none come round.

2.—*Some of the same riddles are posed when* Arendt describes the special burden that genocide, a kind of local extinction, places on our shoulders:

> But as we know today, *murder is only a limited evil.* The murderer who kills a man—a man who has to die anyway—still *moves within the realm of life and death familiar to us*; . . . The murderer leaves a corpse behind and does not pretend that his victim has never existed; if he wipes out any traces, they are those of his own identity, and not the memory and grief of the persons who loved his victim; he destroys a life but he does not destroy *the fact of existence itself.*

This zone of "more" than murder is the zone that extinction reaches and that I suggest is steadily expanding. We, in so many different ways, threaten and impair not just living things but the matrix of living things.

—*Why?*

Now finally a word about what sort of human response might be appropriate.

Perhaps we are heading down the wrong alley when we conceive the problem [as] bigger disaster, requiring and farther and longer reach of compassion, a greater and greater stretch of the ethical imagination. Mind you, the disasters, the hurricane Sandys and droughts and drowning cities will be there. But in seeing all these we may stare past the heart of the matter.

In a word, a more appropriate response, or one that fits the facts of the case better, is something we don't usually associate with politics or public life, and that is something in the vicinity of awe, or a sacred terror. And related to that would be sheer admiration and wonder at the creation.

This is different from ethics and empathy. When something we value is before us, there is a pang we feel that it will be yanked out of existence, be destroyed. This is a very specific, wrenching feeling. But prior to that and per-haps necessary to it is something else. It is wonder and gratitude that this ever came into existence out of nothing in the first place.

This is something more like God's admonition to Job, when, in response to Job's ethical complaints, he steps forward as the creator God, and asks, "Where wast thou when I created Leviathan?" "Where was thou when I laid the foun-dations of the earth?"

Note, though, that this wonder is not for something static and eternal, it is not philosophical wonder of the Greek kind. It is for something changeable, changing, and in fact jeopardized.

Nor is it wonder merely at what presents to the senses; it is specifically for *the source* of all that, the point of origin of incessant renewals of life.

So we are talking about a deeply conservative impulse, but it's conservatism with a difference: it is not, like classical conservatism, a conservatism of the past, a desire to save what went before; it is a conservatism of the future, a desire to permit the entry of what is coming to us out of unshapen future unless we prevent it.

I'll resort to Auden one more time.

And Auden:

> That singular command
> That I do not understand
> *Bless what there is for Being,*
> Which has to be obeyed, for
> What else am I made for.
> Agreeing of Disagreeing

Note

This chapter was originally a discussion paper and is published here in its original form.

CHAPTER 2

THE HUMAN SHADOW

JONATHAN SCHELL

Prologue

> *"Whither is God?," he cried. "I shall tell you. We have killed him—you and I. All of us are his murderers. But how have we done this? How were we able to drink up the sea? Who gave us the sponge to wipe away the entire horizon? What did we do when we unchained the earth from its sun? Whither is it moving now? Whither are we moving now?"*

—Friedrich Nietzsche

A New Order of the Ages

Not long ago, the word *nihilism* meant rejection of all principles, beliefs, morals, and meanings, human or divine. Now modern civilization has become nihilistic in a much more tangible and literal sense of the word: it is poised to annihilate itself physically, bring itself to nothing, as if, without anyone quite noticing what was happening, the old nihilism had supplied itself with a program of action. Once, the destructive powers of human beings exhausted themselves largely within the human order, injuring and destroying human beings and their works. Now they have spilled out of those boundaries and have begun to mutilate, crowd out, devour, deplete, degrade, disfigure, dismantle, and destroy the natural order as well. Since human civilization depends entirely for its continuance on this natural world, it, too, is mortally threatened. The preacher in Ecclesiastes said, "One generation passeth away and another generation cometh, but the Earth abideth forever." It is no longer so.

The new state of affairs has proved bewildering to the human imagination. The particulars of the various depredations, actual and prospective—the blasted plains of Hiroshima and Nagasaki, the spiraling temperatures, the melting ice caps and glaciers, the rising, acidified seas, the dissolving coral reefs, the polluted lakes and streams, the rivers drained of water before they reach the sea, the punishing hurricanes, the killing heat waves, the parched croplands, the falling water tables, the butchered landscapes, the foundering ecosystems, the wholesale extinctions—are grossly and shockingly manifest, having been presented and re-presented for over a half-century in thousands articles, books, films, and unimpeachable scientific reports. Yet the responses on the scale that at first glance seem warranted have not come. With notable exceptions, reactions have ranged from indifference to denial. The annals of preventive action are all but bare.

It has been difficult even to decide which elementary tools of classification to apply to the crisis. It appears to be an event in a narrative that, until recently, we didn't know was unfolding. In which of the traditional stories we human beings have told about ourselves or nature does it find its place? In history— the story of what human beings do and suffer? Not exactly. What we are experiencing cannot be understood merely as the passage from one historical period to another—say, from the post–Cold War era to something else, or even from the modern period to the postmodern period, or from the post-modern period to the next one. In its assault on the natural order, this crisis has clearly transcended merely human doings. Or perhaps it would be better to shun the word *transcend*, with its suggestion of rising above something, perhaps into a spiritual realm or a utopia; for history has not hereby soared into any empyrean, whether of faith or ideals, but rather has broken through downward, into the Earth underfoot, into life's primordial foundations, previ-ously the province not of historians but of geologists, scholars of evolution, and other natural scientists. Not any transcendence but biopolitics—a new fusion of all that is artificial with all that is organic—is born. This is a radical and unexpected reversal whose implications are not easy to absorb, either con-ceptually or practically. For example, in the political thought of ancient Greece, still so important for the self-understanding at least of the West, poli-tics was conceived as that which lifted humans above the organic level. As Aristotle commented, "To have the sensation of the good and the bad and of the just and the unjust is what is proper to men as opposed to other living beings, and the community of these things makes dwelling and the city." To live in a city—to be political—was considered the hallmark of the human as distinct from the bestial. But now politics has, inadvertently, turned its steps backward and downward toward life's biological foundations. It is both burdened with menacing life and charged with the responsibility of physically protecting and preserving it. No longer summoned to dream of splendid goals—to be reached, perhaps, via the long path of "progress" (modernity's version of transcendence)—politics is called on to busy itself with safeguarding its all-too-physical origins, with holding fast to the bare minimum. "To exist"—that is the new utopia. In this sense, a new existentialism has been born.

Seeking to capture our new reality in a word, the climate scientist Paul Crutzen has proposed that the current geological epoch, called the Holocene, which began roughly ten thousand years [ago] with the end of the last ice age, has now been succeeded by a new epoch, which, taking its name from our species, should be called the Anthropocene. His argument is that humankind, which now is the equal in power to tectonic shifts, glaciations, volcanic eruptions, asteroid collisions, and the like, is intervening so forcefully in nature that it has become the dominant influence on Earth even in geological terms. The term has struck a chord, and the geological societies are now discussing whether to make the nomenclature official. The debate revolves around the magnitude of the human impact. Proponents argue that when geologists of the future look back at the record of our time, what they will chiefly find are the consequences of our activity, such as a deposit of radioactivity from the nuclear tests, a layer of plastic on the sea-bottom, changes in soil chemistry from our agriculture, our prodigious population growth, and the mass extinction of other species caused by diverse human activities.

The term has the decisive virtue of shining a clear light on the central feature of the entire ecological crisis, namely, humankind's emergence, unequaled in the annals of any other species in the whole 2.5 billion year history of life, as a decisive actor in the fortunes of life on Earth. (The only near-precedent would be the bacteria that evolved to pump oxygen into the atmosphere 2.4 billion years ago, dooming most of anaerobic life and giving rise to its aerobic successor.) Meanwhile, in the last few decades, scientists have come to appreciate that life as a whole has powerfully influenced the physical conditions of Earth that make life possible, such as temperature and the chemical composition of air, sea, and land. Geology and evolution have merged. That is an old story, but one understood and appreciated only in the last few decades. Crutzen is surely right. Earth, life, and humankind are wedded in a new union. The walls of separation that once divided the human artifice from nature have come down. History has flooded into evolution. And evolution has returned the favor and flooded history. Ecology and economics are fusing. Politics and natural selection are conjoined. Events from now on will transpire on a canvas that weaves together the very short term and the very long term—the daily news cycle and the "deep time" of geology. We will have to respond—and quickly!—to news flashes from the Permian period, 250 million years in the past. Bulletins from the Carboniferous, when the organic carbon deposits that we now loft into the atmosphere were laid down, have particularly urgent import. The tyrannosaurus pokes his nose into the living room. The stegosaurus is in the garden. Their fortunes foreshadow ours. The doings of the ozone layer require headlines alongside the latest political twists and turns. The algae are caucusing, the plankton are heading for their primaries in the ocean stream, and their votes will matter more for what happens in our world than the Iowa caucus in the next presidential election in the United States. (And yet—heaven help us!—the vote in Iowa will also sway the destinies of the algae and plankton.)

> The glacier knocks in the cupboard
> The desert sighs in the bed
> And the crack in the teacup opens
> A lane to the land of the dead.
> (W. H. Auden, "As I Walked Out One Evening")

And yet the new name, for all its aptness, creates risks of new misunderstandings. For with the transformation that is now underway we in fact leave behind conventional geology just as decisively as we leave behind conventional history. The changes now afoot cannot be comprehended merely by ascertaining that the human traces we are leaving in the stratigraphic record are on a scale with those left in the past by natural processes. It will not suffice simply to give the new period a geological name instead of a merely historical one. If that were enough, we could expect the Anthropocene to be succeeded by the next geological epoch in the series, perhaps named after some other species—some Piscenecene or Arachnocene epoch—and that one by another, and so on, as in ages past, time out of mind. But no such confidence is warranted. *Now, unless we humans destroy ourselves, there is reason to believe that that can never happen.* The human rise is like that of no other species. The Anthropocene looks to be irreversible in a way that no previous geological period, no matter how protracted, ever was. The historical and the geological can no more be re-separated into two independent streams than can two rivers that have flowed together. That will not occur even if we humans do finish ourselves off, for in that event it will not be that the streams have parted but that one of them will have dried up. Whether this happens is perhaps the most urgent question before us. (But some would argue that the fate of life on Earth in its totality is more important than the fate of merely human life, and how to weigh these stakes is one more of the questions that our new predicament has placed on the table.) In this world without us, traditional evolution would indeed revive, and the procession of geological ages would resume, though without anyone around to give them names (unless in the fullness of time a new creature, perhaps some gifted fish, evolves to the point at which it can assume the task).

But if on the other hand we save ourselves there is little reason to think the Anthropocene will ever end. For that to happen, humankind would have to shrink its power to down to the point that it *could* no longer destroy itself and other forms of life. But if there is one thing that is beyond the power of humankind it is to do away with that power. We know how to acquire knowledge, but we don't know how to get rid of it, especially the scientific knowledge that makes us powerful. We lack the strength to make ourselves weak. The reason is not mere pig-headedness but something over which we have no control. It is the extraordinary persistence of scientific knowledge in a human order in which knowledge is power. An example is the knowledge that underlay the advent of nuclear weapons, which in this as in so many other respects provide the most illuminating paradigm of the new processes that are the new sources of danger. Humanity can and should abolish nuclear weapons. What it can never abolish are the great advances in physics in the twentieth century—including Einstein's relativity theory and quantum mechanics—that made and will forever make the construction of nuclear weapons possible. The world can be nuclear-weapon-free, but it will always be nuclear-weapon-capable. The same paradox holds true for all the technology that provides the foundation of humanity's power to intervene, for better and worse, in the natural world.

We are left with a curious suspicion. Evolution, with its characteristic time-periods of tens and hundreds of millions of years, seems to us the scene of all in life that is unimaginably long-lasting while by comparison human institutions, measured at the most in centuries and millennia, seem the scene of all that is short-lived, perishable. But is it possible that if history, so far a mere flash in the geological pan, were to be allowed to unfold into future deep time, human achievement, in the form of scientific findings and perhaps other accomplishments yet unknown, would prove to be longer-lasting even than evolution's productions? The suspicion increases when we reflect that, as so many scholars of evolution have recently pointed out, science plays a role in civilization that is highly analogous to the role that DNA plays in evolution. Both provide a foundation of evolving "information" that is passed along from generation to generation, making possible their respective artificial or natural productions. It probably is not a coincidence that just as DNA is the most durable component of natural evolution, science is the most durable component of human civilization. As Richard Dawkins comments, in the evolutionary process, "Genes are the immortals." Might scientific findings be the immortals among human productions? We don't know how far this analogy may extend. What we do know is that the intellectual foundations of our human powers are as deep-rooted as anything in human civilization.

That they are rooted so deep should not be surprising. It was arguably the human capacity to acquire and inherit knowledge that, at the dawn of human civilization, initiated our long accumulation of power, putting us on a collision course with the rest of nature. Civilization first arose in a framework of natural selection but at the same time parted ways with it. A new kind of selection—and a new kind of evolution—was born. The origins of civilization are notoriously murky, but it appears that our ability to pile up scientific findings is not an extension of natural selection so much as a process that runs parallel to it and rivals it. Human control has been won at the expense of natural control. The information in DNA that lies at the heart of every organism on Earth has been steadily pushed aside by the information in the books and computers and all the inherited lore of civilization. The decisions that were made by that mysterious force of evolution, the Selector of Darwinian natural selection, and registered in DNA, have been supplanted more and more by decisions that are made in human councils, or by the invisible hand of collective human activity in default of councils. The orchestra of life, so to speak, now has two conductors, and to the extent that the new, human one is obeyed, the old, natural one is not.

In other words, the deepest of the changes dividing the Anthropocene from the Holocene—a line sharper than any since the first appearance of life on Earth—flows from nothing less than a wholly unprecedented revolution in the way the terrestrial biosphere works and is destined to continue to work from now on. This revolution can be seen as the consequence of a sort of *coup d'état*, a *biocoup*, or, if you prefer, a biorevolution, by humankind, initiating a kind of regime change for life on Earth. If so, then what we face is not so much a new stage in the Darwinian evolutionary process as a progressive exit from that process and the inauguration of a new process. But as in all revolutions it cannot be taken for granted that success in overthrowing the old regime will be followed by success in establishing a new one. Whether this coup, dethroning natural selection and enthroning humanity, will be followed by a new regime viable for either ruler or ruled is anything but clear.

Thus to call the new dispensation the Anthropocene epoch risks misleading us in another fundamental respect. Even as it rightly emphasizes the increase of human powers, it could lead to an overestimation of those powers. The biocoup is an accomplished fact in the sense that, at least since the invention of nuclear weapons, humankind has been capable, if it so chose, of administering unlimited destruction to nature. Humankind is in the position of a great empire that has conquered vast territories and now must decide whether to seek to govern them, and, if so, how. But our powers of destruction of nature are by no means equaled by constructive powers, which are rudimentary.

Above all, we lack sufficient knowledge. The achievements of human knowledge are awesome but the extent of the unknown and the uncontrolled, though by definition hard or even impossible to measure, remains more awesome still. Almost every human intervention into the global life-processes of Earth, whether unintentional or intentional, is more or less a lunge in the dark.

Moreover, attempts to acquire the missing knowledge beyond certain limits are cursed with a difficulty that has no practically or morally tolerable solution. There is a conflict between the nature of scientific investigation and the nature of Earth. Every scientific experiment is a foray into the realm of the unknown. Experimentation is the exploration of the consequences of an action on one specimen of a class of objects—say, atoms or cells—in search of lessons that can then be applied to others of its kind. Indeed, an experimental finding is not usually admitted into the hall of accepted scientific knowledge until additional, confirmatory experiments have been done on sister objects. But the Earth is a singular body in the known universe, a class of one. There are no other planets to which experiments done on life here might be applied. Moreover, from all existing points of view (which for now means all human points of view) the Earth is our all, the necessary foundation of everything we are and everything we hope to be. No stake is higher than this one, because it encompasses all other stakes. This combination of the unknowability in advance of the consequences of any experimentation, the singularity of the object Earth, and the infinity of its value should remove it from all consideration as an experimental object. Since every experiment risks losing or mutilating, perhaps irreversibly, the infinitely precious object we wish to know about, no gain or hope of knowledge or benefit can justify the risk. That we have ignorantly become involved in just such experiments, in the form of such interventions as ozone depletion and global warming, can be no argument for continuing them or inventing deliberate new ones. The experiments that are already underway were commenced largely unwittingly. Had the consequences been known, they should have been prohibited, and perhaps would have been. To now continue them knowingly would be a crime—a crime against Earth, which would also be a crime against humanity. As several scientists have said, "If we conquer nature, we will find ourselves among the defeated."

The disordered relationship between human knowledge and human power can be represented by two tales of science in the twentieth century. The first is the posture we found ourselves in in regard to the survival of other species. Estimates of their number range from a few million to more than a hundred million. Currently, species are disappearing from the planet at a rate estimated at one hundred to one thousand times the "background" rate. The Intergovernmental Panel on Climate Change has stated that by the end of the century, global warming may have doomed a heart-stopping 40 to 70 percent of species.

Genetically speaking, we are in the process of burning down whole libraries of books that we have never read. The purpose of experimentation is to increase knowledge. To destroy, in the pursuit of knowledge or for some other reason, the unknown objects science one day might seek to know is a self-defeat beyond irony or tragedy.

The second tale is the story [of] chlorofluorocarbons and the ozone layer. In 1928, Thomas Midgley, a chemist working for Frigidaire, was charged with finding an improved gas for use in refrigeration. The existing refrigerants were toxic to humans or corrosive to refrigerating pipes or both. Midgley engineered chemical compounds not otherwise found on Earth, chlorofluorocarbons, or CFCs. The gases, which were given the brand name Freon, were later found to be useful as coolants for air conditioning and propellants in spray cans. Not only did CFCs leave pipes in refrigerators unscathed but they seemed unlikely to have any consequences for anything whatever, for they were chemically inert. That is, their chemical bonds were all taken up by their own atoms and therefore could link with nothing, destroy nothing, generate nothing: they could only bounce around the world harmlessly. But then, in the 1970s, teams of scientists, including Crutzen (who thus knew from his own work whereof he spoke when he christened our time the Anthropocene), showed that CFCs did do something, after all. Finding no home in bonds with other chemicals on Earth's surface, where they were produced, they floated into the stratosphere. There, they were bombarded with ultraviolet rays from the sun, whereupon they broke into pieces. The fragments were not inert. They underwent an elaborate series of chemical reactions whose products included chlorine, which had the property of scouring ozone from the stratosphere. A single chlorine atom could destroy one hundred thousand ozone molecules in its career. Later, it turned out that nuclear explosions also depleted the ozone layer.

Ozone, a gas whose properties and role in the atmosphere were fully elucidated only in 1931, two years after Midgley developed CFCs, is critical to the well-being of life on Earth, which cannot thrive in the presence of the high levels of ultraviolet rays that ozone keeps out. It has been said that development of an ozone layer—something that occurred some six hundred million years ago—was a prerequisite for the rise of life on land. Life on Earth without an ozone layer would be stunted and impaired life. In the words of the nuclear weapon effects expert Samuel Glasstone, "life as currently known could not possibly exist, except in the ocean." But sea life is also at risk, because ultraviolent rays at high levels are also ruinous for many kinds of plankton, which are at the base of much of the ocean food chain.

The happy upshot of this tale is that, after an epic political battle, a phaseout of CFCs was agreed upon at a global meeting in Montreal in 1987. Since then, of all the success stories of the environmental movement, this one was surely the greatest.

But that is not the end of the story. Crutzen later realized that another option had been open to Midgley when he hit upon CFCs. Instead of choosing the element chlorine as the foundation of his invention, he might have chosen another inert gas, bromine. Bromine, Crutzen realized, was "one hundred times more dangerous for ozone than chlorine," as he put it in the speech he gave upon receiving the Nobel Prize in 1995. Had Crutzen chosen bromine, the ozone layer would have been almost completely destroyed by the 1970s, well before the destructive effects of chlorine were discovered. As many scientists have noted, when experimenting with the biosphere it is impossible to do only one thing at a time, and some of the many things that happen are ones no one has foreseen. Rarely has this law been better illustrated than by the tale of CFCs and ozone layer. Was ever a chemical less frightening-seeming than this inert substance? Was any ever in fact as dangerous? The discovery of the dangers was an extraordinary episode of serendipity wedded to scientific brilliance. The geophysical experiment came first, the revelation of consequences after. There was no law which stated that the discovery of the destructive effect of CFCs on the ozone layer would precede that destruction. Consequences and knowledge were in a race that the exponents of neither knew they were running. It is easy to imagine that consequences could have won, as they likely would have if Thomas Midgley had chosen bromine. It was a narrow escape for life on Earth—one in which sheer luck played a preponderant role.

The boundary between the Holocene and the Anthropocene, then, is not a line separating one epoch from another, whether historical or geological. It is the point of origin of something far more radical. We can find the appropriate expression on the back of the American dollar bill: *novus ordo seclorum*, a new order of the ages. Like the division into BC and AD of the Christian calendar, this shift sunders all time in two, dividing it into a before and after. A mere age or epoch is destined to end. Like the generations in Ecclesiastes, the dinosaurs came but the dinosaurs passed away. (More than 99 percent of species that have ever appeared have gone extinct.) The Middle Ages arose out of history but then sank back into it. By contrast, a new order of the ages has no necessary end. It is not itself one in a series; it is the start of a series. It signals a change in the composition of any "age" thereafter. History and evolution alike ended as independent processes and a hybrid process mingling both has begun. For this new process (the anthro-evolutionary? the geo-anthropological?) we as yet have no name. However, this is not to say that no solution to our current crisis can ever be found. Just as nuclear weapons can be eliminated even as the capability to build them persists, humans can learn to live un-catastrophically in the Anthropocene order of the ages even as the capability for wrecking nature remains in our hands.

The response to a dilemma that, like this one, is rooted in physical threats (such as ocean acidification) to physical things (such as coral reefs) must begin in science and end in science, which alone can tell us how things stand in the physical sphere. Without authoritative scientific findings diagnosing the sources of harm, their nature, and their magnitude, we would all be wandering in darkness, powerless even to begin to think appropriately. Without authoritative scientific recommendations for preventing damage, we certainly would be unable to act appropriately. Fortunately, the scientific community has stepped forward with an abundance of reliable information of both the diagnostic and the prescriptive types. But between these essential scientific starting and finishing points lies an array of challenges that are not scientific. The most obvious are the political. In any discussion of the environmental crisis, someone is sure to say, "All that is missing is the political will"—will to implement the solutions that scientists and other experts have already mapped out and that, unacted upon, are already overflowing library shelves. But where does political will come from? Where is the lever that moves people to action? Why does the long-endured suddenly become unendurable? Why does a nation suffer oppression and injustice for centuries and then suddenly rise up in revolution to throw it off? Why did one half of humanity, the feminine half, tolerate exclusion from public life from time out of mind and then, in recent decades, insist that its voice be heard and its will felt? Why did Soviet rule, having lasted the better part of a century, suddenly fly to pieces? We must acknowledge that, even after the fact, few questions are less susceptible of convincing answers than these. To predict such moments in advance seems beyond human perspicacity. Nothing is nearer to us than our own will to act—it is within us—and yet nothing, it seems, is harder to find or summon up. And so it has been with the ecological crisis. The political campaigns grind on in a parallel universe, addressing everything but what matters most. The international ecological conclaves follow one after another pouring forth the right words in superabundance but accomplish little. One day, we hope, the igniting spark will fly, perhaps very suddenly, but who can say how or when?

What we can be certain of now is that the political question will be asked and answered in a context that extends beyond the political or even the scientific. Every great crisis challenges and upsets our understanding of ourselves, of the world, and of our place in it, and this is especially true of our ecological crisis, with its threat to our biological foundations. To say it in a word, every new crisis has a new *meaning* and that meaning is never obvious and may sometimes be more elusive even than the adumbrations of the concrete solutions, so that, intellectually, at least, we sometimes figure out *what* we need to do even before we truly know *why*. It may even be appropriate to speak of a crisis of meaning—one in which our actions make sense (or once made sense) only in contexts that have in fact disappeared or are threatened with disappearance. That paradox, I believe, made a notable appearance with the arrival of the threat of human extinction in a nuclear holocaust. This condition was diagnosed by Albert Einstein when he made his remark regarding the nuclear dilemma, "Everything has changed save our way of thinking, and so we drift toward unparalleled catastrophe." Einstein's formula holds good not just for nuclear danger but for the entire ecological crisis, in which our economic as well as political lives are manifestly out of step with the conditions on Earth in which we now actually have come to live. Yet it would be wrong to claim that the search for meaning will conduct us through the fog of confusion and apathy to the missing clear-running well-springs of action. The thinking that discloses meaning is not instrumental—not a means to end. But would it be presumptuous to hope that this search, like Calvinist prayer, which cannot summon divinity but somehow prepares the way for a divine visit, might bring us nearer to the vicinity of those well-springs?

Abraham Lincoln, the champion of the American union, was devoted to the survival of a form of government in the face of a mortal threat to it. In perhaps his most famous saying, he resolved that government "of the people, by the people, and for the people shall not perish from the Earth." No thought that the Earth itself could perish crossed the mind of this mid-nineteenth-century man. But now we are called to frame a similar high resolve precisely with respect to Earth. It's clear that we are certainly *of* Earth —that dust from which we came and to which we all return—and equally clear that we are *by* Earth, brought forth by Earthly life. Can we also be *for* Earth? Will we learn to live at peace with life? Or are we, alone among Earth's progeny, destined to destroy our parent, and, with our parent, ourselves?

This startling, confounding question presses our puzzled species to pose a series of others. What is the place of the living generations in the order of human generations? What standing shall the future generations have among the living? What is the place of humankind in the order of nature? How far has our development, historical and pre-historical, carried us from that order into an "unnatural" realm? To what extent may the solution to our quandary be found precisely in that unnatural realm? By what steps, evolutionary and historical, did the alienation from other life, such as it may be, occur? To what extent is our destructive bent conferred upon us by and inherited from previous evolutionary forms and to what extent is it a strictly human departure from all that? What importance do other species and the ecosystems they form have for us? Finally, *who*, as a species, seen in the light of our newly manifested nihilism, are we? Can a species, like a person, have a character? If so, what is that? What sort of creature is it that readies its own doom? And what sort of creature would we have shown ourselves to be if we so arrange things as to call the cataclysm off? And if we take steps adequate to address the crisis, might other, unexpected positive horizons open up? None of these questions is in itself either a purely scientific or purely practical one, though none can even be asked meaningfully in ignorance of what the scientists have to tell the rest of us. Indeed, one way of describing our task is to say that we have to find out how to live imaginatively and politically in the world in which science has shown that we live factually. This book will be neither diagnostic nor prescriptive; it will offer neither new science nor new plans for action. It will go in search of meaning. Some or most of the questions just raised seem likely to elude any definitive answer, at least for now. However that may be, to address them is the purpose of the pages that follow.

Note

This chapter was originally a discussion paper and is published here in its original form.

CHAPTER 3

THE ANTHROPOCENE AND GLOBAL WARMING

A Brief Update

JAN ZALASIEWICZ

How does the Anthropocene relate to global warming—and vice versa, of course? The simple answer is . . . not at all simply. But closely. Indeed—ever more closely, year by year, as we can see from reports of yet another record-breaking year for global high temperatures, following on from the broken records of the year before that, and of the year before that. Yet, for all that, global warming is not *quite* yet a major component of the Anthropocene, as we continue to enjoy global average temperatures that, so far, have stayed within the normal range for interglacial phases of the Ice Ages. Peak average global temperatures are probably still a fraction of a degree below the peak temperatures of the last interglacial phase, some 125,000 years ago, that is variously called the Ipswichian, the Eemian, the Sangamon, or, more simply, Marine Isotope Stage 5e, when sea level rose to be perhaps five meters higher than it is today.

So, as regards global average temperature and sea level, we are still in familiar territory—familiar, at least for the organisms of the current biosphere, the ancestors of which had to adapt to at least fifty major swings in climate and sea level over the past three million years, and a good deal more lesser fluctuations, many of which took place rapidly—within just decades. That the biosphere took these climate fluctuations in its stride is seen from the fossil record: the Quaternary Period, for all of its febrility of climate, suffered nothing unusual in the way of species extinctions until very late on in its course—some fifty thousand years ago, when the first of a wave of extinctions of the terrestrial megafauna (which were to roll through for much of the subsequent

fifty millennia) took place. What had become normal was the repeated fluc-
tuation of climate (and, as we were later to learn, of greenhouse gases such
as carbon dioxide and methane) between fixed limits throughout this time
span. It was a form of stability of the Earth System, albeit one with such pronounced fluctuations (growing slowly more pronounced through that time
span). The climate fluctuations, of course, did not begin with the beginning of
the Quaternary Period, but extended back for tens of millions of years earlier,
though generally in somewhat more subdued fashion—and even extended
back (though taking somewhat different character) into the warmer greenhouse world that then extended back into the Mesozoic Era, when dinosaurs
lived. Thus, climate change is normal—but it is the nature of its *pattern* that is
significant.

So, if climate is still within the interglacial norm, what has changed?

The changes that may be associated with the Anthropocene, considered
either in a geological context or in the context of Earth System science, where
the concept arose with Paul Crutzen's sudden improvisation of the term in
Mexico in 2000, are not simply detectable traces of human influence of the
environment. Such traces occur from at least the beginning of megafaunal
extinctions, some fifty thousand years ago (for it is likely that human hunting
was the cause of these), and carry on through into the Holocene Epoch, which
began 11,700 years ago, with further hunting and then the yet larger changes
caused by the development and spread of farming across the world. These
changes—and the archaeological record that they have left—may be said to
characterize the Holocene Epoch, rendering it distinct from the earlier inter-
glacials of the Quaternary.

However, throughout all of these early human impacts, and the development of the various human dynasties and kingdoms and nascent nation-
states, the Earth System was essentially stable. A good example of this is the
record of carbon dioxide concentrations in the Earth's atmosphere, as recorded
in air bubbles in polar ice. From a low of about 180 parts per million (ppm), it
rose slowly if irregularly through the next six millennia, reaching a level of
about 260 ppm about eleven thousand years ago. From there, it began to drop
almost imperceptibly, dropping perhaps five ppm by seven thousand years
ago—and then it began to rise, equally slowly, with minor dips and peaks,
until about 1800 CE, by which time it had reached about 280 ppm. This slight
rise *might* (it is controversial) have been due to a drip-feed of carbon dioxide
into the air from the effects of early farming. Whether natural or human-
prompted, though, this slight rise in atmospheric carbon dioxide levels may
have been enough to stave off the beginning of the next glaciation on Earth.
Either way, it provided stability and a continuation of Holocene conditions of
climate and sea level, and so was a key factor in the continued development of
human civilization.

It is the event of the subsequent couple of centuries that constitutes the extraordinary change to the Earth System. Carbon dioxide levels began to rise noticeably, as the Industrial Revolution gathered speed and moved across the world. By the mid-twentieth century they had reached about 310 ppm, and so thirty ppm above normal interglacial levels. Then, with the "Great Acceleration" of the postwar boom in population, energy use, industrialization, and globalization, they began to rise steeply, gaining almost another hundred ppm (the current level is 406 ppm, rising at about three ppm each year). Thus, on a graph, there is a change from Holocene carbon dioxide levels that plot essentially as a horizontal limb to a situation after 1950 where they plot as an approximately vertical (and vertically rising) limb that may be regarded as characterizing the Anthropocene. The rate of rise of this gas is now some 120 times faster than during the (geologically rapid) transition from the Pleistocene to the Holocene, and may be more rapid than at any time in the past sixty or more million years.

This vertical limb of the carbon dioxide graph coincides with an even bigger perturbation of the nitrogen and phosphorus cycles (roughly doubling the amount of reactive nitrogen and phosphorus at the Earth's surface, the former by fixation from the atmosphere using the Haber-Bosch process, and the latter by excavation from limited resources within rock strata). It also coincides with the beginning of the production and global dispersal of metals such as aluminum (with about five hundred million tons now produced), of materials such as plastic (with about five billion tons now produced), and of novel "rocks" such as concrete (with about five hundred billion tons now produced, equivalent to about one kilo for every square meter of the Earth's surface, land and sea). As regards the biosphere, the last seventy years have seen a steep upturn in the rate of extinctions and of transglobal species invasions. And a further development of the biosphere has been the development of the technosphere (a concept developed by Peter Haff), which is now a key player in driving Earth System change—and which is itself evolving rapidly. This together amounts to a global change that has left clear imprints in recent strata globally, and hence may be regarded as representing the Anthropocene—at least within a geological meaning. The Anthropocene as a stratigraphic term continues to be informal, though a case is being prepared (it will take at least a few more years) for its formalization on the Geological Time Scale.

What about climate, though? The change to greenhouse gases in the atmosphere (not just carbon dioxide—the levels of methane and nitrous oxide have also climbed steeply) has clearly begun to change the Earth's radiative balance. Most of the retained heat so far has gone into warming the oceans, which have a greater heat capacity than does the atmosphere—hence there is a lag of decades and centuries before a change in atmospheric composition leads to a change in global average temperature (though this process is beginning), while this process is modulated by various feedbacks to the Earth System, both fast and slow. There is a yet greater lag before the warming leads to significant sea-level rise, as the heat is transferred into melting the vast polar ice masses.

Atmospheric carbon dioxide levels are now similar to those of the Pliocene Epoch some three million years ago (which was a couple of degrees warmer than today, with sea levels likely in excess of ten meters higher than today)—and rising rapidly toward levels seen by earlier and yet warmer epochs. This—both for humans and for the rest of the biosphere—is taking us out of the envelope of the "dynamic stability" of the Ice Age climate fluctuations. And as Earth's geological history tells us that climate is probably *the* primary driver of surface geological and biophysical processes on Earth, the changes to come may take our planet into uncharted and uncomfortable and—not to put too fine a point on it—downright dangerous territory. This is, of course, unless there is a concerted global effort to stabilize and then reduce greenhouse gas concentrations. There is little sign of this happening, and the buildup of these gases is currently accelerating rather than slowing. It will likely be a bumpy ride ahead.

CHAPTER 4

THE EXTRAORDINARY STRATA OF THE ANTHROPOCENE

JAN ZALASIEWICZ

G eologists tend not to think about history, much. Perhaps not even about Earth history. The theme is—one hastens to add—always somewhere there in the background. A planet that bears countless and *detailed* imprints of what has happened to it—to a greater extent, almost certainly, than any other body in the solar system—is fertile ground for such thought. But the sheer scale and complexity of the narrative are a considerable barrier to speculation, whether idle or embarked upon with serious intent. There is simply too much of it. That extraordinarily insightful geologist and paleontologist Richard Fortey once described geology as *everything* that is present or happening on, within, and above the Earth, now and over the past four and a half billion years.[1]

To that substantial compendium of Earthly time, events, and matter, one might now add the Earth's future, something like another five billion years, to the "to do" list—and, just for good measure, all the other planetary bodies whirling around the Sun, which we now know have their own *quite* individual geologies. Then, there are all of the planets and moons orbiting other stars in this galaxy, and that we are just beginning to glimpse; they are about a billion strong, at a conservative estimate. (We may leave from consideration, for the time being, the planetary bodies that will be present in the one hundred billion or so other galaxies within the known universe: it will be some time before our telescopes can be effectively trained on those.) All in all, it is enough to make one's head spin, and a head thus revolving is not ideally placed to undertake cool and considered assessment of planetary history, let alone of the place of any particular, strange, and novel event within it, such as the extraordinary set of processes that we humans have precipitated. So, what is to be done?—and where does one start?

The starting point—to revert back just to this small planet—has traditionally been fragments: small shards of the greater whole that have attracted the attention of some passing geologist, using that last word extremely loosely. It is rather like approaching one of Seurat's pointillist paintings, in a dark room, with one of those fine-beam pen flashlights illuminating just a few of the dots of paint at one time. Examine, then, those one or two splashes of paint—in great detail. Become absorbed by them, to the extent that they become a central pursuit in one's life. Slowly develop a methodology to study them, and a grammar to discuss them (in reality, to consider just a few aspects of the paint specks, which then—by definition—become of critical importance). Suddenly, the flashlight is aimed elsewhere, and a few other paint spots come into view. These are then minutely examined in turn—particularly as regards those chosen aspects, which have now become channels for directed thought. It will be a long time before any sensibly coherent parts of the panorama come into view. It is clear that the paint specks are part of a greater whole, but also that, gazed at intently enough, they have a fascination of their own, and particular aspects—their shape, color, sheen, surface roughness—can become a basis of classification.

The Earth, of course, is somewhat larger than a Seurat canvas. Those dots of information are arranged three-dimensionally. They have not stayed in their relative positions since being put into place, but have been torn apart and reassembled through the unspeakably long tunnel of the fourth dimension. Many have just disappeared. Quite a few have changed color. And they are still being added to.

Nevertheless, focusing on such particulars can make the whole exercise *manageable*. On the larger canvas of the Earth, therefore, some geologists may focus on the chemistry of the crystals in a basalt, others on certain long-dead types of plankton, others still on the patterns that sand grains take up when sped along a sea floor by a current, and yet others on the patterns of space that develop *between* those sand grains (an exercise that is not just the inverse of looking at the sand grains themselves). And so it goes on; for much of the science, the devil really is in the detail (the angel is somewhere there too, of course, to be even-handed about this).

The synthesis, to try to pick out some more general truth from the mass of detail, typically emerges once sufficient detail has been collected to show patterns emerging from what seemed initially to be bewilderingly chaotic. The major steps in this process are now classics of geological history. William Smith,[2] for instance, in the early nineteenth century, demonstrated that rocks and fossils made sense. He showed that the broad patterns of their arrangement—their *superposition*—could be followed from region to region, from country to country. Coal, he said, only occurred in certain layers: if you are a speculator, and sink a mineshaft too high or too low amid the endless strata, you will end up losing your shirt (as many did, before his insights were appreciated).

The most notably coal-rich succession became divided off as the Carboniferous System. This is emphatically a rock unit. Within its several-kilometer thickness, fortunes were made and lost, and miners lived and worked—and all too often died too, as the overstressed rocks exploded in rock bursts around them. To better make money—and to attempt to avoid tragedy—generations of geologists mapped out the exquisitely complex arrangement of the hundreds of individual coal layers within that mass of rock, in ever-greater detail. Even after the coal is exhausted, that knowledge is still needed to predict where the overlying ground may or may not still subside to damage houses, and where sulphurous, iron-rich waters, percolating through old mine-workings, can make their way to the surface to pollute streams and rivers.

This is detailed, practical, absorbing work—a jigsaw puzzle that can take many lifetimes to (only partially) solve. It is work of the here and now. In analyzing the endless detail, the realization that those Carboniferous rocks are a memory of something else entirely—of a world of primeval swamp forests, with amphibians and giant dragonflies, but without flowers, or birds, or mammals—is generally very much at the back of a working mind. Nevertheless, this distant past, now separated off as a segment of time some sixty million years long, which ended some three hundred million years ago, is the basis for the Carboniferous Period. This far-distant period of time may be reconstructed in imagination—but never again touched, or seen, or experienced—from the emphatically tangible rocks of the Carboniferous System.

And so it has developed, this compartmentalization of Earth history. The Cretaceous, to take another example, is separated out on the basis of a kilometer-thick stratum of almost pure chalk (now known to be virtually worldwide), which marked one of the ultimate greenhouse worlds of substantially drowned continents. The Permian was found almost by accident when that grand man of Victorian geology, Sir Roderick Murchison, traveled to deepest Russia.[3] There, he was actually following the rocks of the Silurian and Devonian systems, which he had been instrumental in carving out of the rocks of Wales and southwest England, to see if they had global significance (they did). But while there, in the rock strata around the city of Perm, he noticed strata with fossils in them that were different from those of the underlying and familiar Carboniferous strata—and also very different from those of the overlying Triassic System, by then well known from Europe. By a quirk of ancient geography, ancient Perm had been covered by a shallow sea, teeming with life, while Britain and Europe were a virtually lifeless desert—and so these Russian rocks revealed an entirely new dynasty of life and of Earth's history: the Permian Period, of a long-gone time, the happenings of which may be inferred from the solid rock strata of the Permian System.

The history business has loomed larger within geology in the last few decades, true—not least because enough pointillist-style evidence has accumulated to make sensible inroads into such large questions as the journeys that continents have taken, or the manner in which Earth's climate has changed. Nevertheless, when the Anthropocene was born—in a practical sense, at least—with Paul Crutzen's inspired improvisation at a conference in Mexico just fifteen years ago, the usual procedures were turned upside down.

This new epoch (using the term *epoch* provisionally, for neither formal status nor hierarchical level has been decided) burst upon the scene as a time unit, quite fully formed. It was explicitly a geological time interval, the logic proposed being that the Holocene had ended. And yet, the evidence foremost in mind was of the most insubstantial kind: a change in the air, indeed. And the particular change with which Paul Crutzen's name is indelibly associated—the thinning of circumpolar stratospheric ozone because of tiny amounts of highly reactive human-made chemicals, chlorofluorocarbons—is almost perversely countergeological. While other changes of atmospheric chemistry, such as variations in methane and carbon dioxide levels, can leave an imprint in rock strata—if you look hard enough and devise sufficiently ingenious analyses—this particular change leaves no stratigraphic trace that I am aware of. It is a ghost—albeit one that is convincingly driving significant change in a planetary machine.

Small wonder that it took a while for the geological community—and in particular the formal stratigraphic community (i.e., the one that has the responsibility for maintaining the Geological Time Scale)—to catch on. The idea of the epoch was out there, first among the Earth System science community, monitoring planetary change in real time, and then spreading widely, among the sciences more generally, and among the arts and humanities too.

Only toward the end of the decade was the concept considered as a potential stratigraphic unit, by the Stratigraphy Commission of the Geological Society of London—a national body with no power to change the Geological Time Scale, but nevertheless one that can provide informed comment and opinion. And, on the strength of that opinion (a cautious one, that the Anthropocene "may have merit" geologically), there came the invitation to form the Anthropocene Working Group of the Subcommission on Quaternary Research, a component body of the International Commission on Stratigraphy (the decision-making body that oversees the Geological Time Scale, more technically known as the International Chronostratigraphic Chart)—which itself is under the aegis of the International Union of Geological Sciences, which must ratify any decision.

The process of analyzing the Anthropocene formally is certainly novel, in many respects. The term emerged from an evidence base that was largely observational, of global conditions of the present and recent past. It was not—as is the case with all of the other units of the Geological Time Scale—a synthesis of stratal patterns that became evident after long years of study of the rocks. So, one of the first tasks is to see whether there is, in effect, a stratal unit on Earth that may be systematically recognized and assigned, as a material body, to the Anthropocene Epoch. In the parlance used, this material or "time-rock" unit, parallel to the "time" unit, would be termed an Anthropocene Series.

When considering this material record, it becomes necessary to reconsider one's notion of what geology *is*. For instance, a field geologist mapping the geology of a region would certainly consider the ancient strata below ground as geology. Then, above that, there may be much younger glacial deposits such as boulder clay. Those, too, are geology and are mapped as geological units (they may be the focus of interest of geographical studies, too). Above that, and present almost everywhere, is the present-day soil—and that by contrast is very rarely seen as the province of geology, or shown on a geological map, even though it is very emphatically part of the Earth System and these days considered as a key part of the "critical zone" of this planet (curiously, fossil soils—or "palaeosols"—are considered as geology).[4]

Then, there are those rock and soil deposits transported and reworked by humans around cities, motorway, and railway embankments, in quarries and landfill sites. Here there is an even more curious contrast in geological practice. If the material forms part of waste or rubble layers, then it is typically mapped as strata of "Artificial Ground" on a geological map—and likewise the empty spaces left by the extraction of raw materials such as brick clay and sand and gravel may be mapped as "Worked Ground," and have their own symbol on geological maps (indeed these deposits dominate the surface geology of most urban areas). However, when these materials are in use, in the fabric of homes and factories, roads and railways and airport runways, they are typically seen as outside the realm of geology (though they will reenter it as the buildings are demolished and the road material is ripped up, and these wastes are spread over the ground or dumped within landfill sites).

These are differences only of working practice—and disciplinary territoriality. In reality, the materials, whether within functional human-made constructions or as discarded rubble, represent a new and substantial subsystem within the rock cycle, which, as a large part of the physical structure of Peter Haff's technosphere,[5] is reworking rocks and minerals in new patterns. It is now *very* large in scale. The human-made new rock par excellence is concrete, of which some half a trillion tons have now been made, by far the greater part since the "Great Acceleration" of the mid-twentieth century,[6] and more than half of this (given the current acceleration of the Great Acceleration) in the last twenty years. Most of this new rock is within buildings, dams, roads, embankments (and so technically outside geology as traditionally understood). Within a century, though, most of these constructions are likely to be abandoned or demolished, and so this rock will then be traveling on, somewhere within the planetary rock cycle. Thinking through the Anthropocene necessitates such rethinking and recategorization. Arguably, it is getting the science of geology closer to seeing things as they really are.

There is a similar tension in interpretation when considering minerals. A mineral is, in the standard definition, a solid of more or less fixed chemical composition. According to the International Mineralogical Association, the body that vets and approves new minerals (with about fifty being added each year to the total of some forty-five hundred that have been recognized), a mineral is by definition also natural and not human-made. This is sensible in one way, in that it keeps a pristine view of perhaps the most fundamental set of building blocks of this planet. From another view, though, it masks the scale of the mineralogical reconstruction of the Earth's surface by humans, who have produced pure metals (something that, with a few exceptions such as gold, does not characterize the natural world) in multimillion-ton amounts, who have produced large amounts of other novelties such as tungsten carbide (which forms the ball of your ball-point pen), boron nitride (which is harder than diamond), and many thousands more that continually emerge from the laboratories of the materials scientists. These new materials are minerals in everything but formal definition, and their sudden release onto our planet's surface marks, almost certainly, the single largest change to the Earth's mineral storehouse since the early Proterozoic Era, some two and a half billion years ago, when, in what is now known as the "Great Oxygenation Event," the world changed color, going from the greys and greens of a chemically reducing world to reds, oranges, and browns, as a swathe of oxide and hydroxide minerals appeared.[7]

Even more striking are the "mineraloids," those substances where the chemical composition changes within set limits. The human-made varieties include glass of different kinds, far more pure and transparent than the volcanic glasses obsidian and pitchstone—and that symbol of the modern world, plastic. Plastic in its various forms is essentially a post–World War II phenomenon, now being produced in hundreds of millions of tons per year and—hard-wearing, decay-resistant, light, and easily transportable—is finding its way to the oceans. There, particularly in the almost invisible "microplastic" form (the thousands of fibres, say, that come from your fleece after it has been through a wash cycle), it is transported by currents and within animal bodies and reaches the most distant beaches and the deepest sea floors.[8] To regard these phenomena as part of the fabric of geology is certainly novel—but is defensible, in that it does appear to reflect a genuine departure within the Earth System. However, once accepted as a stratigraphic marker, new materials such as plastics, glass, and ceramics may be straightforward to use—especially the first of these, which is a widely and easily recognizable time marker.

Other aspects of the Anthropocene are more difficult, simply in practical geological terms. The new minerals and rocks of the Anthropocene are components of strata, and one of the prime tasks of a geologist is to fix the location of strata within three-dimensional space, where they intersect with other three-dimensional surfaces (the topographic surface of the landscape, for instance) to produce geological outcrops. Learning this particular skill is often one of the most challenging parts of the life of a student of geology, as it needs several mental images to be manipulated simultaneously inside one's head— and it is often thoroughly counterintuitive. The simplest of horizontal strata may show up as terribly complicated outcrops, if cut through by a network of streams and rivers, for instance. Conversely, severely structurally deformed strata commonly have very simple outcrops (a stratum pushed by mountain-building forces to a vertical position will appear as just a pair of straight, parallel lines, no matter how rugged is the topography that it transects).

Now, Anthropocene strata can be very simple and predictable (the layers of polluted snow on an ice cap, for instance—material that even conventionally is regarded as rock these days. But go to the "urban strata" of a major city and the geometrical complexity can seem limitless, almost fractal.[9] An ancient city such as Rome will have components that are thousands of years old, interlaced with filigrees of material from successive ages and periods of history, now in turn interlaced with new buildings and foundations, roadways, electric pipes, cables, modern sewage systems, fiber-optic cables, and so on. This is the kind of terrain that archaeologists are skilled at analyzing and depicting on plans, but for geologists—who typically have to move rather more quickly over larger areas of terrain, simplifying as need be as they go—it is more challenging to separate out, say, an "Anthropocene" unit (if defined with a mid-twentieth-century boundary, for instance) from a "pre-Anthropocene" one. Simplification of such a three- and four-dimensional patchwork is not straightforward.

One can think of a few partial analogues in nonhuman geology—cave deposits, for instance, and some of the complex structures that form as strata fluidize underground and are injected into both older and younger layers of rock. The challenge for the Anthropocene is novel in scale and extent—but may be resolved, at least partly.[10]

Curiously, there may be less of an issue in considering the extreme youth of the Anthropocene, and in locating its boundary outside those crazily mixed-up urban areas. Most geological epochs, it is true, are much longer than the Anthropocene, almost no matter how it is defined. The five formal epochs of the "Tertiary" period average a little over twelve million years in duration each. But the revolutionary change in duration has already been shown by the Holocene, at a little under twelve thousand years long. This is a full three orders of magnitude shorter than that "Tertiary" average—and the Anthropocene (if defined to start with the "Great Acceleration") is just, so far, some two orders of magnitude shorter than the Holocene. The big time jump in classification has already happened, much earlier in the history of geological time classification (the Holocene was first proposed at the third International Geological Congress in 1885, though its boundary was formally ratified only in 2008),[11] and for eminently practical reasons.

Holocene strata may, for the most part, be wet and sloppy (and smelly, too, quite often) and most unlike the public understanding of a "rock." But they cover a good deal of both the landscape and the sea floor of our planet, being the three-dimensional unit that underlies the two-dimensional surfaces of river floodplains, coastal plains, delta tops, and so on. Take the English Fenland, for instance—some four thousand square kilometers of flat agricultural landscape, a segment of which was one of my first mapping assignments after I joined the British Geological Survey as a rookie field geologist (a bemused one, for I had mostly studied and trained on the eminently hard and ancient rocks of Wales). The Fenland, beneath its surface, was a wedge of muds and silts more than ten meters in thickness, all deposited after the retreat of the last ice age, and with an intricate and delicate internal structure reflecting changes in the position of land and sea, and reflecting too the life and death (almost literally) of spectacular tidal waterways.[12] The Holocene suddenly made the most perfect geological sense—especially as the realization came that the Fenland, huge by English standards, is tiny by comparison with the world-scale examples of Bangladesh, the Louisiana wetlands, the Yellow River delta, and so forth. A mere eleven millennia may be almost nothing at all in the immensity of Earth time, but when it's up close, it can encompass world-shaping forces—and similar logic may be applied to the Anthropocene.

However, a difference appears when the possible futures of these two epochs— the formal and the still informal one—are contrasted. The Holocene—like the Quaternary Period that it is a part of—is essentially defined on the basis of climate. It is defined as starting when the world warmed in emerging from the last glacial phase (just as the Quaternary is defined as starting when the world cooled from the last gasp of Tertiary warmth represented by the Pliocene Epoch). The Holocene is just one interglacial of many in the Quaternary (all the others being lumped together in the Pleistocene Epoch, which makes up almost all of the 2.6-million-year span of the period). In a normal state of affairs, the Holocene as a unit might be expected to terminate within some millennia, given considerations of its "natural length," when the ice would be expected to advance again and cover the currently temperate regions of the

Earth and sea level to fall by a hundred meters or so as the ice caps grow huge once more.[13]

But—and this is a large "but"—the Anthropocene departs from this state and expectation in a number of ways. First, of *all* the characters by which this emerging interval of geological time might be characterized, climate and also sea level are currently among the most trivial and insubstantial ones. True, temperature has risen by a little over a degree, globally on average, in the last century—and this is almost certainly (*quite* certainly, if one shakes off inbuilt scientific caution) due to the human-driven rise in greenhouse gas levels. And sea level has risen by about twenty centimeters over the same time, as the oceans have warmed and expanded, and ice has begun to melt and release water into the sea. But both global temperature and sea level remain within more or less normal interglacial levels. Indeed, temperature would have to rise by about another degree and sea level by more than ten meters to exceed levels seen at peak warmth and sea level of the last interglacial, some 125,000 years ago, when there was no significant human interference in the Earth System whatsoever. It is much more likely than not that these levels of temperature and sea level will be exceeded, later this century and in a few centuries respectively—but this has not happened yet.

It is in a whole variety of other parameters that the Anthropocene is expressed. Some have a long geological pedigree, and there is a sense of planetary déjà vu—such as a marked increase in species extinction and (especially) invasion rates, and changes in sedimentation rates, and changes in greenhouse gas levels, and associated changes in the carbon isotope composition of the global atmosphere. Others, though, include marked planetary novelty, as in the appearance of enormous city-structures that make up the urban strata, and of novel material such as plastics and aluminum and the extraordinary diverse and rapidly evolving technofossils into which they are shaped.[14] And, while some of these new geological constructions will currently only continue to be produced as long as humans persist—the urban strata and technofossils, for instance—in other ways, particularly in the restructuring of the biosphere, we have, in the most fundamental way, changed the course of Earth history.

As Antonio Stoppani realized back in the late nineteenth century (1873), humans are changing not just the present, but also the future.[15] The survivors of the current mass extinction event (like the climate event, in its early stages, but already palpable) and the myriad transplanted (and thriving) invasive species are the ancestors of the life of the future—and hence of the fossils that will be entombed within future strata too. Geologically, the Anthropocene is not just a passing phase—it is *already* a fundamental boundary that might (loosely, and involving a good deal of cross-disciplinary averaging) be assigned an epoch status.[16] Once the climate exceeds Quaternary norms and puts very many species lethally outside their comfort zones, and the mass extinction event unfolds and becomes the "sixth" great such event, as is likely in the next few centuries *even discounting the effects of climate change*,[17] the scale of global change becomes much more reflective of something like a period- or even era-scale boundary.

The nature of the boundary is worth considering. As has been widely noted, there are now many Anthropocenes out there, used for different purposes along different lines of logic in different disciplines. So, one has to draw back the Anthropocene perhaps not quite to its original definition (as a geological time unit, broadly proposed by the atmospheric chemist Paul Crutzen—as well as the lake biologist Eugene Stoermer, who had independently invented the same term a little earlier—within an overall Earth System science context) but to the one that is now the subject of formal inquiry (as a stratigraphical time term that must function both as a unit of Earth history, i.e., a geochronological unit, and as a parallel time-defined material stratal unit, i.e., a chronostratigraphical unit).[18]

The question of the boundary has aroused a good deal of comment, not least as regards the protracted and progressive nature of significant human influence on the Earth, ranging from the beginnings of the extinctions of the terrestrial megafauna, starting as long as fifty thousand years ago as the culturally modern humans sharpened their cooperative hunting skills, through the development and spread of agriculture beginning some ten thousand years ago, to the origin and spread of urbanization a little later. As farms spread and cities slowly grew, a time-transgressive human-altered surface layer resulted, which has been termed the archaeosphere; it has been suggested that this complex, time-transgressive imprint of human influence might equate with the Anthropocene, as its most easily visible reflection.[19] This would be a parallel of archaeological time terms such as Palaeolithic, Bronze Age, and so on, which are of different ages in different regions, reflecting the cultural state of the local human populations (and would also in part parallel the suggestion of the "Palaeoanthropocene" by Foley et al.).[20]

However, this would run counter to a peculiarity of geological time, which is that, at its heart, it is simply time—albeit in very large amounts. A time boundary (whether geochronological or chronostratigraphical) is just an interface in time, of no duration whatsoever—it is less than an instant—between one interval of time (which may be millions of years long) and another. It is inherently synchronous within the domain across which it operates, which is that of the home planet. This "pure" use of time intervals has been drilled into successive cohorts of geological students as part of what used to be called "the holy trinity" of rocks, fossils, and time—which are tightly interrelated, but are separate, and *need* to be kept separate in analysis: broadly, one might say, to keep Earth scientists honest. For rocks can be wildly time-transgressive: think of a stratal unit of beach sand, slowly migrating across wide areas of a landscape as sea level rises over thousands or millions of years—the resultant rock unit will be of different ages in different places. Fossils, now, are often used as hugely effective *proxies* for time—they are still the default tool for dating sedimentary rocks. But they are not the *same* as time; it takes perceptible amounts of time for animal or plant species to evolve from some ancestor—and then yet further time to migrate across whatever part of the world they will eventually come to occupy. So keeping these concepts separate in geology is utterly indispensible—especially as new time proxies, such as chemical and magnetic properties of rocks, have appeared to join the "holy trinity."

The lengths taken by stratigraphers to preserve this separation in boundary definition can appear ludicrous, especially when time overall in geology is measured in millions of years. The end of the Cretaceous Period (and indeed of the whole Mesozoic Era that contains it) and the beginning of the Paleogene Period (and of our current Cenozoic Era) have been placed at the very moment of impact of the asteroid that impacted on Mexico (because it was recognized that the iridium-rich meteoritic debris layer at the "golden spike" global reference locality in Tunisia would have taken some hours, or days, to traverse the intervening distance and fall to the ground).[21] The boundary between the Silurian and Devonian periods has been "placed" in the middle of a "turbidite" layer—effectively a turbulent submarine "avalanche" deposit,[22] which in total probably represents a sediment deposition event lasting less than an hour. An exaggerated precision? Perhaps—but there are times when such precision can be employed in geology. Volcanologists can track "entrachrons" in pyroclastic flow deposits—successive stages in the evolution of these catastrophic and almost (but not quite) geologically "instantaneous" deposits,[23] for instance.

That does not mean that, in normal practice, the carefully tabulated units of the Geological Time Scale can be followed so precisely across the world. The neat colored units of this chart might suggest an equivalent neatness and order in geologically documented nature—but, usually, alas, it isn't so. Take one of the more effective boundaries in the ancient rock record as an example: the boundary between the Ordovician and Silurian periods (or in "time-rock" chronostratigraphic terms, the Ordovician and Silurian systems).[24] This boundary postdates most of the main, tumultuous events that separated the Ordovician and Silurian worlds—a glaciation, enormous sea-level changes, two large pulses of mass extinction, an ocean anoxic event[25]—and places them at the appearance of a particular, distinctive fossil graptolite (a type of extinct marine plankton) species, the "golden spike" being set where this has been documented in the shales of a steep and rather scrambly stream gully named Dob's Linn, in southern Scotland.

This fossil appearance seems to represent the rapid spread of this species across part of the oceanic realm—though whether "rapid" means centuries, millennia, or tens of millennia cannot yet be ascertained. In some parts of that oceanic realm, this species is suspected (but not proven) to have arrived substantially later—by some fraction of a million years—than in that part of the ocean that is now southern Scotland, and the boundary there must be "guesstimated" using other fossil species as guides to time.[26] Once outside the deep sea realm, in provinces that represent the shallow seas of those times, the task becomes yet harder, as the graptolites did not venture much into shallow water, and are rare or absent in these strata—here again, other fossils or other types of evidence need be used to try to constrain the boundary. In strata representing the land surface (lakes, rivers, and the like) of those Ordovician-Silurian times, there are no graptolites at all—and precious few fossils of any kind—and so locating the boundary precisely is currently effectively impossible, and locating it even very imprecisely is difficult. That simply represents current reality in stratigraphy. The kind of error bars involved are rarely quantified, but in these ancient rocks a "precisely located" boundary in strata may have error bars of quite a few thousand years, while the more conjecturally placed boundaries may be plus or minus some millions of years. Some stratal successions have age uncertainties exceeding a hundred million years.[27]

Such error bars in locating boundaries become smaller as the strata become younger, but even in fossiliferous strata of the current Quaternary Period, uncertainties can be many thousands of years, and correlation of the Holocene boundary "golden spike" section for the Holocene (based on chemical changes in an ice core drilled out of Greenland) involves uncertainties—even with five other carefully chosen and closely studied "auxiliary" global reference sections[28]—of up to a couple of centuries.

So, how does a potential Anthropocene boundary measure up to this imperfect reality? The "measurability" (or, in stratigraphical terms, "correlation potential") of the boundary is here the foremost aspect of a geological unit that comes to a geologist's mind: thus, the question "how well can it be defined and traced?" may well come before "how globally important is it?" and "when was the first sign of influence of some major new factor in the Earth system?" And, in terms of the definition of a "stratigraphical Anthropocene," we are dealing with change to the Earth System rather than a change in the extent to which we are recognizing human influence. This kind of definition is seen as a planet-centered, rather than human-centered, phenomenon. So, it is important that the planetary system is *recognizably* changing, and it just happens to be the activities of the human species that are currently the main perturbing force. The Anthropocene would remain just as important geologically, because of the scale of the planetary (and hence stratal) effects, if it had some other cause. Indeed, the concept would then probably be rather easier for humans to comprehend and to react to. Given that, it is important that it is optimally recognizable from the physical evidence. So far, the Anthropocene seems to be effectively distinct—especially if it is "optimally" recognized as the stratal equivalent of the "Great Acceleration" and all that followed.

The mid-twentieth century seems so far to be the best candidate for the boundary, not so much because of its historical importance, but because events around that time left traces that can be detected pretty well everywhere around the globe: the radionuclides from bomb blasts,[29] the fly ash particles,[30] effectively a thin, worldwide—and supremely fossilizeable—dusting of fossilized smoke around the world, the plastic fragments on land and in the sea,[31] the changes in carbon chemistry associated with fossil fuel burning, present in the tissues, bones, and shells of all living things, and plethora of abundant new technofossils (those ball-point pens, for instance). From polar ice layers to tropical swamps, from beach sands to abyssal oozes, there are detectable traces that, with decadal and in places even annual precision, signal such a boundary of pre-Anthropocene below and Anthropocene above: as so defined, one must remember. These are human constructs, albeit ones that try to reflect stratal and geohistorical reality. "The rocks don't lie," says Bryan Lovell, former president of the Geological Society and one who understands the human meaning of "abstract" palaeoclimate studies. He is right, even if rocks don't always speak the easiest of languages. Such an Anthropocene has geological reality, and durability too; that does not necessarily mean that it will or even *should* be formalized in the Geological Time Scale (that will depend just as much on the perceived usefulness of formalization, a knottier problem)—but it does mean that it isn't just a mirage conjured from human overeagerness to see human influence on an ancient planet. More's the pity, perhaps.

The last few pages have, if nothing else, shown something of the stratigrapher's focus—some might say obsession—with details of definition and recognition, and with strata, of course. The wider Anthropocene, as well as its possible significance to human society and culture, or significance as regards, say, the perceived relations between humanity and nature,[32] or as a rallying call to environmental protection (or conversely as counsel of despair), does not loom large here. Indeed, these larger issues are often quite forgotten, as one wrestles with questions such as whether industrial fly ash particles can be consistently distinguished from smoke particles released by natural forest fires, or how long it takes plutonium particles to travel from the atmosphere after a bomb blast to be eventually incorporated into deep ocean muds.[33] This approach may be blinkered, but also seems best placed to exploit the truthfulness of rocks.

The wider use of the Anthropocene is hence beyond the narrative of this essay, or the competence of the writer. But one of the wider inferences drawn seems at odds with standard patterns of geological thought. Thus, the suggestion that, if the deleterious changes associated with the Anthropocene—global warming, biosphere degradation, and so on—are considered as arising from the collective actions of humans, then this hides the fact that some groups of humans (rich ones, in developed countries) drive the process more acutely than others (impoverished communities in developing countries).

This seems to rather oversimplify the logic—perhaps on the assumption that arguments should be oversimplified in geological analysis generally, because of the huge distances of time and space involved. Such an assumption, it seems to me, does not do justice to a geological appreciation of the relation between the complex reality of what happened in the past and the generalizations that need to be made in making interpretations from the patchy evidence available on the ground.

For instance one might say that incoming asteroids cause mass extinctions. Well, there are asteroids and asteroids. For instance, the town of Nördlingen in Germany is sited in a fourteen-million-year-old impact crater. It is a *large* crater—almost twenty-five kilometers across—and the impact would have been totally devastating to the region. Yet, it did not precipitate a mass extinction episode, and the impact is now placed comfortably inside an epoch (the Miocene) rather than providing a catastrophically generated boundary to it.

The end-Cretaceous asteroid impact does represent such a geological boundary. It was a larger body that made a larger crater, true—but it also impacted upon different bedrock geology (limestones including sulphate deposits) that would have vaporized and produced environmentally damaging gases. And it impacted at a time of monstrous, successive volcanic eruptions in the Deccan region of India, which likely was already affecting the fabric of the global biosphere.[34] So, an interplay of these (and other) factors likely contributed to the scale of ecological collapse.

Even with these eruptions—and others of their kind that seem to have triggered other mass extinction events (at the "Great Dying" of the Permian-Triassic boundary, for example)—it is not simply a case of volcanoes as blanket cause. Some volcanic eruptions are (much) bigger than others—and their spacing through time is a major factor in determining how quickly, for instance, greenhouse gas levels might rise, or how much recovery time might have been available between eruptions. The devil is truly in the details (while any remaining angels, here, are keeping at a safe distance).

And so, in trying to understand the processes driving Anthropocene change, it is clear that one must understand the behavior and effect of the human drivers. But—just as with the asteroids or volcanoes—this should not mean that the driving force should necessarily be ascribed to "humanity" per se. It seems clear that there will be large differences in geological motive force (if one puts it this way) in scale and kind between different groups of humans categorized in different kinds of ways and with different levels of detail and perspective. Making comparative analyses would be both useful and fascinating, but the approach should be *at least* as subtle and nuanced as one allows for volcanoes and asteroids. In reality, humans as agents of planetary change are far more complicated and less predictable than are these more ancient forces. Understanding the Anthropocene will not be a small task—particularly when one has the privilege (we should consider it in no other way, because we have no alternative option) of living through the infancy of this new and extraordinary phase of our planet's history.

Notes

This chapter was originally published in Jan Zalasiewicz, "The Extraordinary Strata of the Anthropocene," in *Environmental Humanities: Voices from the Anthropocene*, ed. Serpil Oppermann and Serenella Iovino (London: Rowman and Littlefield International, 2016), 115–132. Reprinted with permission.

1. Richard Fortey, *Life: A Natural History of the First Four Billion Years of Life on Earth* (New York: Vintage, 1999).
2. William Smith was that rare thing, an early-nineteenth-century pioneer of geology who was not "of independent means," but who had to work for a living, as a surveyor. Nevertheless, he is generally credited with inventing the disciplines of lithostratigraphy and biostratigraphy, and he single-handedly produced the world's first national geological map, which covered most of Great Britain. Done by horse and carriage and on foot, it remains an almost unbelievable feat.
3. Michael Collie and John Diemer, eds., *Murchison's Wanderings in Russia* (British Geological Survey Occasional Publication 2, 2004).
4. Daniel de B. Richter and Sharon A. Billings, "'One Physical System': Tansley's Ecosystem as Earth's Critical Zone," *New Phytologist* (March 2015): https://doi.org/10.1111/nph.13338.
5. Peter Haff, "Technology as a Global Phenomenon: Implications for Human Well-Being," *Geological Society, London, Special Publications* 395 (2014): 301–309.
6. Will Steffen, Paul J. Crutzen, and John R. McNeill, "The Anthropocene: Are Humans Now Overwhelming the Great Forces of Nature," *AMBIO: A Journal of the Human Environment* 36, no. 8 (2007): 614–621; Will Steffen, Wendy Broadgate, Lisa Deutsch, Owen Gaffney, and Cornelia Ludwig, "The Trajectory of the Anthropocene: The Great Acceleration," *Anthropocene Review* 2, no. 1 (April 2015): 81–98.
7. Jan Zalasiewicz, Ryszard Kryza, and Mark Williams, "The Mineral Signature of the Anthropocene," in *A Stratigraphical Basis for the Anthropocene?*, ed. Colin N. Waters, Jan A. Zalasiewicz, Mark Williams, Michael A. Ellis, and Andrea M. Snelling, Special Publication 395 (London: Geological Society, 2014), 109–117; Robert M. Hazen, Edward S. Grew, Marcus J. Origlieri, and Robert T. Downs, "On the Mineralogy of the 'Anthropocene Epoch,'" *American Mineralogist* 102 (2017): 595–611; cf. Robert M. Hazen, Dominic Papineau, Wouter Bleeker, Robert T. Downs, John M. Ferry, Timothy J. McCoy, Dimitri A. Sverjensky, and Hexiong Yang, "Mineral Evolution," *American Mineralogist* 93 (2008): 1693–1720.
8. Juliana Ivar do Sul and Monica F. Costa, "The Present and Future of Microplastic Pollution in the Marine Environment," *Environmental Pollution* 185 (2014): 352–364;

Jan Zalasiewicz, Colin N. Waters, Juliana A. Ivar do Sul, Patricia L. Corcoran, et al., "The Geological Cycle of Plastics and Their Use as a Stratigraphic Indicator of the Anthropocene," *Anthropocene* 13 (2016): 4–17.

9. Ford et al. discuss the geological practicalities of dealing with urban strata in Jon R. Ford, Simon J. Price, Anthony C. Cooper, and Colin N. Waters, "An Assessment of Lithostratigraphy for Anthropogenic Deposits," *Geological Society, London, Special Publications* 395 (2014): 55–89.

10. Ricky L. Terrington, Érika C. N. Silva, Colin N. Waters, H. Smith, and Stephen Thorpe, "Quantifying Anthropogenic Modification of the Shallow Geosphere in Central London, UK," *Geomorphology* 319 (2018): 15–34.

11. Mike Walker, Sigfus Johnsen, Sune Olander Rasmussen, Trevor Popp, et al., "Formal Definition and Dating of the GSSP (Global Stratotype Section and Point) for the Base of the Holocene Using the Greenland NGRIP Ice Core, and Selected Auxiliary Records," *Journal of Quaternary Science* 24 (2009): 3–17.

12. Dinah M. Smith, Jan Zalasiewicz, Mark Williams, Ian Wilkinson, Martin Redding, and Crane Begg, "Holocene Drainage of the English Fenland: Roddons and Their Environmental Significance," *Proceedings of the Geologists' Association* 121 (2010): 256–269; Dinah M. Smith, Jan Zalasiewicz, Mark Williams, Ian P. Wilkinson, James C. Scarborough, Mark Knight, Carl Sayer, Martin Redding, and Steven G. Moreton, "The Anatomy of a Fenland Roddon: Sedimentation and Environmental Change in a Lowland Holocene Tidal Creek Environment," *Proceedings of the Yorkshire Geological Society* 59 (2012): 145–159.

13. Chornis P. Tzedakis, E. W. Wolff, L. C. Skinner, V. Brovkin, D. A. Hodell, J. F. McManus, and D. Raynaud, "Can We Predict the Duration of an Interglacial?," *Climate of the Past* 8, no. 5 (September 2012): 1473–1485.

14. Jan Zalasiewicz, Mark Williams, Colin N. Waters, Anthony D. Barnosky, and Peter Haff, "The Technofossil Record of Humans," *Anthropocene Review* 1 (2014): 34–43; Jan Zalasiewicz, Mark Williams, Colin N. Waters, Anthony D. Barnosky, et al., "Scale and Diversity of the Physical Technosphere: A Geological Perpsective," *Anthropocene Review* 4, no. 1 (2017): 9–22.

15. Antonio Stoppani, *Corso di geologia*, vol. 2., ed. G. Bernardoni and G. Brigola (Milan, 1873).

16. See the discussion of biosphere transformation in Mark Williams, Jan Zalasiewicz, Peter K. Haff, Christian Schwägerl, Anthony D. Barnosky, and Erle C. Ellis, "The Anthropocene Biosphere," *Anthropocene Review* 2, no. 3 (June 2015): 196–219, https://doi.org/10.1177/2053019615591020; and the summary of Anthropocene evidence in Colin N. Waters, Jan Zalasiewicz, Colin Summerhayes, Anthony D. Barnosky, et al., "The Anthropocene Is Functionally and Stratigraphically Different from the Holocene," *Science* 351, no. 6269 (2016): aad2622, DOI: 10.1126/science.aad2622; and Jan Zalasiewicz, Colin N. Waters, Mark Williams, and Colin P. Summerhayes, eds., *The Anthropocene as a Geological Time Unit: A Guide to the Scientific Evidence and Current Debate* (Cambridge: Cambridge University Press, 2019).

17. Anthony D. Barnosky, Nicholas J. Matzke, Susumu Tomiya, Guinevere O. U. Wogan, et al., "Has the Earth's Sixth Mass Extinction Already Arrived?," *Nature* 471 (March 2011): 51–57.

18. For an exploration of how these two parallel time-cataloguing systems are related today, see Jan Zalasiewicz, Maria Bianca Cita, Frits Hilgen, Brian R. Pratt, André Strasser, Jacques Thierry, and Helmut Weissert, "Chronostratigraphy and Geochronology: A Proposed Realignment," *GSA Today* 23, no. 3 (2013): 4–8.

19. Matt Edgeworth, Dan deB Richter, Colin Waters, Peter Haff, Cath Neal, and Simon James Price, "Diachronous Beginnings of the Anthropocene: The Lower Bounding Surface of Anthropogenic Deposits," *Anthropocene Review* 2, no. 1 (April 2015): 33–58.

20. Stephen F. Foley, Detlef Gronenborn, Meinrat O. Andreae, Joachim W. Kadereit, et al., "The Palaeoanthropocene—The Beginnings of Anthropogenic Environmental Change," *Anthropocene* 3 (2013): 83–88.

21. Eustoquio Molina, Laia Alegret, Ignacio Arenillas, José A. Arz, Njoud Gallala, Jan Hardenbol, Katharina von Salis, Etienne Steurbaut, Noël Vandenberghe, and Dalila Zaghbib-Turki, "The Global Boundary Stratotype Section and Point for the Base of the Danian Stage (Paleocene, Paleogene, 'Tertiary,' Cenozoic) at El Kef, Tunisia—Original Definition and Revision," *Episodes* 29, no. 4 (2006): 263–273.

22. Anders Martinsson, ed., "The Silurian-Devonian Boundary: Final Report of the Committee of the Siluro-Devonian Boundary Within IUGS Commission on Stratigraphy and a State of the Art Report for Project Ecostratigraphy," *International Union of Geological Sciences*, series A, no. 5 (1977): 347.

23. Richard J. Brown and Michael J. Branney, "Internal Flow Variations and Diachronous Sedimentation Within Extensive, Sustained, Density-Stratified Pyroclastic Density Currents Flowing Down Gentle Slopes, as Revealed by the Internal Architectures of Ignimbrites on Tenerife," *Bulletin of Volcanology* 75, no. 7 (2013): 727.

24. Michael Melchin, Peter M. Sadler, and Brad D. Cramer, "The Silurian Period," in *The Geological Time Scale 2012*, ed. Felix M. Gradstein, James G. Ogg, Mark Schmitz, and Gabi Ogg (Amsterdam: Elsevier, 2012), 526–558.

25. Jan Zalasiewicz and Mark Williams, "The Anthropocene: A Comparison with the Ordovician-Silurian Boundary," *Rendiconti Lincei—Scienze Fisiche e Naturali* 25, no. 1 (2014): 5–12.

26. Tatjana N. Koreń and Michael J. Melchin, "Lowermost Silurian Graptolites from the Kurama Range, Eastern Uzbekistan," *Journal of Palaeontology* 74 (2000): 1093–1113.

27. E.g., Ryszard Kryza and Jan Zalasiewicz, "Records of Precambrian-Early Palaeozoic Volcanic and Sedimentary Processes in the Central European Variscides: A Review of SHRIMP Zircon Data from the Kaczawa Succession (Sudetes, SW Poland)," *Tectonophysics* 461 (2008): 60–71.

28. Walker et al., "Formal Definition and Dating."

29. Colin N. Waters, James P. M. Syvitski, Agnieszka Gałuszka, Gary J. Hancock, et al., "Can Nuclear Weapons Fallout Mark the Beginning of the Anthropocene Epoch?," *Bulletin of Atomic Scientists* 71, no. 3 (2015): 46–57.

30. Neil L. Rose, "Spheroidal Carbonaceous Fly Ash Particles Provide a Globally Synchronous Stratigraphic Marker for the Anthropocene," *Environmental Science and Technology* 49, no. 7 (2015): 4155–4162; Graeme T. Swindles, Elizabeth Watson, T. Edward Turner, Jennifer M. Galloway, Thomas Hadlari, Jane Wheeler, and Karen L. Bacon, "Spheroidal Carbonaceous Particles Are a Defining Stratigraphic Marker for the Anthropocene," *Scientific Reports* 5, no. 10264 (2015): 1–6.

31. Ivar do Sul and Costa, "Present and Future of Microplastic Pollution," 352–364.

32. E.g., Dipesh Chakrabarty, "Climate and Capital: On Conjoined Histories," *Critical Inquiry* 41, no. 1 (Autumn 2014): 1–23; Bruno Latour, "Face à Gaïa," in *Les Empécheurs de Penser en Rond* (Paris: La Découverte, 2015).

33. Waters et al., "Nuclear Weapons Fallout," 46–57.

34. Jane Qiu, "Dinosaur Climate Probed," *Science* 348 (2015): 1185.

CHAPTER 5

THE ANTHROPOCENE DATING PROBLEM

Disciplinary Misalignments, Paradigm Shifts, and
the Possibility for New Foundations in Science

KYLE NICHOLS AND BINA GOGINENI

Geoscience Background and Dating Issues

In the last several decades the field of geology has gone through a disciplinary
realignment. Traditionally, geology has meant the study of the lithosphere, or
the solid Earth (e.g., solid rock, soil, sediments, molten rock beneath Earth's
surface). However, as *geologists'* research started to examine how the lithosphere interacts with the hydrosphere,[1] the atmosphere,[2] and the biosphere,[3]
the narrow definition of geology no longer seemed appropriate. Thus, the
broader terms *geosciences* and *Earth science* started to replace *geology* as the
common disciplinary term in academic settings. While the terms *geology, geosciences*, and *Earth sciences* are used interchangeably, there is a distinct difference in content and in temporal frameworks. We make the distinction that
geology is the most traditional subset of the broader *geosciences* (or Earth
sciences).[4]

Geoscientists research across a broad range of temporal and spatial scales.
The traditional geologists—such as igneous and metamorphic petrologists as
well as stratigraphers—engage with the deep Earth and deep time. Stratigraphers in particular have long occupied the privileged role as keepers of the
geologic timescale because the stratigraphic record has traditionally been the
sine qua non for understanding the Earth's history. Analyzing the geologic
record contained in rock, sediment, and ice to identify the boundaries for each
of the geologic time periods, stratigraphers traditionally work in temporal
scales that are much longer than the human timescale: stratigraphic records in

rock preserve geologic change over millions to billions of years, a scale too vast to register human impacts that have occurred over the course of mere decades and centuries (or, at most, millennia). Often stratigraphic data are recorded in deposits that accumulate slowly (generally less than 0.1 mm per year)[5] and are difficult to measure on decadal or centennial timescales.[6] Such slowly accumulating and difficult-to-measure geologic deposits stand in contrast to the rapidly accumulating and short-term-measurable carbon dioxide in the atmosphere. Climatologists, who measure such recent changes in the atmosphere; glaciologists, who measure climate changes recorded in ice; and geomorphologists, who study Earth's surface change, engage directly with the temporal scales and spatial scales affected by human activity.

In short, the relevance of human activity to all geoscientists is not equal. The most sensitive portions of the Earth System—the atmosphere, the hydrosphere, the biosphere, and Earth's surface—record human impacts in real time. The least sensitive portion of the Earth System, the vast majority of the lithosphere, stores deep-time changes in geologic deposits subsequently deciphered by stratigraphers. It therefore makes sense that the empirical studies recording the increase of carbon dioxide in the atmosphere,[7] and similarly consequential changes in the surface of the lithosphere,[8] the hydrosphere[9] and the biosphere[10] on the decadal to centennial timescale, would have been impertinent to traditional stratigraphers. But it is curious that such studies, well known to other geoscientists, were never fully considered in concert until Paul Crutzen, an atmospheric chemist, and Eugene Stoermer, a biologist, concocted the category "Anthropocene."[11] Why did it take so long for geoscientists, first, to recognize that humans had become a driving force in the Earth System and, second, to name that unprecedented phenomenon as a geologic time period?

The explanation has to do with the history of science and with the nature of scientific research. In terms of the history of science, it is important to note that until recently geology, by definition, meant the deep-time study of nature. Deep time was what set geology apart from all other sciences. When, in the middle of the twentieth century, new specialties dedicated to shorter-term study of Earth emerged or migrated from other fields such as geography, they were not wholly accepted into the bastion of geology. Traditional geology banished them to their various marginal silos, where they languished in isolation until very, very recently (as betokened by the nascent conglomerate term *geosciences*). Because of the extreme siloing of these various specialties studying recent Earth, they remained out of touch with one another's findings; and so, they were unable to conclude that, indeed, humans were forcing the Earth's entire system. Even within a given silo, there wasn't, until very recently, enough research accumulated to conclude definitively that humans were a dominant force on the Earth. Perhaps surprisingly, these subfields did not resist their ghettoization by traditional deep-time geology; rather, they happily burrowed down into their specialties, steering clear of any politics that might come with an un-siloed, controversial, and morally implicating category like the "Anthropocene." It was only very recently that radical scientists like Crutzen and Stoermer began to transgress the taboo of politics in science. Notably, these research partners who coined a geological age came from outside the geosciences entirely. Within the geosciences, climatologists and glaciologists tend to be the most politically radical (for example, James Hansen, the former head of the NASA Goddard Institute for Space Studies); this is because they have the most detailed records of climate over the last eight hundred thousand years, and because ice is the element most sensitive to climate change.

A second challenge that had faced researchers was the difficulty in distinguishing short-term trends from "noise" in the data.[12] When geologic deposits, such as ice, recorded Earth System change on decadal to millennial timescales, researchers did not always recognize it. To illustrate: During the course of the twentieth century, scientists' understanding of how rapidly climate could change progressively shifted by orders of magnitude. In the 1950s, climate scientists looked for trends on the order of tens of thousands of years, so they dismissed any trends on shorter timescales as noise. In the 1970s, climate scientists discovered that trends on the millennial timescale were robust, but they dismissed trends on shorter timescales as noise. In the 1980s, climate scientists identified robust trends on centennial timescales and, by the 1990s, on decadal or shorter timescales.[13] Technological advances have allowed for the progressive refinement of these data sets. But in the meantime, the coarseness of data sets thwarted researchers from discerning between noise and short-term trends.

A third challenge related to the difficulty distinguishing between noise and short-term trends is one common to all science: scientists only see what they are looking for or what seems plausible at the time. A signal or pattern in the data might easily be overlooked when it is irrelevant to the hypothesis that a researcher is testing. Alternatively, even when a researcher is looking for a particular signal or pattern, she or he might dismiss it if what she or he sees is incommensurate with prior knowledge. To use the example from earlier: if one is taught that climate change cannot possibly happen on an order of magnitude less than what has been established by science, then one is unlikely to study, let alone recognize, patterns that are smaller in magnitude. To use another example, one concurrent with the climate scientists' undertakings discussed earlier: a range of geoscientists developed high-resolution data sets that contained all of the data necessary to confirm recent climate change. But they dismissed the significance of the short timescale trends of the data vis-à-vis climate change for either one of two reasons: (1) they were testing hypotheses that did not pertain to climate change; or (2) they did not allow themselves to interpret short-term trends in such a way that challenged prior understandings of the temporal scale of change. Here, we can see how prevailing scientific assumptions constrain researchers in their assessment of the plausibility of their own interpretations. Weart explains:[14] "Over the decades, many scientists who looked at tree rings, varves, ice layers, and such had held evidence of decade-scale climate shifts before their eyes. They easily dismissed it. There were plausible reasons to dismiss global calamity as nothing but crackpot fantasy."

It took forty years for the conventional understandings to shift to allow scientists to permit and interpret this evidence of decadal climate change. Again, Weart puts it well: "If the conventional beliefs had been the same in 1993 as in 1953—that significant climate change always takes many thousands of years—the short-term fluctuations in ice cores would have been passed over as being meaningless noise."[15]

In sum, short timescale research in the geosciences is always initially constrained by previous research informed by larger temporal scales.[16]

In addition to the difficulty of interpreting data "against the grain" of conventional understandings (e.g., how rapidly climate could change), there is a related difficulty pertaining to data collection itself. Methodological norms constrain data collection procedures, with the effect of sometimes obscuring signals in the data. For example, if the methodological norm is to collect sediment at, say, five-centimeter intervals, that increment may amalgamate one hundred years of deposition, which would make decadal resolution impossible. In fact, data collection protocols inform hypotheses: sometimes the questions appropriate to a particular temporal scale are not even being asked. What's more, one's method of collecting the data *now* (as in the five-centimeter-increment example) can even make it impossible to reexamine them *later* to address different questions informed by subsequent advances in science. That is why the data repository is so important. If one collects the data in as refined a way as possible, then the noise will seem excessive now, but there's the chance for later scientists to reexamine it in light of new understandings and possibly to discover patterns unperceived by prior researchers. The discovery of progressively shorter timespans for natural climate change is one of these new understandings that have required climate scientists to shed methodological norms and to mine old data for new patterns.

While all three of these issues—silos, noise, conventional understandings and methods— confront science across the board, they have been particularly pitched vis-à-vis the short-timescale geosciences. Below all of that is a profound epistemological constraint that explains the delayed recognition of humans as a driving force in the Earth System and the consequent identification of the "Anthropocene": for most of the past 300 plus years, geologists have thought of humans as separate from the natural system.[17] This axiom informed traditional geology so profoundly that all conventional understandings, methods, and research questions derived from it. In fact, one might say that the normative distinction between traditional geology and the more recent short-scale geosciences is that the former takes as axiomatic the separation of humans from nature, whereas the latter do not. To recognize such a thing as the "Anthropocene" is to reject the fundamental axiom of traditional geology. Hence the controversy over the nomenclature. It is remarkably incongruous that those who take as axiomatic humans' separation from Earth (the stratigraphers) should be tasked with identifying if and when humans began to alter nature profoundly. The background assumptions of stratigraphers necessarily disincline them from recognizing anthropogenic impacts. Far better to assign those geoscientists less wedded to the notion of humans' separation from nature. These short-timescale geoscientists, too, are epistemologically constrained—not only in the more detailed ways mentioned in the preceding pages, but also in their inevitable inheritance of many methods and tools originally developed with traditional geology's background assumption that humans are separate from nature. Indeed, this very belief may occasionally nag even researchers who work on the human timescale.

All of this said, these shorter-timescale geoscientists tend to strive to supersede, and even to upend, traditional geology's assumption of humans' separation from, and irrelevance to, nature. We might think of this distinction in terms of the third-person versus first-person point of view on the world, as parsed out by Akeel Bilgrami. The Newtonian orthodoxy effectively disenchanted nature such that it became an object to be studied by human observers (i.e., scientists) necessarily detached from it.[18] Furthermore, it wasn't just scientists performing their studies but humans in general that were seen as separate from nature. As the prestige of science and the metaphysics of objectivity and detachment grew—with far-reaching implications for political economy (colonialism, extractionism, and the like)—nature came to be *exhaustively defined* as that which could be objectively studied by science.[19]

"Anthropocene"-oriented scientists are now challenging that long-standing view of nature, demonstrating instead humans' inextricability from nature. Ironically it is the detached, third-person point of view held by traditional science/geology that has brought the other geosciences to this more engaged, first-person view of the world: the damage of four-hundred-plus years of humans' detachment-enabled domination of nature has forced some scientists to see that we humans were never separate from nature and that it places normative demands/constraints on us. Such recognition has induced scientists (particularly short-timescale geoscientists) to take an unprecedented perspective on nature and their own work. Once science has accepted that humans are part of its object of study, it requires new approaches in that study, including moral imperatives—i.e., a scientific response to the normative demands of nature. What that might look like remains to be fully explored and articulated; but already scientists prompted in the first place by ethical considerations are working within the parameters of the scientific method. Moral imperatives—the desire to study recent Earth history in order to mitigate its damage—have, in fact, driven most short-term geoscientists to choose their specialties and their particular research projects. Thus, the very idea of objective, machine-like scientists whose own humanity plays no role in their study becomes problematic when it comes to the "Anthropocene": humans studying their own impacts on a nature with which they understand themselves to be in reciprocal relation cannot possibly maintain the objectivity traditionally associated with purely descriptive sciences like stratigraphy. And we consider the partially "first-person" approach in these geosciences to be a good thing because it has resulted in a wealth of new knowledge about recent Earth history—new knowledge that is not yet uncontroversially recorded in the stratigraphic record but that can allow humans to act now to mitigate damage.

What Is the "Anthropocene"?

In lieu of traditional geologists, an atmospheric chemist, Paul Crutzen, and Eugene Stoermer, a biologist, identified and coined the term *Anthropocene*. This term's subsequent popularity among a broad spectrum of humanist and social science scholars has, in any case, now required geoscientists, and specifically stratigraphers, to address the usage and feasibility of a geologic time period invented by others.

The definitions of the "Anthropocene" are ambiguous on multiple levels. First, there are inconsistencies among the proliferating rhetorical definitions of the term. Second, there is inconsistency between the "Anthropocene" Working Group's (AWG)[20] own rhetorical definition and its official definition, which will delineate the starting date, and the total length, of the "Anthropocene" in the geologic timescale. That the institution tasked with bringing together these two registers—rhetorical and official, i.e., consequences and chronology—is unable to reconcile them highlights the fundamental incongruity of the discourse at large. The nonstratigrapher stakeholders are interested in consequences, the stratigrapher stakeholders in chronology. Yet, the one register without clear reference to the other has no value.

Though there is no unified rhetorical definition, many generally state that the "Anthropocene" is the time when humans act a global and geological force. Most of these rhetorical definitions do not make reference to when humans *started* to have an impact; they simply take for granted that somehow we have ended up in the midst of a new period, the "Anthropocene." The working (rhetorical) definition of the AWG for the "Anthropocene" is "the present time interval, in which many geologically significant conditions and processes are profoundly altered by human activities."[21] However, like all other rhetorical definitions, the AWG's rhetorical definition does not explicitly state what is needed to identify the *commencement* of the "Anthropocene," a necessary component of the official geologic definition of the "Anthropocene."

If the AWG's rhetorical definition *implicitly* refers to the beginning of the "Anthropocene," then even that rhetorical definition is not consistent with itself. The commencement of the "Anthropocene" cannot be defined as both the time when geologically significant *conditions* (read, the geologic record) are profoundly altered by human activities *and* the time at which many geologically significant *processes* are profoundly altered by human activities. This is because conditions and processes are asynchronous at the human timescale; processes must precede conditions. Such a distinction between processes and conditions is not significant in *deep time* because large yet necessarily admissible temporal uncertainties surrounding the beginning of a new geologic time period would obscure the lag-time from processes to conditions. However, the distinction between processes and conditions is important when discussing the geology at the *human timescale*.

If the "Anthropocene" is officially designated as a *geologic* epoch, then the starting date will be determined in accordance with *geologic* and more specifically *stratigraphic* principles.[22] Such a process seeks to identify synchronous, global change (or a change in condition) in the stratigraphic record: a Global Boundary Stratigraphic Section and Point (GSSP), informally known as the "golden spike." However, for such a change to be recorded, a threshold in the natural system must be crossed. Thresholds in nature are rarely crossed when a force commences; rather, there is a lag-time before the cumulative effects of the driving force push the system into a new regime (much like the "tipping point" in the climate change discourse). Thus, the defining of a new geologic time as the *crossing* of a threshold is a temporal mismatch with the general understanding of when human activities *started* to have a significant impact. Herein lies one of the fundamental confusions of the "Anthropocene" for geoscientists. Does the "Anthropocene" begin with a significant change in condition or a significant change in process? The difference between change in condition and change in process is not trivial. If the "Anthropocene" is formalized, the starting date will have broad implications for the multidisciplinary and interdisciplinary perspectives that are central to the discourse and that are often mentioned as important by members of the AWG.

Insight into how stratigraphers, and in particular the AWG, are thinking about the commencement of the "Anthropocene" is offered by Jan Zalasiewicz, the convener of the AWG, in a *Generation Anthropocene* podcast.[23]

Zalasiewicz asks the rhetorical question "Where do we draw the boundary?" and then answers it by saying, "It won't be at the first sign of human influence, it will be at what seems to be the best marker layer to separate the world of the present and the future from the world of the past." Zalasiewicz's statement suggests that the AWG is not concerned with the time when human activities started to have a profound impact on processes; rather, they are looking for an event horizon or marker in the stratigraphic record (i.e., a change in condition), a GSSP or "golden spike." In an article published in 2015 by twenty-six of the thirty-eight members of the AWG,[24] the authors suggested a date of July 16, 1945—the detonation of the first atomic weapon—as the commencement of the "Anthropocene" time period. Such a date would seem to fetishize a global and synchronous marker in the stratigraphy; but it completely misses the mark because it is hard to justify geologically that this date "separate(s) the world of the present and the future from the world of the past." On the other hand, this date does represent a "golden spike" in geopolitics. Herein lies an interesting departure from traditional geology and stratigraphy. Rather than using inductive analysis where stratigraphers find a signal in the geologic record and seek to explain its occurrence, the AWG uses deductive reasoning by assuming that there must be a radionuclide signal in the stratigraphic record. The article does not present data that show a significant change in the geologic record from July 15, 1945, to July 17, 1945. History, not stratigraphy, informed the hypothesis in the first place.

While the date July 16, 1945, may not meet the GSSP criteria, it is in line with some of the geologic time boundaries defined by a Global Standard Stratigraphic Age (GSSA, a numerical age assigned when there is a dearth of stratigraphic evidence). Indeed, Waters et al.[25] mentioned that a GSSA is arguably preferable to a GSSP for the "Anthropocene," due to the high resolution of dating techniques that thereby obviates the stratigraphic synchronicity needed for a GSSP. Although the GSSA is more commonly used in the geologic time periods that preexist the stratigraphic record, it also has been used to define younger geologic time boundaries. The current epoch, the Holocene, was first defined by a GSSA (ten thousand radiocarbon years before ~1950 CE) before stratigraphic evidence was agreed upon to establish a GSSP at ~11,700 years before 2000 CE.[26] If, per Waters et al., the "Anthropocene" can be defined as a GSSA, and if, per the AWG's suggestion of July 16, 1945, geological dates can be chosen in consultation with historical information, then why not choose a starting date that has broader implications for humanity and is more in line with the time when "geologically significant . . . processes [were] profoundly altered by humans"?

There are several other suggested starting dates for the "Anthropocene" that, as well, include some combination of geologic knowledge and prior historical/cultural knowledge. These proposed starting dates range from several millennia, to several centuries, to several decades ago. Some argue that humans have impacted the Earth System since the Holocene and even before. Thus, they argue that the "Anthropocene" should include the entirety of the Holocene (e.g., Smith and Zeder,[27] Certini and Scalenghe).[28] Others, like climatologist Bill Ruddiman, argue for an early Holocene start date[29] based on the hypothesis that humans changed atmospheric carbon dioxide and methane levels as early as seven thousand years ago. Other potential start dates for the "Anthropocene" (in order from oldest to youngest) include 1610 CE, a low value in the carbon dioxide record called the Orbis spike, which represents the change in the Earth System as a result of colonization of the Americas (Lewis and Maslin, 2015); 1800 CE, which represents the Industrial Revolution;[30] 1950 CE, which represents the "Great Acceleration";[31] and 1964 CE, which represents the peak of radionuclide fallout from atmospheric atomic weapons testing.[32] Each of these proposed dates, based on a combination of geological and non-geological information, bears different implications for the meaning of the "Anthropocene," such as where to place responsibility and how to move forward politically and economically. One could imagine different outcomes of UN Climate Change Conferences if a formally defined "Anthropocene" started with man's control of fire (i.e., all of humanity bears responsibility) rather than, say, in 1610 CE, when empires extracted resources from colonies and created a highly unequal distribution of resources and technology that launched us to four hundred ppm of carbon dioxide in the atmosphere.

Paradigm Shifts

Interestingly, the "Anthropocene" discourse that intersects the geosciences with the humanities/social sciences could initiate paradigm shifts in the geosciences and possibly the sciences in general. The "Anthropocene" discussions call attention to some of the fundamental assumptions of science. The first paradigm shift challenges the assumption of the dispassionate and unbiased scientific process. One of the curious and unexpected developments in the interdisciplinary discussions is that science itself has been shown to be "impure" or nonobjective. The very roots of modern science have been shown to be imbricated in worldly alliances (i.e., with the mercantile, political, and cognitive elite of the seventeenth century; see Bilgrami).[33] Even today, science is politically implicated. To take a dramatic example: The National Atmospheric and Space Administration toned down the implications of climate change research by James Hansen, a radical climatologist who has since been arrested several times during climate change demonstrations. Though less overtly involved with political ideology than controversial figures like Hansen and Crutzen, even ordinary scientists'—especially geoscientists'—choices of specialty, specific research interests, and background assumptions are informed by politics and ethics. The science of the "Anthropocene" unsettles the metaphysics of "objectivity" that has, since the Enlightenment, pervaded all sectors of public life, resulting in the widespread alienation from nature.

Science, as we were taught, advances through detached observations and repeated experiments. However, science is not detached from humans who design experimental/data collection methods and interpret the data; both activities are informed by their training, experience, and existing knowledge.[34] The traditional stratigraphic process of simply identifying a marker in the geologic record that represents a threshold or tipping point is no longer sufficient, given what we know from human history. Stratigraphers do not need to determine inductively what caused a change in the stratigraphy (if one is to be found); human history has already recorded the causes. Even the AWG unwittingly suggested a start of the "Anthropocene" based more on historical knowledge than on stratigraphy. As acknowledged by Jan Zalasiewicz in the *Generation Anthropocene* interview discussed before, "When we consider the "Anthropocene" as a potential formal system, we have to consider whether it works or not geologically, we have to consider its use, not just for geologists but its use also for much wider sections of society. This is very new in geology." Zalasiewicz is hinting at a paradigm shift. However, the full impact of this statement is compromised when Zalasiewicz, only a few minutes prior to the previous statement, responded to the question, "How are you able to keep politics out of the 'Anthropocene'?" by answering, "I don't think we will, I think it will come and I think baggage will arrive with it as well. I think partly we will deal with it in classic geologic maneuver of simply ignoring it as best as we can." The contradiction that Zalasiewicz expressed is common among geoscientists who wish to remain impartial but realize that the science has an important impact on society, and vice-versa. *If* the AWG formalizes the "Anthropocene," *and* it acknowledges that human culture largely informs the definition of the "geologic age of humans," then a paradigm shift in science will certainly have started.

A second paradigm shift of the "Anthropocene" is its reversal of the centuries-long epistemological trajectory in science that diminished the importance of humans in natural processes. This trajectory started over three hundred years ago with the Copernican notion that the human environment (Earth) is not the center of the universe (i.e., geocentric compared to heliocentric cosmology).[35] Later, Darwin disabused humans of their exceptionalism vis-à-vis nature by suggesting that they are just one branch of the evolutionary tree. So, it is nothing short of a paradigm shift for "Anthropocene" geoscientists to acknowledge that humans occupy a special role in the Earth System, driving significant changes in all of its portions: the atmosphere, the biosphere, the hydrosphere, and even the surface of the lithosphere. In other words, geoscientists are becoming *more* anthropocentric. Ironically, this shift is opposite to that of the humanities/social sciences and is therefore worth noting. In the humanities/social science, the discourse centers on becoming *less* anthropocentric (e.g., Mbembe) as an outcome of the realization that humans cannot control nature.[36] To put a finer point on it: the "Anthropocene" represents the humanities' paradoxical realization of humanity's current extreme vulnerability to nature as a result of its past extreme domination of it. In any case, through the "Anthropocene" discourse, the geosciences and the humanities/social sciences are converging upon more common ground than heretofore.

Broader Implications

The significance of when the "Anthropocene" started extends beyond a stratigraphic marker to interdisciplinary concerns. We argue that the interdisciplinarity of the debate is good for science, the geosciences in particular, and for the humanities and social sciences as well. The necessity acknowledged by Jan Zalasiewicz to have the definition of the "Anthropocene" be good for "much wider sections of society" opens an avenue to bridge the geosciences with other disciplines, including history, philosophy, politics, international law, and economics (e.g., Biermann et al.,[37] Chakrabarty,[38] Robin,[39] Brown and Timmerman,[40] Vidas et al.).[41] Indeed, geoscientists know that human forcing of the Earth System goes beyond carbon dioxide emissions (the most common Earth ill associated with the "Anthropocene"), and includes also the landscape changes created by extraction of resources such as deforestation or mining that were hallmarks of colonial empires.[42] Acknowledging such practices seems to be another a blind spot for geoscientists, arguably because the training of many geoscientists is dominated by Western countries. Yet, historical knowledge helps geoscientists link cause and effect in the recent natural history of the Earth System. Prohibiting historical knowledge in Anthropocenic inquiry is like putting blinders on scientists seeking to link cause to effect, the very goal of science.

Formalizing an interdisciplinary start to the "Anthropocene" will have important implications going forward. For example, an early-Holocene date could potentially underrepresent the importance of current crises such as global warming, acidification of the oceans, sea-level rise, and regional as well as global pollution that happened much later than the rise of agriculture. The 1610 CE date[43] would likely signify the importance of colonialism and its extractionist practices on the Earth System. Such a date could put squarely on the shoulders of the former empires the responsibility of bearing the weight of mitigating and adapting to global climate change, among other environmental damage. However, this date is controversial.[44] Specifically, the carbon dioxide marker used by Lewis and Maslin[45] to define the 1610 CE date is refuted by the AWG.[46] The AWG claims that the carbon dioxide marker associated with the 1610 CE date is not outside the natural variability of the record. In other words, it is just noise in the data. The 1800 CE date would suggest that technology designed to increase profit, specifically fossil-fueled machines, is the cause for the "Anthropocene" (as suggested by the term *Capitolocene*).[47] Similarly, the July 16, 1945,[48] date, the 1964 CE peak radionuclide fallout date,[49] and the 1950 CE Great Acceleration date would represent the *culmination* of four hundred years of technological, economic, and political practices oriented toward the full conquest of nature, rather than the *beginning* of human forcing on the Earth System.

Centuries of using nature to benefit societal interests, combined with population growth, led to the compounding impact of humans on the Earth System. In an effort to support a mid-twentieth-century start date, Waters et al.[50] present many data sets that show a significant change in the mid-twentieth century. However, most of these data sets are based on production (aluminum, black carbon, concrete, nitrogen fertilizer, and plastic) and emissions (nitrous oxides), rather than deposition in the stratigraphic record. Waters et al.[51] also present stratigraphic data such as carbon dioxide, methane, and carbon isotopes stored in glacial ice that show changes caused by human activity, but these changes started in the eighteenth and nineteenth centuries. The only stratigraphic record that they suggest shows a change in the mid-twentieth century is the fallout of radionuclides produced by atomic weapons testing. Of course, such a distinct marker would be due to the absence of atomic weapons before 1945, so it would seem to be a clean candidate to mark the beginning of the "Anthropocene." However, the radionuclide production and emissions data do not identify the beginning of human forcing, but rather some subsequent moment in time well into the anthropogenic forcing of the Earth System. It should be noted that even invoking the production and emission data bends the requirements for a stratigraphically defined "Anthropocene." Even if we consider those data as part of the stratigraphic record, they do not address the commencement of human forcing, which is more important to interdisciplinary, and even geologic, research and thought. After all, the goal in geology is to determine when and why events happened so as to understand the implications going forward: i.e., to be descriptive *and* prescriptive.

Indeed, the AWG has repeatedly mentioned that the geologic definition of the "Anthropocene" has broader implications and should work for other disciplines as well.[52] Further confirming the AWG's interdisciplinary commitment, four of the seven authors in the editorial introduction of the first issue of *The Anthropocene Review*, a journal founded as "transdisciplinary," are on the AWG.[53] However, members of the AWG have repeatedly identified the mid-twentieth century as the best candidate for the geologic definition of the "Anthropo-cene" (e.g., Wolfe et al.,[54] Waters et al.,[55] Zalasiewicz et al.,[56] Steffen et al.,[57] Waters et al.,[58] Zalasiewicz et al.,[59] Waters et al.),[60] a date that they suggest abides stratigraphic convention, but one that ignores the interdisciplinary/transdisciplinary nature that they so often invoke. Historically, delineating a point in time in the stratigraphic record worked well when humans were not around to determine the cause of the change in the Earth System. However, since humans are here to record and witness the change in the Earth System, the official definition should involve more than just a point in the stratigraphy and should provide a deeper understanding of why the change in the Earth System happened.

There are many reasonable ideas in the humanities and social sciences that can be drawn on to define the start of the "Anthropocene." Nevertheless, if the "Anthropocene" becomes a formalized *geologic* time period, it will most likely be defined by stratigraphic principles. However, the narrow protocols used to define older geologic ages are not sufficient in the face of the abundant infor-mation that geoscientists have from these other disciplines. The complement-ing of stratigraphic records with cultural/historical knowledge is born of necessity for two reasons: because simple empirical measurements of the human timescale recorded in geologic deposits are no longer sufficient now that we know a myriad of human activities not yet recorded in geologic depos-its are still central to, and even dominate, natural processes; and because the implications of an officially defined "Anthropocene" could be far-reaching.

While some stratigraphers prefer to establish a geological marker (aka "golden spike") independent of additional historical/cultural knowledge,[61] there has yet to be consensus on the geological marker that is global and syn-chronous. We argue that such a stratigraphically defined marker would not likely represent the best understanding of the "Anthropocene" because once one is established it will be too late to be useful for "wider sections of society." A future stratigraphic layer that would be both global and synchronous would only exist after the sea level rises several meters and deposits sediment over several million square kilometers of newly submerged coastal land; this would devastate poor nations and small island nations even more than developed countries. Such an extreme scenario of prioritizing the golden spike highlights two problems of science that arise from the constraints of disciplinary con-vention. The first is the notion of the dispassionate data collector immune to the implications of the science and the second is the limitation imposed by unduly strict protocols for defining a geological age. These geological ages established by humans matter only for our understanding of Earth. But now, the constraints we have ourselves imposed on such dating seem to be limiting rather than enhancing our understanding of Earth. The traditional geological specialties are descriptive because their processes are so slow that human timescale doesn't matter. The prescriptive geosciences study shorter-timescale events that are relevant to human life. Some examples include geomorphology (landslides and floods), seismology (earthquake prediction), and volcanology (volcanic eruption prediction). There's an incongruity between assigning the standards of the most descriptive geosciences (i.e., geology) and the establishment of a geological age whose main value/pertinence is its prescriptive quality. Such an age would seem to be most relevant to, and most easily measured by, the more prescriptively oriented (yet still descriptive) geosciences: the short-timescale geosciences. If stratigraphy is going to adhere to its strict protocols for defining geological ages, then the "Anthropocene" should not, for all of the reasons discussed here, be designated an official geologic period. However, if stratigraphers do officially instate the "Anthropocene" as a "new kind of geology," then they must take into account additional nonstratigraphic evidence, such as historical and cultural information and other data from the short-term geosciences. Doing so would serve not only geology but also humanity in the best possible way. In other words, in order for stratigraphy to be prescriptive, the causes of past events and the implications of current events need to be considered in concert.

Science, without question, is valuable and useful to society (after all, how would we understand the magnitude of climate change if we did not understand the geologic record?) but it is limited because of biases and blind spots that are exposed by the "Anthropocene" discourse. The humanities and the social sciences have correctly pointed out some of these. The humanities/social sciences also understand that culture and politics influence research. Moreover, they can offer new ideas that can be formed into testable hypotheses. However, science is still using the same models that have been orthodox since the scientific revolution despite having been exposed by the "Anthropocene." A purely stratigraphically defined "Anthropocene" *would officially* recognize man's impact on the Earth System; however, such a stratigraphically defined "Anthropocene" would negate many of the humanities/social science perspectives that perhaps matter more to mitigating and adapting to the effects of humans on Earth's system. It is the time for scientists to think differently about the different protocols of science. The AWG has already started thinking more broadly, but the interdisciplinary knowledge is still lacking in the process.

Notes

This research received no funding from the public, private, or nonprofit sectors. A subsequent, peer-reviewed version of this article was published as "The Anthropocene's Dating Problem: Insights from the Geosciences and the Humanities," *Anthropocene Review* 5 (2018): 107–119.

1. E.g., oceans, rivers, lakes, groundwater.

2. E.g., gases such as nitrogen, oxygen, carbon dioxide, water vapor.

3. E.g., plants, animals, microbes.

4. For the sake of streamlining, we will use *geosciences* although *Earth sciences* could just as well be used.

5. Heiko Hüneke and Thierry Mulder, eds., *Deep-Sea Sediments*, vol. 63 (Amsterdam: Elsevier, 2010).

6. Except for glacial ice, which can have depositional rates that are often higher than those of sediment.

7. NOAA Earth System Research Laboratory, www.esrl.noaa.gov/gmd/ccgg/trends/.

8. Roger LeB. Hooke, "On the Efficacy of Humans as Geomorphic Agents," *GSA Today* 4 (1994): 217, 224–225. Humans are the dominant global geomorphic force changing Earth's surface, even greater than glaciers and rivers.

9. Tsung-Hung Peng and Wallace S. Broecker, "The Utility of Multiple Tracer Distributions in Calibrating Models for Uptake of Anthropogenic CO_2 by the Ocean Thermocline," *Journal of Geophysical Research: Oceans* 90, no. C4 (1985): 7023–7035; Nicholas R. Bates, Yrene M. Astor, Matthew J. Church, Kim Currie, John E. Dore, Melchor González-Dávila, Laura Lorenzoni, Frank Muller-Karger, Jon Olafsson, and J. Magdalena Santana-Casiano, "A Time-Series View of Changing Ocean Chemistry Due to Ocean Uptake of Anthropogenic CO_2 and Ocean Acidification," *Oceanography* 27 (2014): 126–141. E.g., chemistry of the oceans.

10. David W. Schindler, "Detecting Ecosystem Responses to Anthropogenic Stress," *Canadian Journal of Fisheries and Aquatic Sciences* 44, no. S1 (1987): s6–s25; Yann Hautier, David Tilman, Forest Isbell, Eric W. Seabloom, Elizabeth T. Borer, and Peter B. Reich, "Anthropogenic Environmental Changes Affect Ecosystem Stability via Biodiversity," *Science* 348 (2015): 336–340. E.g., changing ecosystems.

11. Paul J. Crutzen and Eugene R. Stoermer, "The Anthropocene," *Global Change Newsletter* 41 (2000): 17–18.

12. Spencer Weart, "The Discovery of Rapid Climate Change," *Physics Today* 56, no. 8 (2003): 30–36.

13. Weart, 30–36.

14. Weart, 30–36.

15. Weart, 30–36.

16. Weart, 30–36.

17. Joanne Vining, Melinda S. Merrick, and Emily A. Price, "The Distinction Between Humans and Nature: Human Perceptions of Connectedness to Nature and Elements of Natural and Unnatural," *Research in Human Ecology* 15 (2008): 1–11.

18. Akeel Bilgrami, *Secularism, Identity, and Enchantment* (Cambridge, MA: Harvard University Press, 2014); Libby Robin, "Histories for Changing Times: Entering the Anthropocene?," *Australian Historical Studies* 44 (2013): 329–340.

19. Akeel Bilgrami, "The Political Possibilities in the Long Romantic Period," *Romanticism and Politics* 49, no. 4 (2010): 533–552.

20. Anthropocene Working Group (AWG) is a subcommittee of the International Commission on Stratigraphy that is charged to determine if and when the "Anthropocene" began. It comprises thirty-eight members, most of whom are geoscientists: thirteen from the United Kingdom, nine from the United States, nine from Western Europe, and one each from Australia, Brazil, China, Canada, Kenya, Poland, and South Africa.

21. "Working Group on the 'Anthropocene' (AWG)," Subcommission on Quaternary Stratigraphy, http://quaternary.stratigraphy.org/workinggroups/anthropocene/.

22. Colin N. Waters, Jan A. Zalasiewicz, Mark Williams, Michael A. Ellis, and Andrea M. Snelling, eds., "A Stratigraphical Basis for the Anthropocene?," *Geological Society, London, Special Publications* 395 (2014): 1–21.

23. Generation Anthropocene podcast, "Masters of the Anthropocene Boundary" (2013), www.stanford.edu/group/anthropocene/cgi-bin/wordpress/wp-content/uploads /2013/02/AnthropoceneWorkingGroup.mp3.

24. Jan Zalasiewicz, Colin N. Waters, Mark Williams, Anthony D. Barnosky, et al., "When Did the Anthropocene Begin? A Mid-Twentieth Century Boundary Level Is Stratigraphically Optimal," *Quaternary International* 383 (2015): 196–203.

25. Waters et al., "Stratigraphical Basis," 1–21.

26. Mike Walker, Sigfus Johnsen, Sune Olander Rasmussen, Trevor Popp, et al., "Formal Definition and Dating of the GSSP (Global Stratotype Section and Point) for the Base of the Holocene Using the Greenland NGRIP Ice Core, and Selected Auxiliary Records," *Journal of Quaternary Science* 24 (2009): 3–17.

27. Bruce D. Smith and Melinda A. Zeder, "The Onset of the Anthropocene," *Anthropocene* 4 (2014): 8–13.

28. Giacomo Certini and Riccardo Scalenghe, "Is the Anthropocene Really Worthy of a Formal Geologic Definition?," *Anthropocene Review* 2 (2014): 177–180.

29. William Ruddiman, "The Anthropocene," *Annual Review of Earth and Planetary Sciences* 41 (2013): 45–68; William Ruddiman, Erle C. Ellis, Jed O. Kaplan, and Dorian Q. Fuller, "Defining the Epoch We Live In: Is a Formally Designated 'Anthropocene' a Good Idea?," *Science* 348 (2015): 38–39.

30. Paul J. Crutzen, "Geology of Mankind," *Nature* 415 (2002): 23.

31. Will Steffen, Wendy Broadgate, Lisa Deutsch, Owen Gaffney, and Cornelia Ludwig, "The Trajectory of the Anthropocene: The Great Acceleration," *Anthropocene Review* 2 (2015): 81–98.

32. Simon L. Lewis and Mark A. Maslin, "Defining the Anthropocene," *Nature* 519 (2015): 171–179.

33. Bilgrami, *Secularism*.

34. Bruno Latour and Steve Woolgar, *Laboratory Life: The Construction of Scientific Facts* (Princeton: Princeton University Press, 1986).

35. Simon L. Lewis and Mark A. Maslin, "Defining the Anthropocene," *Nature* 519 (2015): 171–179.

36. Achille Mbembe, "Decolonizing Knowledge and the Question of the Archive" (2015), http://wiser.wits.ac.za/system/files/Achille.pdf.

37. Frank Biermann, K. Abbott, S. Andresen, K. Bäckstrand, et al., "Navigating the Anthropocene: Improving Earth System Governance," *Science* 335 (2012): 306–307.

38. Dipesh Chakrabarty, "Postcolonial Studies and the Challenge of Climate Change," *New Literary History* 43 (2012): 1–18.

39. Robin, "Histories for Changing Times," 329–340.

40. Peter G. Brown and Peter Timmerman, eds., *Ecological Economics for the Anthropocene* (New York: Columbia University Press, 2015).

41. Davor Vidas, Ole Kristian Fauchald, Øystein Jensen, and Morten Walløe Tvedt, "International Law for the Anthropocene? Shifting Perspectives in Regulation of Oceans, Environment and Genetic Resources," *Anthropocene* 9 (2015): 1–15.

42. Naomi Klein, *This Changes Everything: Capitalism vs. the Climate* (Toronto: Alfred A. Knopf Canada, 2014); Daron Acemoglu, Simon Johnson, and James A. Robinson, "The Colonial Origins of Comparative Development: An Empirical Investigation," National Bureau of Economic Research: Working Paper No. 7771, June 2000, 69.

43. Lewis and Maslin, "Defining the Anthropocene," 171–179.
44. Members of the Anthropocene Working Group (AWG), "Colonization of the Americas, 'Little Ice Age' Climate, and the Bomb-Produced Carbon: Their Role in Defining the Anthropocene," *Anthropocene Review* 2 (2015): 117–127; Clive Hamilton, "Getting the Anthropocene So Wrong," *Anthropocene Review* 2, no. 2 (2015): 102–107, https://doi.org/10.1177/2053019615584974.
45. Lewis and Maslin, "Defining the Anthropocene," 171–179.
46. AWG, "Colonization," 117–127.
47. Donna Haraway, "Anthropocene, Capitalocene, Plantationocene, Chthulucene: Making Kin," *Environmental Humanities* 6 (2015): 159–165.
48. Zalasiewicz et al., "When Did the Anthropocene Begin," 196–203.
49. Lewis and Maslin, "Defining the Anthropocene," 171–179.
50. Colin N. Waters, Jan Zalasiewicz, Colin Summerhayes, Anthony D. Barnosky, et al., "The Anthropocene Is Functionally and Stratigraphically Different from the Holocene," *Science* 351, no. 6269 (2016): aad2622, DOI: 10.1126/science.aad2622.
51. Waters et al., "Anthropocene Is Functionally," aad2622.
52. Generation Anthropocene podcast, "Masters of the Anthropocene Boundary"; Waters et al., "Anthropocene Is Functionally," aad2622.
53. Frank Oldfield, Anthony D. Barnosky, John Dearing, Marina Fischer-Kowalski, John McNeill, Will Steffen, and Jan Zalasiewicz, "The Anthropocene Review: Its Significance, Implications and the Rationale for a New Transdisciplinary Journal," *Anthropocene Review* 1 (2014): 3–7.
54. Alexander P. Wolfe, William O. Hobbs, Hilary H. Birks, Jason P. Briner, et al., "Stratigraphic Expressions of the Holocene-Anthropocene Transition Revealed in Sediments from Remote Lakes," *Earth-Science Reviews* 116 (2013): 17–34.
55. Waters et al., "Stratigraphical Basis," 1–21.
56. Jan Zalasiewicz, Mark Williams, Colin N. Waters, Anthony D. Barnosky, and Peter Haff, "The Technofossil Record of Humans," *Anthropocene Review* 1 (2014): 34–43.
57. Steffen et al., "Trajectory of the Anthropocene," 81–98.
58. Colin N. Waters, James P. M. Syvitski, Agnieszka Gałuszka, Gary J. Hancock, et al., "Can Nuclear Weapons Fallout Mark the Beginning of the Anthropocene Epoch?," *Bulletin of the Atomic Scientists* 71 (2015): 46–57.
59. Zalasiewicz et al., "When Did the Anthropocene Begin?," 196–203.
60. Waters et al., "Anthropocene is Functionally," aad2622.
61. Whitney J. Autin and John M. Holbrook, "Is the Anthropocene an Issue of Stratigraphy or Pop Culture?," *GSA Today* 22 (2012): 60–61.

CHAPTER 6

DISCIPLINARY VARIATIONS ON THE ANTHROPOCENE

Temporality and Epistemic Authority
Response to Kyle Nichols and Bina Gogineni

NIKOLAS KOMPRIDIS

1.   While not directly related to the theme of this workshop, the focal concern of this paper impinges on the issues of nature and value, perhaps a little more urgently and a little more directly.

This is my second encounter with this paper, and I am impressed by the exemplary collaboration between the humanities and the sciences it represents and prefigures. I am also impressed by the illumining analysis of the Anthropocene debates in the geosciences that Gogineni and Nichols provide, and the concomitant proposal for greater integration of and greater cooperation between the humanities and the sciences, at the level of both problem description and problem solving.

I am very sympathetic to this proposal since it is more or less identical to one I have been making in my own writings on the Anthropocene. The Anthropocene not only names a new geological era that shatters all Holocenic assumptions about the climatic stability of the Earth and its living systems; it also reveals the entwinement of human history and natural history, of environmental justice and social justice. Under such circumstances it is not enough for academics to carry on their activities as they have done so in the past. The Anthropocene calls for more than an urgent shift of scholarly attention; it calls for an unprecedented degree of cooperation and collaboration across the disciplines of the sciences and the humanities. This entails much more than overcoming the chasm between the cultures depicted in C. P. Snow's famous paper. Snow's depiction of the "two cultures" is truly undercomplex, but the chasm is nonetheless real. I would not underestimate how much work scholars in the sciences and in the humanities must do (on themselves) to bridge their disciplinary and cultural differences if they are to work together to fashion a sufficiently coherent and sufficiently effective response to the various challenges of the Anthropocene. It would require a process of reciprocal elucidation, and a readiness to engage with and be altered by other modes of knowing.

Nichols and Gogineni arrive at a like-minded conclusion from a very compelling and fascinating story about the internal conflicts within the geosciences, conflicts that are played out and indeed induced by the disciplinary challenge of defining the Anthropocene. This is the entirely new problem of how to bring the "human" into the object domain of the geosciences, where it had no place before, an object domain that has become unsettled and far less stable in the "age of the humans." This problem is entangled with another: the problem of identifying when exactly the Anthropocene begins—at what point precisely do we enter into this new geological age?—which they describe as the search for the "golden spike."

2.   I would propose a different title for the paper, which frames the problem somewhat differently, a framing that I think is truer to Gogineni and Nichols's analysis: "Whose Temporality? Which Scales?"

Gogineni and Nichols reveal a *conflict of temporalities* within the geosciences between the study of "deep time" (millions/billions of years) and the study of human-sized temporalities (thousands and hundreds of years). The former makes geology distinctive, but it also renders human impact and concerns ultimately irrelevant. This conflict between geoscientific disciplines whose temporal unit is deep time and those whose temporal unit is human-sized exposes how far apart stratigraphers can be from scientists who focus on changes in the hydrosphere, the biosphere, and the Earth's surface. The differences between their respective temporal orientations make the problem of defining (through integration of the "human") and dating the Anthropocene intractable.

A challenge is emerging to the primacy traditional geology accords to deep-time explanation, from which it acquires its own epistemic authority, but at the cost of withholding recognition of the relevance of the human-sized temporalities studied by their geosciences colleagues—e.g., geomorphologists, climatologists, and glaciologists.

The disciplinary conflict of temporalities that Gogineni and Nichols expose underpins the answer to one of their key questions: "Why did it take so long for geoscientists, first, to recognize that humans had become a driving force in the Earth System and, second, to name that unprecedented phenomenon as a geologic time period?"

And their answer is that the deep-time bias of traditional geology toward human-sized temporalities slowed down significantly the identification of the Anthropocene: "if one is taught that climate change cannot possibly happen on an order of magnitude less than what has been established by science, then one is unlikely to study, let alone recognize, patterns that are smaller in magnitude." "In sum, the short-timescale research in the geosciences is always initially constrained by previous research informed by larger temporal scales."

Geosciences that study changes in tree rings, varves, and ice layers (in human-sized temporalities) have been far more sensitive to the geological force of human activities and interventions in the Earth Systems. But their knowledge value was not sufficiently acknowledged by the deep-time bias of their geoscience colleagues.

Thus, we have had a "delayed recognition of humans as a driving force in the Earth System, and the consequent recognition of the 'Anthropocene.'"

> The normative distinction between traditional geology and the more recent short-scale geosciences is that the former takes as axiomatic the separation of humans from nature, whereas the latter do not. To recognize such a thing as the "Anthropocene" is to reject the fundamental axiom of traditional geology. Hence, the controversy over the nomenclature. It is remarkably incongruous that those who take as axiomatic humans' separation from Earth (the stratigraphers) should be tasked with identifying if and when humans began to alter nature profoundly.

Gogineni and Nichol's account of this epistemic conflict within the geosciences gives us a much better and clearer understanding of what is at stake over the definition of the Anthropocene. Their account also makes salient what is at stake in the question of who *owns* the concept of the Anthropocene.

Which epistemic community should make the call about its definition and its starting point?

Should a single epistemic community be given exclusive epistemic authority over the meaning of the Anthropocene and its causality? At what epistemic and practical cost? "A purely stratigraphically defined 'Anthropocene' *would officially* recognize human impact on the Earth System; however, such a stratigraphically defined Anthropocene would negate many of the humanities/social science perspectives that perhaps matter more to mitigating and adapting to the effects of humans on Earth's system."

3.   I want now to look at what Gogineni and Nichols have to say about what is at stake in the "when did it begin?" question: "Does the 'Anthropocene' begin with a significant change in condition or a significant change in process? The

difference between change in condition and change in process is not trivial. If the 'Anthropocene' is formalized, the starting date will have broad implications for the multidisciplinary and interdisciplinary perspectives that are central to the discourse and that are often mentioned as important by the members of the AWG [Anthropocene Working Group]."

The issue is whether the commencement of the Anthropocene is determined strictly by geological methods (stratigraphy) or is determined by cooperative convergence of both historical (humanities) and scientific considerations.

At stake are also the "practical value possibility" of collaboration and integration of the epistemologies of science and the humanities and the generation of a systemic and comprehensive response to the multiple and hypercomplex challenges of the Anthropocene.

Which is why Nichols and Gogineni ask the following question regarding the dating issue: Why not choose a starting date that has broader implications for humanity and is more in line with the time when "geologically significant processes were profoundly altered by humans"?

This would then end the conflict over and between temporalities: there would be some kind of reasonable accommodation of the two, and thereby also the conditions for collaboration between and integration of humanities and natural science perspectives.

Now we may need to concede that we will never arrive at a noncontroversial starting date for (or definition of) the Anthropocene. However, Gogineni and Nichols help us see that what matters is the kind of controversies we can live with and accept, and *why* it is that we can live with them and accept them. Leaving the definition and dating of the Anthropocene to a single geoscience discipline (stratigraphy) that privileges one temporality above all others not only arrogates far too much epistemic authority to that discipline; it also impedes the collaboration and integration of *all* the relevant disciplines necessary to meet the challenges of the Anthropocene. We can surely live with unavoidable controversies over dating and definition; but we don't have to accept the undemocratic and epistemically unjustifiable exclusion of key epistemologies and perspectives from the community of inquiry whose object of urgently needed study is the "age of the humans."

4.   A couple of criticisms and suggestions. First, the suggested paradigm shift to another kind of science not rooted in the metaphysics of "objectivity" is described too quickly; it needs more careful development and justification if it is to be persuasively argued. Similarly, the second paradigm shift that is to come from the acknowledgment that "human culture largely informs the definition of the 'geologic age of the humans'" needs to be spelled out more suggestively, and with more detail. Finally, Gogineni and Nichols need to say

more about how the recognition of humans as part of the geosciences' object of study can alter geosciences that work with human-sized temporalities. Deepening our understanding of the implications of the geoscientific recognition of "the human" as a legitimate and necessary object of study would make their case all the more compelling, and all the more fascinating.

And, just a quick question, to close off my response to this paper: Just what would "a scientific response to the normative demands of nature" look like? Scientists don't normally respond to normative demands of any kind. The normative is not part of the domain of objects that science studies, or a property of the objects it studies. Even if such a response were possible, and, indeed, necessary, just what would it actually involve and require?

CHAPTER 7

VALUE AND ALIENATION

A Revisionist Essay on Our Political Ideals

AKEEL BILGRAMI

I will start with a broad philosophical claim—made with far more brevity than it deserves—about one aspect of the nature of values and their relation to agency, which I hope to exploit in the subsequent discussion on the themes of alienation and our political ideals.

There is a superstition of modernity, which declares that nature contains no properties that are not countenanced by the natural sciences. By superstition, I mean that we take it for granted without knowing when or how it was proved, or how it helps us to live better.

On the assumption that the natural sciences do not study values, I want to resist this superstition by assembling some relatively familiar points (having to do with practical reason and the nature of human agency) that, when assembled in just this way, might make it seem both natural and right to conclude that the world (including nature) contains value properties that make normative demands on us.

I would hesitate to say that such a conclusion, suggested by considerations such as those I present, is properly called *realism* about values. Thus, for instance, those who seek moral realist conclusions in order to repudiate a moral relativism will find nothing on offer in my remarks to promote their cause. For all I say, there might be substantially *differential* perception of the value properties in the world by subjects from different cultural backgrounds, and should that ring alarm bells about a relativism in the offing, there will be nothing in what I say in this paper that could hush those alarms. I am not a relativist, but that is for reasons that will not be detectable in the points I am about to assemble. And relativism apart, even some of the other standard themes and interests surrounding questions of realism, such as the mind-independence of the reality at stake, will not be addressed by me, and in fact, it might appear to many that there is far too much mind-dependence in the conception of values that I present. I don't myself think that the sort of mind-dependence I allow does anything to cancel or even qualify the claim that values are genuine and irreducible properties *in the world*, but I realize that there are realists who will, even so, find the claim, in these respects, falling measurably short of what they conceive realism to be. There is nothing to do but to acknowledge that that might be so and proceed.

Let me begin with a familiar distinction that philosophers (Sartre, but Spinoza before him) make in detecting an ambiguity in an utterance that a subject might make: "I will do . . ." This could be a statement of intention or a prediction. That distinction might be elaborated as follows. When one predicts that one will do something, one steps outside of oneself and looks at oneself as the object of behavioral and causal and motivational tendencies, one looks at oneself as another might look at one, and so this is sometimes called the "third-person" point of view on oneself. But when one intends to do something, one is asking, "What should I do?" or "What ought I to do?," one is being an agent, in the practical sense, not an observer of oneself, one is a practical subject rather than an object, and that is why this is described as a "first-person" perspective on oneself. (Caveat: this vocabulary invoking the first and third person to make the distinction may be misleading since it is a philosophical distinction and not a linguistic one having to do with the first- and third-personal pronoun. So it may be better to make the distinction as one between an agentive—or engaged—and detached point of view, but I may not always be careful to do so.)

We can have both these points of view on ourselves, but not simultaneously since the one crowds the other out. The more important point, for the purposes of this paper, is that, given its nature, if we did not ever possess or occupy the first-person point of view on ourselves, then (to the extent that we can even find that idea coherent) we would have abdicated our practical agency. This is a crucial point, which I will ask you to keep stored in mind till it is deployed in the argument a little later. I'll repeat the point by putting it in its reverse: If all we had were a third-person perspective on ourselves, we would have abdicated our practical agency. (Caveat: In making this point, I say "practical" agency, and mean it. I am certainly not denying that in "I predict that I will . . . ," the first occurrence of "I" is the "I" of agency. It's just that when one observes or studies oneself, one's *attitude* toward oneself is one *not* of practical engagement but rather, if anything, of merely theoretical—or what Kant called "speculative," the Latinate etymology of which is to "look from a vantage point as if from a watchtower"—engagement and so, in the practical sense, *detached,* despite the presence of that agency. And that is why the second occurrence of "I" marks not a subject or agent, but an object, an object of a detached observational gaze, often perhaps with explanatory purpose in mind.)

Now, though, as I've presented them, these are two distinguishable points of view on *oneself*, it is natural to think that there ought also to be a similar distinction that holds for perspectives we have on the *world*. We can have a detached perspective on it, a perspective of study (as is paradigmatically found in natural science, though that is just one very highly systematic and regimented form that that perspective takes), and we can have a perspective of agency on the world, one of responding to it with practical engagement rather than with detached observation and explanatory purpose.

So, these contrasting points of view one has can apply to oneself *as well as* to *the world*. Let's, then, consider the *latter* and ask the question, what must the world be like, what must the world contain, such that it moves us to such practical engagement, over and above detached observation and study? If the world prompts such engagement, it must contain elements over and above those we observe and study from a detached point of view. A natural answer to the question is that over and above containing the facts that natural science studies it contains a special kind of fact, evaluative facts and properties, or more simply, it contains *values*; and when we perceive them, they put normative demands to us and activate our practical engagement. Values, being the sort of thing they are, are not primarily the objects of detached observation, they engage with our first- rather than our third-person point of view on the world.

Thus if we *extend* in this way *onto the world* a presupposition of the fundamental distinction between intention and prediction (the presupposition of two contrasting perspectives that one can have *on oneself*), we get a conception of values that is not something *we* always generate with our mental tendencies, our desires and moral sentiments (as Hume and Adam Smith would have it), and then "project" onto the world (a long-favored metaphor), but they are properties that are found in the world, a world of nature, of others who inhabit nature with us, and of a history and tradition that accumulate in the relations among these, and within which value is understood as being "in the world." It is not as if sympathy and moral sentiments are left out of this picture, but sympathy and moral sentiments, on this picture, are our responses to the normative demands that we apprehend in our perceptions of the evaluative properties of the world. This understanding of values (sometimes in recent years attributed to Aristotle, by John McDowell), therefore, seems something that can be motivated by the most commonsensical considerations about what is distinctive about practical agency and its unique point of view.[1]

But motivating some thought or conclusion may merely make it seem natural, something one could come to believe without strain; it does not yet approach the prestige of an argument.

I think we can go a step further toward that, if we ask—and attempt to answer—the following question, intended as a *challenge* to the view: Why can't agency consist in nothing more than the fact that we try to fulfill our desires, intentions, and so on? True, there is a first-person point of view that is activated and exercised in agency, but why can't it simply be exercised merely in our efforts to satisfy our desires and fulfill our intentions? Why do I insist that *agency* comes into play only when our desires (and moral sentiments) are *responding to the callings of something external, the evaluative properties in the world*? Why can't our characterization of desires have them be *self-standing*, independent of the need for such an external calling?

To answer this challenge, I want to look a little harder at the relationship between desires and agency.

Gareth Evans in a brilliant passage said illuminatingly that questions put to one about whether one believes something,[2] say, whether it is raining outside, do not prompt us to scan our mental interiority to see whether we possess that belief; they prompt us to look outside and see whether it is raining. That is to say, one not only looks outside when one is asked, "Is it raining?" but also when one is asked, "Do you believe it is raining?"

I'd like to extend this insight and ask, is this true of questions put to one about whether one *desires* something? When someone asks one, "Do you desire x?," are we prompted to ponder our own minds or are we prompted to consider whether x is desirable? Of course, x's desirability may not always be perceptible to us, if x is not sensibly present to us, but I am assuming that something like perception is basic and that when we try to *imagine* the desirability of x or just even reflect on its desirability, these are capacities that depend on some background of past perceptions, if not of x itself, then perhaps ingredients of x or things sufficiently approximating or analogous to x, and so on. So, to return to the question, "What does one do when one is asked whether one desires x?," I think the answer is that though there may be special sorts of cases where we might ponder our own minds (say, on an analyst's couch), for most ordinary cases I think, we would simply consider x's desirability. This suggests that our desires are often presented to us as having *desirabilities* in the world as their objects, which, putting subtle differences aside, is the idea of value properties in the world, as intended by the Aristotelian picture of values recently revived by McDowell.

But, since it is an argument we are seeking to provide, we should persist a little more, and ask, even if there is room for extending Evans's point about beliefs to many desires in this way, why is it so *important* to find it extendable in this way to desires? What makes it seem a philosophical *necessity* that we do so?

Perhaps the best way to answer this question is to look critically at what would seem to follow, if one thought my extension of Evans's point to be wrong.

If one thought that a question about whether one desires something never requires one to consider its desirability but rather to step back and consider by scanning our minds whether we desire it, then that would suggest that our desires were presented to us in a way that was accessible to us only when our angle on ourselves was—to use my earlier term—in the *third* person. And here we need to go back to the discussion of the two different points of view, not on the world, but on *ourselves*. If it is indeed true, as we said in that discussion, that practical agency is present in the possession and exercise of the *first-*person rather than the third-person point of view on ourselves, then it would seem that to exercise our practical agency, the appropriate way to experience our desires *cannot* always be to experience them as desir*ed* by us. What alternative way *is* there, then, for our desires being present for us? And the answer it would seem would have to be by perceiving the desirability of what is desired. In short, if our desires were always experienced by us only as desir*ed*, our very agency would be in jeopardy; and it would only be restored if we found a way of saying that our desires are sometimes given to us *not* as desired and therefore available only to the agency-threatening third-person point of view but rather in the first-person point of view when we are perceiving the desirability of what we desire, where no stepping back is involved. It is only when we experience them in this latter form that we are in the perspective of the first person that carries our practical agency. In other words, when we step back and register that we desire something, that desire cannot induce the agentive aspect of our mentality. It sits there as the leaden object of our detached gaze. And so the challenge before us, with its claim that our agency consists in just our possession and pursuit of self-standing desires, is exposed as insufficient because it is only when our desires are not self-standing but given to us via the desirabilities that are given to us in our perception that desires will prompt in us the responses of our practical agency.

If in order to be practical agents we must be able to experience our desires not merely directly as elements located in our own mentality but indirectly via the perception of values or desirabilities in the world, what does this suggest about the nature of value properties in the world and their relation to our states of mind such as desires (and moral sentiments)?

I want to get to this relation between states of *mind* (desires) and properties in the *world* (values) indirectly by first looking at the nature of some emotions.[3] Aristotle had said of emotions such as anger that are directed at others that they *presuppose* certain beliefs about others in the subjects who undergo emotions. Thus, for instance, he said that my anger at someone presupposes the belief that he has done me harm.[4] This may seem like common sense but in many cases I think it goes against the phenomenology of our emotions such as anger and their relation to the world and our beliefs about the world. In many cases of anger, the word *presupposes* is the wrong description for the relation between the anger and the relevant belief, suggesting as it does that the belief that someone has done one harm *is all in place first, and then the anger wells up.* Rather the relation is more like the one we find in some accounts of pain (physical pain). On these accounts, pain is a way of perceiving parts of one's body. So take, for example, a toothache. We may perceive a tooth tactually, by touching it with a finger. We may go to a mirror, unfurl a lip, and perceive it visually. But we may also, more internally and involuntarily, perceive it by— and here we may run out of the right logical grammar—by "paining" it. And emotions are often similar. Often when we are angry with someone, it is not as if the belief that he has done one harm is in place before the anger, as the idea of "presupposition" suggests. One's anger is often *a way of perceiving that he has done one harm.* And, I am claiming that the right view of the relationship between our desires and the value properties in the world is something like that. I had said earlier that to be practical agents we must often perceive values or desirabilities and *in doing* so experience our desires indirectly via that perception of desirabilities, rather than directly as when we step back and scan our desires as elements of our mentality. And what that suggests is that desires are our ways of or conduits for perceiving the value properties in the external world.

This shows that we need to be a little careful when we say that we perceive value properties in the world that make a normative demand on us. It is not as if the value properties that prompt our desires are a source of normative promptings that are entirely alien to the kind of sensibility we have in *having* desires. Creatures (frogs, say) who lack desires of the sophisticated, discriminating sort that we possess will see darkness in the world where we perceive values.

At any rate, quite apart from these points about the exact nature of the relation between our desires and the value properties in the world, the overall and basic point of importance is that if there was no complex relationship of the kind I have been sketching that our desires bear to value properties *without*, we put into doubt the possibility of practical agency *within* us. It is often thought (in some writings of the *verstehen* and hermeneutical tradition, for instance) that the world around us is disenchanted but we are special and enchanted by the agency and intentionality and value-forming faculties we possess. If what I've been saying so far is right, that thought is shallow. There is no understanding whatever that is meant by our being special in these senses if the world itself were not possessed of value properties, just what the superstition I began with denies. (Caveat: Though I won't argue for it here, this talk of the world being "enchanted" with value properties should carry no implication whatever of revivalist ideas of a world laden with intentional vitalism for which Bruno Latour and Jane Bennett have recently argued in their writings on the environment. The rhetoric of "enchantment" is irresistible if *dis*enchantment is defined upon the superstition I have been resisting, but that rhetoric should not mislead us into unnecessary attributions of intentionality where it does not exist. To say that the world contains value properties attributes no intentionality to those properties, or any vitalist element.)

A way of summing up would then be to say that, if I am right in these arguments so far, ethics is primarily a perceptual discipline: the world, the perceptible world we inhabit, over and above containing the properties that natural science studies, contains value properties (or more simply values) that we perceive as making normative demands on us to which our practical agency responds. To stress this primacy of moral *perception* is not at all to deny the importance of ethical *deliberation*. Rather, deliberation is a secondary sophistication. For instance, it occurs when there are conflicting perceptions of value in the world or it occurs when we find that our practical agency falls short of or falls afoul of the normative demands we perceive in perceiving the value-laden layout of the world. In the former case, we do the usual ranking and weighing that go into much moral deliberation. In the latter case moral deliberation takes a reflective and self-critical form. But all these are sophistications at second remove. The primary framework of ethical life is a perceptual one.

I want to now turn to presenting—again in much broader stroke than it deserves—some rudimentary intellectual history from within which I will generate a dialectic about the theoretical nature of our political ideals and goals and their possible revision, a dialectic in which the metaethical conclusion I have just argued for over these last many pages will eventually return to play a facilitating role.

* * *

It is a large curiosity, indeed a perversity, that the political Enlightenment, as soon as it articulated its two great ideals of liberty and equality, proceeded with theoretical and methodological developments that put them in irresoluble tension with each other. There are many theoretical sources of this tension and I will briefly mention only two.

One source is too well known and well mined to bear much more than the most minimal mention, and that is the linking of the notion of *property* to a notion of the personal liberty that its ownership bestows on one, a liberty that is carried in a "right" and therefore enshrined in the law of the land. How the possession of private property, when seen in these terms, undermines equality in the economic sphere (and therefore in other spheres) has been the subject of extensive commentary, and Marx was, of course, only its most famous and most powerful critic. As I said, this topic is so well mined that it would be to belabor the obvious to say anything more about it at just this stage.

Less explicitly and familiarly mined is another feature, which I will call the *incentivization of talent*. It is the most natural thing in the world to think that someone's talent should be acknowledged as *hers*, and that it is *she* who should be praised and rewarded for its productions. We don't—at any rate, not in the past many centuries—simply praise and reward the zeitgeist for such productions. That would be a failure to respect someone's individuality. And with the entrenchment of a conception of liberty as a form of *individual* self-governance, one's liberty to reap the fruits of one's talent became quite as prominent and taken for granted as the other liberties. Notions such as "dessert" thus also get linked to one, among other, rights possessed by individuals. This goes so deep in our thinking that it is likely to be considered a hysterical egalitarian ideologue's artifice to deny it. Denying it seems to fly in the face of our intuitive understanding of what it is to be an *individual* (rather than just a symptom of the zeitgeist in embodied human form); it violates what we conceive to be the *liberty* of an individual to reap the praise and rewards of the exercise and efforts of his talents, not to mention the liberty of *others to enjoy* the productions of these efforts at their most excellent because they are *incentivized* to be as excellent as they can be. But, like the liberty attaching to possession of property, this way of thinking of liberty as attaching to talent also promotes social and economic inequality. This second feature of liberty attaching to the idea of talent is less structurally central to our culture than the liberty that is tied to property, but it perhaps goes even deeper psychologically and the dichotomy it generates with equality is, therefore, more subtly troubling; and it seems just as impossible to overcome.

There are several other features of the political philosophy we have inherited that one could summon to present the tension between liberty and equality, and I have mentioned only these two just to give a completely familiar sense of how far such thinking has gone into our sensibility, how entrenched it is in the very way we deploy these terms, and how, therefore, it would seem almost to change the semantics of the terms if we were to think that the tension could be removed or resolved. That is to say, if we managed to see them as not being in tension, it would only be because, as Thomas Kuhn might have put it, we have changed the *meanings* of the terms *liberty* and *equality*, not because we have produced an improved theory or politics *within* the framework of the Enlightenment. Within that framework things are, on this score, unimprovable. In other words, what I mean by framework here is perhaps one of the (diverse) things that Kuhn meant by his term *paradigm* and, if so, clearly we need to shift to another framework if we are ever going to remove the tension between these two notions. In such a new framework, neither "liberty" nor "equality" would *mean* what they mean in the framework of Enlightenment thought, no more than "mass" in Einstein's physics meant what it meant in Newtonian mechanics, if Kuhn is right.

How might such a shift in framework be sought? Here is a proposal. Let's, as a start, usher the very ideals of liberty and equality off center stage. If this is to disinherit an entire tradition of liberal thought of the Enlightenment, so be it. Once these exeunt, we need to replace them on center stage with a third, more primitive concept, that is to say, a concept more fundamental to our social and political life than even liberty and equality. And this is to be done with the idea that "liberty" and "equality" may *subsequently* be introduced once again—by the back door, as it were—but now merely as necessary conditions for the achievement of this more basic ideal that occupies the central position. So when they are reintroduced, there is reason to think that these terms may have undergone substantial revision in their meaning, and thus may not any longer express concepts that are at odds with one another.

We need, then, to fasten on an appropriately more fundamental concept. To be more fundamental than concepts such as liberty and equality, which have been so central to our theoretical understanding of politics, it would have to track something not necessarily older and more traditional in our political understanding so much as something that speaks more immediately to our experience and our ordinary lives, and I suggest that a natural concept and ideal on offer for this role is that of an *unalienated life.*[5]

What is that?

The term *alienation* has had a remarkable elaboration in the early work of Marx, and though that remains very much in the back of my mind as I proceed now, I will not, in these remarks to come, be speaking at the level of specificity that Marx does, tying the phenomenon centrally, as he does, to the experience and consciousness of the population of wage-laborers in particular. I will instead step back and speak of alienation at a much higher level of generality. Moreover, as Marx himself and many others discuss it, alienation takes many forms and has many aspects, but, given the level of generality at which I am addressing the issue, I will focus only on one, in my view very central, form and aspect that it possesses.

Here is how I've allowed myself to think of it.

It's hard to say what an "unalienated life" is in any direct statement that elaborates a definition, so I will come to it indirectly by trying to uncover the thinking, the mentality, that underlies the sources by which liberty (as it got attached to property and talent) grew to be in tension with equality in our standard orthodox liberal traditions.

One can convey the thinking in a variety of ways, but let me do so by a brief consideration of one prominent strand in the intellectual history by which the attaching of liberty to property, in particular, came to be seen as a form of moral and political achievement, elevating what was happening hitherto by predatory force in the enclosures movement into a form of right and rationality. In the early modern period one particular social theory argued with clarity and with the force of the great intellect of its propounder for the political *rationality* and therefore the historically progressive *necessity* of transforming nature not merely as a site of living and livelihood, but as a resource for a specific form of gain, what in our time we call agribusiness as against mere agrarian living. (The claimed rationality is repeatedly invoked to this day as the basis for further claims that the rest of the world, including Europe's erstwhile colonized lands, would have to inevitably adopt these rational political and economic arrangements as a historically progressive and therefore *necessary* form of development. Alas, even so humane an economist as Amartya Sen has recently done so, an indication of how widespread the grip of this liberal assumption about political rationality is.)[6]

This theory was contractualist in conception. There are, of course, several powerful strands in social contract theory—Hobbes, Locke, Rousseau, and Rawls and Nozick in our own time, to name only the most well known. On the questions I have just raised, the strand that is relevant is the one that owes to John Locke.[7] I will focus on this strand. The point of all social contract theory, whether Lockean or any other, is to establish that in an originary scenario described as "the state of nature" (in a later idiom and theory, "the original position"), which is a prepolitical condition, *freely chosen consent* by a people to certain principles or arrangements to live by immediately transforms those people into citizens and the state of nature into a polity—but it only does so if the consent to those principles and arrangements is demonstrated to be rational in a very specific way: the principles and arrangements freely consented to must make these people *better off* as citizens than they hitherto were as mere people, prior to polities, in a state of nature. Thus, there are two conditions that must be met to ensure the rationality of the outcome. The principles and arrangements contracted into must make us *better off* and our implicit consent to them must be seen as a *freely* made consent.

In the strand I am concerned with the canonical scenario has it that were someone in a state of nature to come upon a stretch of land in the common and fence it and register it at an elementary form of office that they set up for this kind of registry, the land becomes *his* (obviously, no other pronoun is apt given the time in which the theory was first propounded). Suppose then that this is done by some of the people and they each keep faith with (a somewhat fuller statement of) the first general requirement I mentioned earlier—namely, this can only be done if no one is made worse off and at least some are made better off than they were in the state of nature—a requirement that they then elaborate further by adding the following crucial clause: if those who had thus claimed ownership of the land were then to hire others who had not done so, at wages that enabled them to live better as well, then this too would be an arrangement that is justified since they too are in fact better off than they were in the state of nature.

Such was the explicit claim of the Lockean ideal of the social contract (roughly an argument from Pareto-improvement), which went on to become the cornerstone for certain political principles and arrangements that came to be called liberalism in which, among other things such as free speech (except for atheists, Catholics, and so on), private property and wage labor were seen as progressive advances justified by the mutual advantage or amelioration of *all* concerned (or, in the limiting case, amelioration for some and no resulting *dis*advantaging of anyone else).

When one asks the question of what in the historical context was motivating the articulation of such a contractualist theory, the answer has to be that the theory philosophically consolidated the system of enclosures, which, as I said, had been practiced by brute force for many decades up to that point, and in doing so it prepared the ground for it to become a form of right with law and governance to back it up. The point was to present the political principles and arrangements which justified the system of enclosures as a moral and political achievement since it was based on a form of rational and freely chosen consent that generated a compact that created a citizenry in its inhabitation of nature and its relation to the sources of livelihood and production, and the polity to sustain it in these relations.

Marx's twenty-seventh chapter of *Capital*,[8] which presented in critical detail the predatory nature of such primitive accumulation in general, but also of the enclosures in England in particular, had its premonitional anticipation in the widespread protest against the enclosures among some of the radical groups during the English revolution who predated Locke but whose protest on behalf of the quite different ideal of the collective cultivation of the existing commons can be seen as seeking to preempt the claim to rationality in Locke of such an implicit consent that he had attributed to all in the originary scenario of a state of nature. Let me then construct a specific counterargument against the Lockean contract and attribute it (completely anachronistically, of course, since these dissenters predated Locke) to these dissenters as the theoretical source of their protest and as proposing instead an *alternative notion of consent*. So, someone like Winstanley could have been heard as anachronistically saying:

The entire contractualist scenario as you have presented it generates an opportunity cost. An opportunity cost is the cost of an avoided benefit paid for making a certain choice. That avoided benefit is the collective cultivation of the commons that is prevented by the choice to privatize the land in your initial step in the scenario. Once the step is taken, it is true what you say that those who were hired for wages are better off than they were in *the state of nature* but they are not better off than they *would have been* if the land had not been privatized in the first place and if there was a collective cultivation of the commons instead.

The criticism is based on a relatively simple counterfactual. But despite its simplicity, its theoretical effect is complex and interesting because, as I said, it proposes a quite different notion of consent than the one that Locke assumes. Consent must now be viewed as a more complicated act than Locke understands; it should be viewed as follows: Whether someone can be said to have consented is not necessarily to be viewed as this tradition proposes but rather to be viewed as what he or she *would choose in antecedently specified sorts of conditions that do not obtain*—in which case the entire Lockean tradition of thought may be assuming that we have implicitly rationally consented to something that we in fact have not.

If in this way we shift the focus of this imagined dispute between the pre-emptive Winstanley and Locke to a dispute about which notion of implicit consent is at stake in the social contract, a further issue opens up about the nature of freedom and coercion. Suppose that Locke were to respond by saying: "I have offered a perfectly good notion of implicit consent and I see no reason to accept yours." Winstanley's response would then presumably have to be, "If you ignore my counterfactual and insist that the sense of consent you have on offer suffices in the contractarian scenario and that everyone has indeed implicitly consented in that sense, then I will have to point out that the implicit consent you have attributed in particular to those who are hired to work for wages was *coerced* by a condition that they could not avoid, their nonpossession of the land (in the face of the possession of the land by others). *My* alternative notion of consent was articulated with the view to establishing that that nonpossession (in the face of possession by others) should be seen as avoidable. Thus, your insistence on your notion of consent, even despite the assertion of my counterfactual, brings out in the open that possession of the land by some and not others is a coercive condition in which the latter has to 'consent' in *your sense* of the term. And so the contractualist tradition presents a *coerced* implicit consent fraudulently as a *freely chosen* implicit consent."

I had said earlier that a great deal of social theory presented the developments in political economy in the early modern period as advances in political *rationality*, and that it was this claim of rationality that was invoked as the basis for recurring further claims over the centuries down to our own time that the rest of the world should adopt these rational political and economic arrangements as a historically progressive and therefore *necessary* form of development. But, if I am right, the premise of the argument, i.e., the claim to the rationality of the contractual ideal that philosophically rationalizes historical developments (initially) in England, depends on two things: (a) on what is consented to making one better off than one was in the state of nature and (b) the consent being freely made. And if the criticism attributed as implicit in Winstanley's dissenting stances is correct, *these two conditions cannot be satisfied jointly.* The counterfactual notion of consent offered by Winstanley's implicit criticism makes clear that the first requirement has not been met, and if you simply deny the counterfactual notion of consent, the Lockean notion of consent fails to meet the second requirement that the consent be freely chosen.

Does this argument suffice to undermine Locke's contractualist conclusion? It would be premature to think so since Locke's account has been implicitly fortified in more recent times, by theoretical developments in the liberal tradition that I am addressing.

I have argued that Locke's argument for privatization of property, when fully laid out, has a rebuttal in an argument from opportunity cost that can be attributed (anachronistically) to the radical Puritan dissenters. The rebuttal I attributed made counterfactual use of the ideal of a *collective cultivation of the commons.* This was essential to the argument. But it is not as if the long tradition of liberal theory which that rebuttal was targeting (a tradition in which Locke was only a very early influential founding figure) did not deploy *further* conceptual resources to try to undermine this ideal of the commons that was essential to my efforts. So, I now want to address one central strand of such resources that updates the Lockean picture in *contemporary* liberal assumptions with a fortifying argument that, I believe, can only be rebutted by invoking considerations having to do with human subjectivity and alienation.

Perhaps the most standard resource that liberal theory relies on is the idea and argument behind what has come to be called the "tragedy of the commons," which in its more sophisticated form is formulated in terms of a multiperson prisoners' dilemma. The idea is to raise an intractable problem for any ideal of cooperative life, such as collective cultivation of the commons (or in our own time the pressing concern about the preservation of the environment, fresh air, say, being the relevant common). The intractable problem that is supposed to arise is that individual human psychology at its most rational is required to behave in ways that undermine the collective by failing in the cooperation needed to keep it going. This is because the collective ideal asks the individual to contribute resources (sometimes restraint may be a negative form of contribution of resources, when the goal is, say, to prevent overuse or overcultivation of the common) that produce a benefit that is shared by and therefore divided over the whole collective while the cost is borne by each individual. If everybody does what is required of him or her, of course everyone gains. But since one is in the epistemic dark about whether others are contributing their bit of the resources demanded of them, one is constantly filled with qualm that one's contribution would be wasted if others don't do their bit. In such an understanding of the collective ideal—which is pervasive in liberal frameworks of social thought—some individual commoner who decides not to cooperate therefore is always at an advantage since the gains of noncooperation will accrue all to oneself and they will be immediate as well as assured, whereas the gains from cooperating will be divided over the whole group and will be long-term and, as just said, uncertain because the cooperation of others cannot be taken for granted. Noncooperation for him, as an individual, would thus be rational. But the common cannot survive if each individual commoner does this individually rational thing. It is doomed. Thus the tragedy. So privatization is a better bet. This very briefly expounded argument sums up a widespread outlook of liberal political economy that had developed over a very long period since Locke and was summarized in a very well-known paper by Garrett Hardin called "The Tragedy of the Commons." It might be thought of as a game-theoretic updating of Locke so as to deal with the anti-Locke argument from opportunity cost, which I had made on behalf of the radical dissenters.[9]

In recent years, Elinor Ostrom has responded to this argument of the tragedy of the commons by claiming that it should be seen not as an argument for privatization of the land of the commons, but rather as an argument for the regulation of the commons by policing, detection, and punishment. (I am focusing only on one lesson intended in her work *Governing the Commons*, which is an impressively detailed and effective antidote to the scare tactics of an impending tragedy. The seven principles of common pool resource governance that she lays down are drawn from close empirical work done in a number of commons in several different continents, which show the specter of tragedy to be highly exaggerated.) Now, it would be foolish to deny that regulation via policing, detecting, and punishment of noncooperation is a good thing and should be deployed to avoid a tragedy that owes to noncooperation. And no one has presented a more convincing alternative picture to the prevailing liberal consensus around Hardin's paper than Ostrom.

But what I'd like to do is to ask what a philosopher is bound to ask: What are the underlying lessons for the *mentality* that must underlie any alternative approach to the commons, if we are not to be browbeaten by the argument from tragedy? The reason for this is that, quite apart from the well-known difficulties of detecting many nonobvious forms of noncooperation, the real problem is that the *same multiperson prisoners' dilemma–style argument* can be raised for *why anyone should cooperate with policing and punishment* if he can get away with *not* cooperating—by offering bribes, for instance, or making mafia-style threats against those who detect and police or those who cooperate with the policing and detecting, or, in political cultures that congratulate themselves on having outgrown a culture of bribes and threats, by loopholing the law. The deeper and underlying flaw in the "tragedy of the commons" idea lies in its basic way of thinking and that can't be rectified by solutions like "regulation" and policing, solutions that are vulnerable to the same strategy of argument that generated the "tragedy of the commons" in the first place.

Without probing beyond the idea of regulation and policing and punishment, however laudable and necessary such regulation is, and without seeking an alternative *outlook* to the liberal notion of rationality that is found in the tradition from Locke down to Hardin, it will rightly be said that Locke, anticipating this argument of the tragedy, was correct to see his version of the implicit consent of all contractors (possessors and nonpossessors of land) as rational, indeed *even freely chosen* if the contractors had an implicit or tacit understanding of the looming threat of the "tragedy of the commons." By contrast, it will be said, the counterfactual-based notion of implicit consent that I have put in the mouth of Winstanley and the dissenters, which vitally depends on the possibility of a collective and cooperative cultivation of the comments, is hobbled precisely by the fact of a looming "tragedy" of the commons, unless a more fundamental critique is provided of the thinking involved in the very formulation of the argument that there is a tragedy that looms.

It is just here that the ideal of an unalienated life begins to surface in relevance to identify the deeper flaw.

So I ask again, what is this ideal of an unalienated life? And I am a little better situated to spell that out (though it will have to be brief in a paper that is already quite long), keeping in mind the eventual task I have set myself—of defending the counterfactual-based notion of consent of the dissenters against Locke's notion of implicit consent and against the particular way of thinking that generates the tragedy of the commons that would sustain and fortify Locke's idea of such a consent to a privatized economy.

* * *

The first thing to note is that the term *an unalienated life* even as I have wielded it is ambiguous. One sense of it is the unalienatedness that came with the sense of *belonging* that was made possible by the social frameworks of a period prior to modernity. All political and social theorists, including Rousseau and Marx, have tended to agree that whatever the defects of societies prior to modernity were, alienation was not among the defects. But the point remains that, as is well known and widely acknowledged by the very same theorists, the unalienated life of those earlier times was deeply marred by the oppressive defects in those societies. (To say "feudal" to describe that form of oppression would be merely to use a *vastly* summarizing and somewhat misleading category that we have all been brought up on.) It is precisely those defects that the sloganized ideals of liberalism, Liberty and Equality, were intended as *directly* addressing. And I have argued that since the methodological and theoretical framework within which those two concepts were then developed made it impossible to so much as conceive how they could be jointly implemented, we should no longer see them as something to be *directly* deployed, but rather as indirectly deployed—merely as necessary conditions for the achievement of a quite different (directly deployed) ideal—thereby transforming the concepts of liberty and equality. Now, if the achievement of an ideal of an unalienated life were to bring in its wake—*indirectly*—conditions of liberty and equality (however transformed), it is bound to be very different from the unalienated life that is acknowledged to have existed in times prior to modernity because the conditions in which it existed then were also acknowledged to be acutely *lacking* in, precisely, liberty and equality. Thus, given this rudimentary conceptual dialectic, what we need to show is how a new framework that breaks out of the dialectic would solve for *three* things at once—a transformed notion of liberty and equality, as I have said from the outset, *but also it would now seem a transformed notion of the unalienated life*. So, this is to be conceived as a holistically *triangular* transformation—we overcome a certain conceptual-historical dialectic and in doing so *together and at once transform all three concepts* that feature in the dialectic.

To do this vastly ambitious thing, we have to ask, first, what can be retained of the general idea of social "belonging" of an earlier time in any revision of the idea of an unalienated life for our own time. We know from the other elements of the dialectic that the social belonging of an earlier time was marred by the defects of a lack of liberty and equality, but we also know from what I have said that the attempts to directly overcome those defects were, in turn, marred by the fact that liberty and equality flowered in conception within a social framework in which a highly *individualized* notion of individual liberty that attached to property and talent made for liberty's conceptual incoherence with equality. *That* was the fundamental source of the shortcoming of the liberalism that emerged out of the standard political Enlightenment. So it would seem logical then, given this entire dialectic, to conclude that a concerted and triangulated transformation of all three notions would have to find its first hook, find its initial roots, in *individual* liberty being conceived in nonindividualistic terms. It is only such a notion, I think, that could be of a piece with the ideal of an unalienated life.

Liberty, in its broadest and most intuitive sense, is the idea of self-governance, the power to make the decisions that shape the material and other aspects of our lives. If so, it would seem then that to transform the notion of liberty in the way that we have just seen as being required, we would have to envisage each individual as approaching these decisions not primarily with her own interests in mind but the interests of everyone in society. Now, the last few words of that last sentence express something utterly familiar, a cliché, a piety. The critique of self-interest has long been with us. Moreover since my goal is to show the shortcomings of the argument from the tragedy of the commons, which is manifestly based on individual rationality as conceived in terms of individual self-interest, it is not all that interesting to just *say* that one is opposed to self-interest. But I am *not* merely saying that one should be opposed to self-interest. I am hoping that I am saying something more interesting because I think it is far less well known and hardly at all theoretically developed that what such a critique of self-interest amounts to *is the construction of a notion of liberty.*

Why is that little known and developed? I repeat: because individual self-governance (i.e., liberty) has for so long been viewed in individualistic terms. But what is it to have a *non*-individualistic conception of *individual* self-governance? Just to be clear: it is *not* group or collective self-governance, which is a different notion (interesting in a different way) but not relevant to liberty that is felt and exercised by individuals (as indeed is alienation). Rather, it is something like this. And here I return to the metaethical conclusion of the first section that I have held in reserve.

The world (both the natural and the social world) I have said contains value properties that present us with perceptible normative demands to which our (practical) agency must respond. If this is so, then, as Aristotle first suggested implicitly in the various ingredients that go into his notion of phronesis, we have to *see the world right*, to *see correctly* what its normative demands are. The point is essentially phenomenological (a further reason not to see this part of the argument as merely being a repudiation of self-interest). And to be unalienated, I believe, is for our agency to be in responsive sync with these perceptually presented normative demands coming from the world as we apprehend it, when we apprehend it correctly. That is what it is to be at home in the world—to be sensitive in our perception to the demands that the value elements in the world make on us and to respond suitably with our agency to those demands. And the centrally important point then is that for one to *see* these demands of the *world* for what they are, *one's own orientation* to the world in viewing the normative demands of its value-laden layout has to be primarily not through the lens of one's individual point of view but from a larger point of view.

Consider a physical analogy that needs to be extrapolated to the social—how when one (an individual) drives a car one orients oneself perceptually to the demands of the road ahead not from the point of view of *one's own individual body* but from the point of view of something larger than one's individual body, that of the *whole car*. This is obvious when one considers how when (say, before one starts the car to drive) one turns to talk to someone sitting next to one in the car, one's orientation to the world is from the point of view of one's own individual body onto the world (to the person next to one). It is only when one turns away and looks ahead to the road to drive that one's orientation toward the world is not at all from the point of view of one's individual body but rather navigating and responding to the demands of the world from a larger point of view, the point of view of the car. If that transition did not happen, one would crash the car. In a sense, then, with the proper extrapolation of this bodily analogy to the social, the tragedy of the commons is akin to the tragedy of such a car crash.

In the dialectic I have set up in this paper, the idea of such a larger orientation when extrapolated from this physical or bodily example of the car to the *social* has a very significant outcome. Even though it may involve the mentality and agency of individuals, because they each exercise their *liberty* or self-governance in perceiving and agentively responding from the point of view of something larger, the social or collective orientation of each individual to the world (seeing the world's demands from the point of view of the collective) is bound to *internally* cohere with equality in its outcomes. For equality would on this picture not be seen as something *extra or further* that is conceptually configured as something to be navigated in terms of a trade-off with liberty, but rather as *built-into* the deliverances of the exercise of *liberty* itself, when the exercise of liberty is the exercise of a mentality in this form of unalienated agentive responsiveness to the normative demands of the "world," properly perceived along the lines I've just briefly sketched. (I want to make clear what should in any case perhaps be obvious, that my theoretical motives here are not to deliver equality itself—no mere philosophical argument could do that—but to remove the chronic trade-off relations between these two ideals that are found in the liberal tradition that I am trying to reconfigure and transcend.)

If all three notions, liberty, equality, and the unalienated life, are triangulated in this way together, we have a notion of unalienatedness that is not the same as the one of premodernity with its absence of liberty and equality, and we have a notion of liberty that is not generative of inequalities, unlike in the liberal framework, where it is individualistically conceived (in the form that attaches to talent and property), but rather a notion of liberty that is nonindividualistically conceived in the way that I have just outlined. In this new framework or paradigm, it is quite impossible to even so much as raise the difficulties that lead to the tragedy of the commons. To even so much as have the qualm, "Would my efforts and contributions to the collective cultivation (or restraint from overcultivation) be wasted if others don't also contribute?" *is already to be thoroughly alienated*, by the lights I have set up in the ideal of the unalienated life I have just presented. If this is right, then it should follow that if one pursues the path of this dialectic I have set up about how to reconfigure a trio of ideals, the tragedy of the commons far from fortifying Locke simply exposes how Locke's notion of consent has no superior rationality to my imagined Winstanley's alternative notion of consent.

Let me come to the significance of this point about the unalienated life from a different angle—by seeking an extension of Wittgenstein's idea of a "basic action." Wittgenstein, after the discussion of a celebrated example, had concluded that an individual's *basic* intentional action is an action you do and there is no other action you do in virtue of which you do it. In the example, he asked: What is the difference between my waving to a friend and my arm going up? He had in mind to criticize philosophers who say that the intentional act of waving (to a friend, say) is to be considered as follows. There is something more conceptually *basic*, the physical arm going up (or my moving my physical arm). Then an intention (or further intention) *accrues* to this basic element and that accretion conceptually transforms it into a waving to a friend. By contrast, he insisted the right view is that what is basic is the *waving to a friend*. Full stop. That's it. It (the intentional act, the waving) has *no conceptual breakdown* into other more basic elements and accruals upon them.

Wittgenstein, in formulating the notion of a basic action, was concerned with the idea of individual *intentionality*. And I am asking whether there might be an extension of Wittgenstein's thinking here, regarding the notion of a basic action, that reflects an idea of individual sociality. One might even, straining what Wittgenstein calls "logical grammar," say this: Wittgenstein was interested in individual intentional action, and my effort to extend his idea of a basic action is to find it relevant also to individual social action. (The last three words seem so odd to us only because our conceptual frameworks are impoverished by the fact that the idea of an unalienated life is not—where I've been struggling in this paper to place it—on *center* stage.) A phenomenology of the normative demands coming from the value properties in the world by *seeing* them from an orientation larger than our own, I am suggesting, might help with the extension. That was the point of the bodily analogy with seeing the world from the point of view of the car rather than one's own individual body. If the social version of that phenomenology were in place, an individual's liberty-constituting or self-governing actions would not emerge from a more basic form of thinking that is present in the formulation of the tragedy of the commons argument. The tragedy of the commons argument, the very idea of such a tragedy, is only intelligible within a framework that assumes that if there is any idea we have of the social, *it is built up via accruals attaching to a more basic form of thinking*; and it is this form of thinking that is then pervasively taken for granted and becomes the basis for the argument that leads to the tragedy, just as the idea of the intentional (in the view that Wittgenstein repudiates) is built up on accruals attaching to something more basic.

This implies that to even have the qualm, "What if I were to do my bit and other's didn't?" would be akin to saying that the arm movement is having a qualm about the waving to a friend, an incoherent idea if ever there was one. I mean this quite literally. In an unalienated life and society, that qualm (what if I did my bit and others didn't?) which underlies the rationality that fuels the tragedy of the commons is *literally* as incoherent and unintelligible as the arm movement being said to have a qualm about the waving. And nothing short of establishing that incoherence captures the ideal of an unalienated life. To be unalienated is to be free of a certain malaise, but since that malaise gets a rather abstract description, to be "unalienated" itself must be understood in relatively abstract terms. It is not to have sympathy or feel fraternity with others or to show solidarity toward others (see note 5), good though it is to have and do that—not all good things are the same good thing! Rather, it is to be free of a way of thought in which—when we make the decisions we make in governing ourselves as individuals in the exercise of our liberty—we do not just find it wrong but we cannot even find it *so much as intelligible* or coherent to have the qualm in question that leads to the tragedy, suggesting that game-theory is itself a higher-order symptom of our alienation; and to be free of that way of thought is simply (well, "simply" is an ostentatious bit of rhetoric here) the other side of a fitting phenomenology of value. To *see* the world and its value properties that make normative demands on us aright and to overcome alienation are not two things but one.

Notes

1. Hume and Adam Smith are the classic accounts that eventually rest the basis of values in desires and moral sentiments. McDowell's account of Aristotle is most clearly presented in John McDowell, "Virtue and Reason," *Monist* 62, no. 3 (1979): 331–350. McDowell himself has interestingly developed Aristotle's view in his own terms in subsequent papers via criticisms of Mackie, Blackburn, and Philippa Foot. See the essays in section 2 of McDowell, *Mind, Value, and Reality* (Cambridge, MA: Harvard University Press, 2001). On one crucial point, though, McDowell fails to see the implications and significance of taking the Aristotelian view, which, I think, undermines the coherence (I repeat coherence, not truth) of all ideas of the supervenience of value properties on nonevaluative natural properties. This has serious consequences for how to understand the relationship between value and agency, which is crucial to the argument I am making in this essay. For an exchange on this subject, see the essay by Akeel Bilgrami in *McDowell and His Critics*, ed. Cynthia McDonald and Graham McDonald (Hoboken, NJ: Wiley-Blackwell, 2006), and the reply by McDowell, and chapter 5 of Bilgrami, *Self-Knowledge and Resentment* (Cambridge, MA: Harvard University Press, 2006).

2. Gareth Evans, *The Varieties of Reference*, ed. John McDowell (Oxford: Oxford University Press, 1982).

3. The point I am about to make about emotions is inspired by P. F. Strawson's claim about something like resentment in his remarkably innovative essay "Freedom and Resentment." P. F. Strawson, "Freedom and Resentment," in *Freedom and Resentment, and Other Essays* (London: Methuen, 1972). There he argues that it is through reactive attitudes such as resentments that we apprehend that there are *agents* present in the world.

4. In any edition of Aristotle, this may be found in *De anima* 1.1.403a16–32, where he distinguishes between what the natural scientist and the dialectician say about rage. He further takes up this point of view in *Rhetoric* 2.2.1378a31.

5. There is a great temptation here, though a glib one (owing to a preexisting trio of ideals within the Enlightenment), to suggest "fraternity" instead of "the unalienated life" as the more basic ideal for what I want to say. The temptation should be resisted for a wide variety of reasons that I will not present here. Those reasons would also show why the notion of sympathy in Hume and Adam Smith (which is meant to be part of a picture of values that stands opposed to the picture presented in the first section) is quite the wrong notion with which to theoretically develop the notion of an unalienated life. I hope to develop those reasons in a sequel to this paper.

6. For example, even so humane an economist as Amartya Sen in his "Prohibiting the Use of Agricultural Land for Industry Is Ultimately Self-Defeating," *Indian Telegraph* (Kolkata, July 23, 2010), said against those who protested the "eminent domain" dispossession of the peasantry of rural Bengal and other tribal foresters of the commons in various parts of the country: "England went through its pain to create its Londons and Manchesters, India will have to do so too."

7. John Locke, *Second Treatise on Government* (1689; Nicholson Press, 2013).

8. Karl Marx, *Capital: A Critique of Political Economy*, vol. 1 (1867; New York: Penguin, 1976).

9. Garrett Hardin, "The Tragedy of the Commons," *Science* (1968).

CHAPTER 8

EQUALITY AND LIBERTY

Beyond a Boundary
Response to Akeel Bilgrami

SANJAY G. REDDY

Akeel Bilgrami has written a profound essay, employing a powerful philosophical imagination to reach to the heart of our contemporary dilemmas and difficulties. He argues that the way out of these dilemmas and difficulties, which beset our politics, society, economy, and ecology, is to see them differently, by adopting a new conceptual lens, which involves accustomed terms being understood in an unaccustomed way, not so much to resolve the problem as to dissolve it.

The concept of dissolving a problem rather than (re)solving it came to my attention some time ago through study of Piero Sraffa, who, like Ludwig Wittgenstein, appears to have been greatly influenced by this idea, possibly introduced to them both as young men through their encounter with a then prominent textbook of classical mechanics by the German physicist Heinrich Hertz, *The Principles of Mechanics* (1899).[1] Hertz's textbook emphasized the ability to resolve problems through a change of perspective, permitting even apparently complex problems to be addressed.[2] It may be argued that Sraffa and Wittgenstein were each, in different ways, attempting to apply a similar method. The method aims to generate clarity where there is obscurity by finding a vantage point from which one may substitute a simple description for a complex one, without any loss of pertinent information. In doing so it is often found that it is not necessary to put effort into solving an apparent problem, since the change of description has caused it to become a nonproblem (i.e., to be *dissolved*). Bilgrami has adopted a similar method.

Addressing the "possessive individualist" perspective on individual free-dom most closely associated with Locke, Bilgrami points out the contradic-tion between the understanding of personal freedom in this tradition and interpersonal equality. The Lockean answer to egalitarian claims emphasizes that persons are made better off under a regime of individual and very proba-bly unequal property rights than in the state of nature. Bilgrami notes, citing Winstanley, that this idea neglects another counterfactual, which involves alternative forms of productive organization, including the formation of worker collectives that might, in theory at least, make ordinary workers better off than they would be under a private property rights system. However, that proposal in turn meets another response from the Lockean "possessive indi-vidualist," which is that such alternative forms of productive organization must necessarily fail, due to the existence of collective action problems ("free-rider problems") that are ostensibly inescapable in the presence of "rational" agents. Bilgrami responds to this objection with a novel argument that meets it on philosophical ground, insisting that no such thing is mandated by ratio-nality. Insofar as Locke's argument is based on an empirical counterfactual (concerning the improvement constituted by society over the state of nature) an alternative empirical counterfactual, if it could be shown to be more con-vincing, might seem to be perfectly adequate to the task of overturning his argument. However, Bilgrami, in suitable philosophical fashion, wishes to provide a conceptual and not *merely* an empirical overturning, as important as the latter may be for cementing the case for an alternative conclusion. As a result, he argues that the so-called free-rider problem is no problem at all, philosophically speaking, since a proper understanding of our ends and of the reasoning that should go with them would lead us in an entirely different direction: not seeing the free-rider problem as a conceptually ever-present problem potentially empirically avoided (as demonstrated by Elinor Ostrom and others),[3] but rather seeing it as a nonproblem generated by a defective way of seeing. In Bilgrami's perspective, seeing from the larger point of view is mandated by nonalienated rationality, a perspective in which equality and lib-erty may be fully reconciled. The choice of such a perspective dissolves the apparent tension between the concepts, by generating a "holistically triangu-lar transformation," which changes our understanding of all three concepts. As a result, the problem is turned into a nonproblem, and a fundamental and complete solution is made possible, avoiding any need for a weighing or trading-off.

Within the alienated perspective from which the prisoner's dilemma or the free-rider problem is present, the argument for "enlightened self-interest" is in part instrumental: that seeing from the larger perspective allows everyone to be made better off. Consider Bilgrami's two alternatives to the state of nature, the Lockean scenario of the establishment of a private property system (P) and the cooperative management of the means of production (C). As he rightly notes, the Lockean argument is that a private property system makes everyone better off (as least materially speaking) as compared to a state of nature (S); speaking anachronistically, this is to claim that P generates a "Pareto-improvement" over S and thus provides a social contract argument for P. Now, what if cooperative management (C) also makes *everyone* better off than in the state of nature (S), as Bilgrami imagines? There are three possibilities, namely (i) C makes some better off than in P and others worse off, (ii) C makes everyone better off than in S but worse off than in P, and (iii) C makes everyone better off than in S and better off than in P. If (iii) is true, then no one would have a reason to make an argument for P (and Locke's argument becomes one for C rather than P). If (ii) is true, then no one would have reason to make an argument for C (and Locke's argument for P goes through in an extended form). The conceptually interesting and important case is (i). One might also add that it is empirically plausible (we may think of cooperatives of sheep farmers as against large landed estates on which sheep farming takes place). In this case, the argument for universal consent for C over P must turn not on narrowly self-interested considerations but on seeing from the larger point of view enabling agents to see that they have reason to choose C over P despite the possibility that it may be contrary to their (narrowly and inappropriately conceived) self-interest.[4] From this perspective, the importance of Bilgrami's decision to focus on the *conceptual* argument becomes clear. Only by doing so can one potentially make an argument for C even in cases where doing so involves an element of material sacrifice for some. Of course, C and P are polar conceptual abstractions and more generally one may wish to consider alternatives that include a cooperative component as against those that are purely privatist. However, the extent to which C promises a benefit to the majority as compared to P is an empirical question that would have to be assessed when analyzing empirical choices for a political or economic system, even from the "larger point of view." The choice between systems must inevitably be shaped by both facts and values. Understandings of the kinds of reasoning that are available to and mandated of us given our natures and our historical circumstances are crucial here. The perspective of nonalienation does not provide the basis of an irrefutable argument but rather a point of view through which we might hope to find a resolution, or "overcome" a prior "conceptual dialectic." If the reconciliation of freedom with equality required merely that we recognized the greater *efficiency* of a system embodying the latter (i.e., its ability to do better by everyone) then no philosophical work as such would be required to do so. It is because more is required that the perspective of nonalienation might be thought to provide the key to dissolving and thus solving the problem.[5]

Nonalienation may be sufficient to dissolve the problem of collective action and thereby that between freedom and equality, but is it necessary? It is imaginable that a larger point of view might be arrived at through other means, e.g., enlightened self-interest arguments or perspectives of obligation which broaden the reasons that an individual is permitted to recognize without substituting, as Bilgrami wishes to do, a nonindividualist conception of how individuals might approach the exercise of liberty. On a conception of rationality that centers on having good reasons to do what one does rather than mechanically and instrumentally pursuing "payoffs," the free-rider problem is not mandated.[6] Here and elsewhere, as we shall see, more pluralism may be possible in the resolution of the dilemma than initially recognized.

The example of driving a car illustrates beautifully Bilgrami's point that one may see from a larger point of view, and indeed that it may be necessary to do so, in order to engage in practical action. On the other hand, it also illustrates the possibility, and indeed the necessity, of moving between perspectives—larger and smaller ones, so to speak—in order to occupy the distinct roles that make up the aspects of practical agency (we may think of moving between driving, turning on and off the car radio, and speaking to passengers in the car). Even if we accept Bilgrami's argument for a larger point of view in "dissolving" apparent problems of collective action, we may also recognize that this sort of "moving between" may play an important role in our lives. It is possible, and indeed proper and necessary, for persons to recognize distinct perspectives, for instance, in order to occupy the different roles that they have (e.g., as individuals or as members simultaneously of communities of different scales and types). Condemnable though sectarian tensions may be, distinct roles and obligations beneath the level of the community as a whole provide part of the richness of existence and cannot be viewed only as manifestations of "alienation." The perspective of nonalienation may provide a way of "dissolving" the so-called free-rider problem but it leaves important issues unresolved, including that of how it can apply across all social scales, roles, and configurations of identities, when these may give rise to distinct legitimate appeals that are not always easily reconciled. We may think, for instance, of the citizen's problem of how much to contribute to provision of public goods in the city, county, province, and country in which she resides and the world as a whole. The idea that the perspective of nonalienation eliminates the tension between levels or forms of concern may appear to be at least hopeful and at worst metaphysical.

On Bilgrami's account, to even admit the problem of the free-rider is already to occupy an alienated consciousness. The role of his analogy to the Wittgensteinian concept of a basic action here is to bring out a way of seeing in which the nonalienated point of view is complete and freestanding. Appealing as this purism is, it may not be tenable. The very fact that we are engaged in the investigation of how to reconcile freedom and equality, and that we feel the need to concern ourselves with the presumed free-rider problem, demonstrates that we are already "fallen." To view a "nonalienated" point of view, from which the free-rider problem and the conflict between freedom and equality dissolve, as such (a point of view) is not a corrupting compromise, but a recognition of who we are and where we are situated. The idea of a basic action is that it is not further reducible to subsidiary descriptions, but is simply that. In this view, only one description (the unitary one, of the waving of the hand) is tenable. In a contrasting, mathematical metaphor, a change of basis may provide more than one description of the same object. No one description is uniquely correct, even though some may, when describing a specific object, be simpler or indeed more appropriate to certain purposes or otherwise have reasons to recommend them, as compared with others.[7] The idea of an appropriate choice of perspective turning a problem into a nonproblem survives the recognition of this plurality.

Bilgrami's argument is framed by another important contrast, between values being present in the world and our living in the world in accordance with certain values. He would like to argue that values are best seen in the former light and thus have an objective character. If we have reason to see from the larger point of view, this is therefore in the nature of things rather than ourselves. The reason for Bilgrami to argue this point (with which he begins and ends the paper and thus frames the argument as a whole) is to provide additional force for seeing "from a larger point of view" by viewing it as nothing but a consequence of "seeing the world right." However, as meritorious as such a perspective might be—about which there is forceful disagreement—the argument for seeing from a larger point of view (and thus for negating the free-rider problem and choosing C over P) would not seem to depend upon it. The values in question and the reasons that go with them can be present in a "value-laden" world or in our perception of and response to that world (or in the relation between the world and ourselves) while having a similar directive force. Bilgrami points to the example of the gap-free relation between desiring an object, finding an object to be desirable, and deeming that the object possesses desirability in order to make his case for "value properties in the world." On another view, however, this is not a proof of the case for value properties inhering in the world but rather a demonstration of the functional *interchangeability* of distinct ways of speaking, some of which are more roundabout than others.

A famous example of the relation between perception and evaluation is provided by that of the snake and the rope, in which how an object is perceived and how it is evaluated by a weary traveler on a path are integrally connected: only by changing one's perception of the object can one change one's evaluation of it, and these are tied in ways that are integrally linked to who and what we are, as Bilgrami too acknowledges.[8] (If we conceive of the snake as potentially dangerous, it is because we are the kinds of beings we are.) The example also illustrates the centrality of perception to the relation between person and world. To treat the world as simply having certain objective features and associated normative demands can be to miss the role that *we* have in ascribing certain features to the world, even when perception has the quality of unreasoned immediacy that is involved in assessing whether the object before one is a rope or a snake. This is a point that is crucial when applied to Bilgrami's interpretation of the demands of an "unalienated" response to the social world. On his own account, premodern unalienated worldviews were not necessarily compatible either with liberty or with equality, let alone both (he recognizes that the reconciliation of liberty and equality that is called for may not have been realized in any prior social order—except perhaps for "primitive communism"). However, in Bilgrami's view the modern expression of an unalienated worldview in the exercise of liberty must "cohere with equality in its outcomes." Why this is necessarily so is not wholly evident, but if it is so it must be because how we might experience an unalienated worldview, and thus exercise our liberty, is the product of our social character, constituted over historical time in ways that shape our perceptions and our evaluations. Yet, the role that we have in responding to the world is an active one, not reducible to such determinations. The supposition that a modern unalienated consciousness leads to an exercise of liberty compatible with (indeed conducive to) equality, apart from all historical experience, seems in this respect to be more of a hopeful possibility, for which we might at best believe there is some evidential warrant (for example, in the form of our living in an "age of democracies"), than a deduction. An inquiry into the practical or psychological conditions in which a modern form of nonalienated consciousness could emerge and thrive, even as a frame that we inhabit occasionally, is necessary to give content to the idea, but such an exercise is also necessarily speculative.

The distinct metaethical doctrines that Bilgrami contrasts—respectively, that values are present in the world and that they are present in our perception of or response to the world—may both be compatible with his proposed "triangular" method of resolving the tension between liberty and equality. His argument for seeing from a larger point of view is that the resulting change of perspective is simply the right way of seeing. The example of the Wittgensteinian basic action is meant to underline this idea. It is as a by-product that this shift in perspective also removes an apparent contradiction between substantive values, by making the unalienated exercise of liberty conduce to equality. The case for adopting the "larger point of view" does not therefore appear to depend on a metaethical premise about the nature of values. The case for presenting these ethical and metaethical perspectives together may have to do with the aesthetic or the argumentative value of doing so. If so, this ought to be elaborated, lest a plausible ethical perspective be made hostage to a less plausible metaethical one.

The specific issues raised here are but qualifications. Adequate discussions of our modern predicament ought henceforth to take note of Akeel Bilgrami's ambitious and insightful argument.

Notes

I would like to thank the NOMIS Foundation and the participants in its workshop on Nature and Value as well as Cosima Crawford for notes of the discussion.

1. For some related discussion, see Ajit Sinha, *A Revolution in Economic Theory: The Economics of Piero Sraffa* (New York: Palgrave Macmillan, 2016). I am obliged to the author for mentioning the relevant biographical background concerning Sraffa and Wittgenstein, based upon archival research.

2. Heinrich Hertz, *The Principles of Mechanics: Presented in a New Form*, trans. D. E. Jones and J. T. Walley (London: Macmillan, 1899). Hertz analyzed how various apparently distinct principles in classical mechanics might be interpreted as applications of a single Fundamental Law ("Every free system persists in its state of rest or of uniform motion in a straightest path") through appropriate redescription or change of variables that give rise to "derived equations" which substitute for the original ones.

3. Although Ostrom has provided perhaps the best-known empirically grounded response to the standard view that collective action must necessarily collapse due to the "free-rider problem," there are multiple theoretical and practical reasons to conceive of the propensity to failure as exaggerated. In general, public goods provision problems can have multiple equilibria as the extent of an individual's contribution is dependent on the perceived likelihood of others making contributions, even when narrowly conceived self-interest is the motive. Moreover, in experimental contexts, public goods provision is much greater than expected on the basis of the conventional theory, even without repetition or other features that might be expected to generate greater propensity to cooperation. Elements of noninstrumental and instrumental reasoning both appear to play an empirical role in contributions to public good provisioning, although the former are hardly recognized in the conventional theory. For more on these issues, see Sanjay Reddy, "Externalities and Public Goods: Theory or Society?," *Reading Mas-Colell* (blog), Institute for New Economic Thinking, November 19, 2015, www.ineteconomics.org/perspectives/blog/externalities-and -public-goods-theory-or-society.

4. Those who insist on interpreting the "larger point of view" in a framework amenable to conventional economic analysis might think of it as being a perspective in which the individual is guided by the "marginal social benefit" rather than the "marginal private benefit" of an action, i.e., by the sum of "private" benefits rather than the agent's privately experienced benefit alone.

5. The echoes of Marxian reasoning are rather obvious here and elsewhere, although Bilgrami's canvas is wide.

6. A Kantian approach, mandating cooperation in the presence of supposed collective action problems, provides one example of such a conception of reasons to act. For an early effort to explore the impact of Kantian reasoning on the critical assessment of solution concepts in fame theory, see Shaun Hargreaves-Heap and Yanis Varoufakis, *Game Theory: A Critical Introduction* (London: Routledge, 1995).

7. This is the idea at the heart of Sraffa's investigation, embodied in his conception of a "standard system": Piero Sraffa, *Production of Commodities by Means of Commodities* (Cambridge: Cambridge University Press, 1960).

8. "Creatures (frogs, say) who lack desires of the sophisticated, discriminating sort that we possess will see darkness in the world where we possess values." In conversation, Bilgrami has also referred to the necessity to refer in this respect to our "species being."

CHAPTER 9

EXPERIMENTING WITH OTHER PEOPLE

JOANNA PICCIOTTO

This essay urges a reconsideration of Max Horkheimer and Theodor Adorno's influential attack on the "factual mentality" of Francis Bacon and his followers. We've all heard some version of the argument: by divorcing meanings from things, the new science of early modernity converted an intrinsically valuable Creation, brimming with symbolic resonance, into raw material for the depredations of "instrumental reason." Now "what human beings seek to learn from nature is how to use it to dominate wholly both it and human beings. Nothing else counts."[1] Within a "disenchanted nature," nothing else could. The argument is such a familiar part of our intellectual landscape that its claims seem not just routine but necessary: without them, our major narratives about modernity would collapse. I think that this is the best thing that could happen for our understanding of early modern experimental science and its legacies, on which critiques of instrumental reason must finally depend.

In a recent contribution to this tradition of critique, Akeel Bilgrami has blamed Bacon and his self-appointed heirs, "the ideologues of the Royal Society," for reducing nature, "without remainder," to natural resources.[2] Counterexamples to this claim abound. The Royal Society's first major publication, John Evelyn's best-selling *Sylva: A Discourse of Forest-Trees*, addressed the problem of deforestation. Evelyn, a founding member of the Society, also published a jeremiad about air pollution, suggesting remedies in use today.[3] The experimentalist inheritance is a variegated one, visible in ecological thought, conservationism, and political economy (all organized by "the needs of desiring bodies") and in projects now firmly identified with the aesthetic.[4] When we mount arguments against the creep of reductionist logic and the market regulation of domains once organized by norms, we inevitably draw on a discourse of stewardship that early modern Baconians developed.

The notion that responsibility for the present environmental crisis rests with long-dead intellectuals who stripped nature of its symbolic meanings has undeniable appeal. The demands such a diagnosis places on us are so extravagant as to be, practically speaking, weightless. Belief in the inherent meanings of things can't be revived by sheer force of will. If we are prepared to distinguish more sharply between the realms of meaning and value, however, we can easily recognize that a division between symbols and bodies entails no devaluation of the latter. Indeed I want to suggest that it was precisely by sacrificing the category of exogenous meaning that early modern experimentalists created opportunities for nonhuman creatures to be recognized as fellow subjects rather than merely animated symbols.

This claim builds on another: the emergence of objectivity as a regulative ideal of investigative practice did not involve an understanding of science as "value-free." Stephen Gaukroger has recently described the process of modernization as one through which all cognitive values came to be shaped around scientific ones at the very same time that science was reinventing itself as a value-free enterprise: hence the contradiction at modernity's core. As Gaukroger asks, "How can something that is value-free realize human ideals and aspirations?" The answer of course is that it can't. What happens instead is that "scientific, technological, and economic goals replace—rather than realize—more traditional political, social, and cultural ones."[5] As a conceptual analysis this is unexceptionable, but as history it is inadequate. The story of objectivity's emergence is a circuitous one, but it was shaped throughout by religious and moral values.

It was the renewed interest in innocence as a state of epistemological privilege, stimulated by Martin Luther's lectures on Genesis, that gave moral authorization to the ideal of estranged or defamiliarized perception.[6] The ideal of objectivity emerged from the conviction that natural investigation was an expression of humanity's created rather than its corrupted nature. By cultivating epistemological modesty, studiously "unprepossessed" investigators hoped to reenact Adam's innocent encounter with the creatures. Imagined as both the primal scene of discovery and the first religious service, Adam's naming the creatures according to their natures was expanded into a model for "the experimental life."[7] In taking up this life, Bacon's self-defined followers aspired to fulfill the offices of a priesthood that began with the first man: hence the topos of man as "priest of nature," whose importance to seventeenth-century practitioners was asserted by Harold Fisch long ago.[8] Experimentalist efforts to purify and redeem fallen perception were part of a spiritual discipline.

The welter of projects pursued under the sign of innocent curiosity began with experiments in immersive yet estranged perception, revealing the alien within the fold of the everyday. In the arts of description that testified to these experiments, we can recognize not only the prehistory of the modern ideal of objectivity but modern understandings of the aesthetic function as well, before these become associated with divergent, even conflicting, projects. In the efforts of experimentalists "to increase the difficulty and length of perception" (to borrow the famous formulation of the Russian Formalist Victor Shklovsky), the aesthetic function was put to productive use. The effort to recover the insights of innocence was cultivated in full awareness of the investigator's susceptibility to transformation by the objects of his attention.[9] This, indeed, was largely the point.

As Gaukroger observes, a major reason for the extraordinary success of science in the early modern West, particularly in Northern Europe and Britain, was its alliance with religion. It's worth remembering, in this connection, that in the late sixteenth century and for most of the seventeenth century the English word *experiment* and its cognates appeared most often in religious contexts: the religious usage predated and largely governed the scientific use. As Peter Harrison has shown, this usage generated a series of oppositions between experiential knowledge and mere dogma, oppositions that organized devotional and scientific practice in the period and entangled them.[10]

In what follows I will focus on one knot in this tangle: physico-theology, a blend of natural history and Christian apologetics dedicated to revealing evidence of the Creator's wisdom and benevolence "in every *fibre* of a *Plant*, in every *particle* of an *Insect*, in every *drop* of *Dew*."[11] Physico-theology's distinctive twist on the design argument was to appeal not just to a lawful order in nature—a traditional staple of natural theology since Cicero—but to examples of adaptive design directed to purposeful ends. In spite of Newton's disapproval, this would remain its signature emphasis throughout its long ascendancy.[12] And it was very long. Physico-theology towered like a cloud formation over the eighteenth century, continually precipitating new literary kinds, from loco-descriptive nature poetry like James Thomson's *The Seasons* to the Boyle lectures, which shaped popular interpretations of nature until the time of Darwin (whose work was a triumph of physico-theological reasoning rather than a departure from it).[13] Some sixty years before the Bridgewater Treatises, we find meditations on adaptive design becoming central to the course of study that John Wesley laid out for Methodist lay preachers, along with readers of his Christian Library. Wesley's *Survey of the Wisdom of God in the Creation: A Compendium of Natural Philosophy in Two Volumes*, originally published in 1763, had by 1777 swelled to five. Its epigraph, taken from the first couple's prayer in *Paradise Lost*, is a full-throated endorsement of the idea that physico-theological contemplation is a revival of innocent devotion: "These are thy glorious Works, Parent of Good, / Almighty! Thine this Universal Frame, / Thus wondrous fair! Thyself how wondrous then!"

This literature "magnified" (i.e., praised) creation by revealing not what it meant, but how it worked. Through physico-theology, the reading public was weaned from an emblematic understanding of nature and inducted into an understanding of nature as a vast machine: an "Oeconomy" of awesome complexity, of which each creature was a working part. Making nature imaginable as a system in which the human must find a place consistent with the activities and entitlements of other creatures, physico-theology challenged readers to surrender the assumption—at once traditional and modern, perhaps even natural or reflexive—that the value of the nonhuman world depends on the symbolic or practical work it can perform for us.

The category of people will be my focus in what follows, because this literature is interested in estranging and thereby expanding it. I will suggest that by sustaining experiments in provisional identification with "people" up and down the chain of being, this literature inducted its readers into a new understanding of personhood, predicated not on human particularity but on the common physical grounds of lived experience—grounds from which no creature could be excluded. In doing so, it offered its readers an alternative to the anthropocentrism that shaped hermeneutic approaches to nature and that continues to govern its careless development today.

In Defense of Feeble Philosophy

Despite the grand claims I want to make for it, no one has had anything good to say about physico-theology for well over a century.[14] The admittedly meager scholarship on this literature is an unbroken record of condescension and insult. Neal Gillespie sums up the frustration of generations of readers when he asks, "How could so many capable naturalists allow themselves to be persuaded by such feeble philosophy?"[15] One only has to quote the texts to imagine the derision. William Derham invokes the Goldilocks principle on almost every page of his *Physico-Theology*: just as the Earth is "neither too full, nor too empty," so the human body is neither "too Pygmean, nor too Gigantick."[16] Windy tautologies abound: reptiles, we learn, are remarkably well equipped "to perform all the Offices of their Reptile State" (223). Dozens of pages are dedicated to showing that eyes are perfectly designed for seeing. It's hard to imagine such otiose "demonstrations" creating many converts. Aside from physico-theology's shaky argumentation, readers have long complained about its sheer tediousness. The argument from design hardly requires the multiplication of examples these writers provide; all that is really needed is an articulation of its premises. Once you've reached the five hundredth example of natural contrivance, it's unlikely that another one will make any persuasive difference. John Ray's declaration that "design doth so clearly appear" in nature, "that he must needs be very stupid that doth not discern it, or impudent that can deny it," suggests as much: once it's articulated, the argument barely has to be made.[17] But what generations of intellectual historians have treated as a set of inexplicably bad arguments might better be thought of as a liturgy guiding an exercise, requiring intellectual, imaginative, and affective work by the reader. The suasional structure offers a framework for a meditative practice.

In the seventeenth century, reading nature's book could still be a hermeneutic enterprise. The peacock was a walking emblem of vainglory; the turtle, always at home, literally embodied domestic virtue; the pelican could still be read as a living sign of Christ's sacrifice. Investigators who drew inspiration from Bacon worked hard to purge natural history of this lore, as the fruit of fallen learning (that is, error). They took particularly sharp aim at its anthropocentric bias. The preface to Francis Willughby and John Ray's *Ornithology* famously attempts a kind of exorcism: "we have wholly omitted what we find in other Authors concerning . . . *Hieroglyphicks, Emblems, Morals, Fables, Presages*, or ought else appertaining to . . . any sort of Humane Learning: And present [the reader] only with what properly relates to Natural History."[18] Willughby and Ray weren't simply trying to erect a disciplinary division—they were focused on the total desymbolization of the world.

While instrumental reason is a reassuring enemy for scholars in the humanities, a number of literary theorists have invited us to consider the violence that underwrites "instrumental meaning": the forcible imposition of meaning onto bodies that guarantees the otherness of a creature or thing will be effaced at the moment of its inclusion into the system, generating a world in which nothing is alien, nothing irrelevant to our interests and needs. Within the thoroughly curated worlds of allegory, we can be confident that everyone encountered by the hero will relate to his own person: Despair, Envy.[19] The same confidence was possessed by readers of nature's book with interpretive habits shaped by the bestiary tradition. The great achievement of physico-theology was to destroy this confidence, by dismantling the world that sustained it and offering different satisfactions in its place. Rather than parsing a world of signs pointing to a central protagonist, readers of this literature were invited to contemplate the strangeness of alternative centers of experience or selves: the other worlds experienced by other subjects. In shedding their emblematic meanings, nonhuman creatures were able to acquire the status of subjects. Rather than representing a virtue or vice, or a single aspect of a person, they gained a shot at full personhood. Paradoxically, it was when they were stripped of their "characters" that nonhuman creatures could become people. The book of nature, long read as an allegorical romance, became a novel.[20]

The change is registered in contemporary usage. Thomson's *The Seasons* presents us with "happy People, in their waxen cells," "plumy People," "Swallow-People," "houshold feathery People," "soft fearful People" (these with "woolly sides"), and "unseen People" that inhabit algae and drops of water (brethren of Alexander Pope's "green myriads in the peopled grass").[21] Throughout *The Seasons*, the logic of this usage is upheld by other collective nouns that embrace human and animal worlds indifferently: *tribes, inhabitants,* and *nations,* the latter coming in "busy," "tuneful," "nameless," "quivering," "plumy," and "furry" varieties.[22] Despite some earlier precedents, this extended use of the word *people* only hits the mainstream in the later seventeenth century, and it's clear that for Thomson it still carries the feel of an exciting discovery.[23]

Thomson's animal descriptions are sometimes described (or, rather, dismissed) as anthropomorphic, but such critiques beg the question, since the question these descriptions raise with some urgency is precisely what counts as a person. A related objection, that these animal-people are sentimental types, depends for its force on a widespread sense that there is something "emblematically sentimental" about sympathizing with animals at all. As Tobias Menely observes, one of the projects accomplished by the devaluation of sentimentality is the policing of "the border of human community, a border disrupted by the cross-species sympathies widely promoted within sentimental culture."[24] Of course, it's fair to question how disruptive such border-crossing really is. It was probably easier for many eighteenth-century readers to sympathize with one of Thomson's "feathery People" than with a beggar, prostitute, or slave. As today's advertisers are well aware, animals exhibit no race, class, or cultural identity that might interfere with their status as identificatory objects.[25] The bears that promote Coke and Charmin are *people,* subjects unspecified as regards class or identity, rather than *persons,* a word that "emphasizes the plurality and individuality of the referent."[26] It stands to reason that this individuating word, in its singular or plural form, appears nowhere in *The Seasons*. Thomson is attracted to general or indefinite designations for persons because he is interested in stretching the boundaries of the category.

It's important to keep in mind, in this connection, that the effort to drain nature of symbolic meanings drew inspiration from a restorative state-of-nature myth about a stripped-down Everyman, an avatar of everybody and no one in particular. I have observed that animals exhibit no race, class, or cultural identity, but in early modern culture, Adam was the single *human* subject that could carry the burden of representing humanity prior to its descent into personal difference. The first man was the iconic representative of the Royal Society; the organization was founded in large part to repair the intellectual losses caused by the unequal distribution of knowledge across classes.[27]

In physico-theology's primal scene, both Adamic man and the creatures he contemplated are thus linked by a common feature, or featurelessness: as candidates for identification, they don't exclude anyone in particular. Which is not to say that it is easy to identify with them, or that such identifications can ever result in identity; like any meditative practice, physico-theology describes a project that does not end.[28]

William Ashworth has suggested that the dismantling of the emblematic worldview began with the discovery of new world specimens; like creatures revealed by the microscope, these flora and fauna presented themselves to their new witnesses entirely naked of associations.[29] There is a way in which these creatures resembled their observers, who were themselves shorn of traditional attributes associated with intellectual authority. Since these observers did not ground their claims in Pliny, Aristotle, or Aquinas, "their authority rested solely on their status *as* observers."[30] Some of the most exciting encounters between animals and humans in this period were thus encounters between denuded "people." The idea that the ideal observer was an innocent observer—in effect, a stripped-down person—facilitated the practice of observation as a cooperative enterprise, the proper vocation of the undifferentiated humanity that Adam represented. But in order to revive the scene of naming as an encounter between minimally defined persons, natural philosophers had to reinvent both themselves and nature from the top down. They had to investigate creation not as individuals with passions, prejudices, and preconceptions but as "indifferent" representatives of an Adamic office.[31]

In parallel fashion, the literature of physico-theology presents its readers not with a parade of familiar characters but with agents negotiating an environment. Rather than interpreting the creatures, readers are invited to imaginatively inhabit them: to identify with unfamiliar forms of sentience engaged in more or less familiar life-struggles. In contrast, bestiaries often arranged animals under the headings of easily identified virtues or vices. The crocodile, for example, was a familiar villain in the bestiary tradition, his tears an emblem of insincere repentance, while the whale was a symbol of Satan's deceit: unsuspecting sailors who mistook his back for an island met with disaster when he reared up. Readers of English literature may be grateful to this bit of lore for generating one of Milton's most celebrated similes, but no self-respecting physico-theologist would have any truck with such an anthropocentric interpretation of events.[32]

Derham's appreciative description of the sting of a wasp offers us a very different perspective on animals that do humans harm. Furnished with "small, sharp, bearded Spears . . . somewhat like the Beards of Fish-hooks," it is "so pretty a piece of Work" that one cannot help but admire it: "By means of this pretty Mechanism," Derham is delighted to find, "the Sting . . . is able" not merely "to pierce and sting us," but to remain "lodged deep in the Flesh" (240).

Derham's attitude is entirely approving. It is not that he adopts an impersonal view directly. Instead, he takes up a temporary residence in the perspective of the wasp, a creature that, like every other, has a "need of Security" and desire for "*Self-Preservation*" (239, 238). For the duration of Derham's description, her ends become our ends, and we are relieved to find that her lack of protective "Armature" is compensated for by such elegant and effective weaponry (239).

This is the key principle of physico-theology: in order to identify each creature, the Adamic investigator has first to identify *with* it. To magnify God's workmanship, one must adopt the perspective of every creature he has made. One does not attain the objective view by trying to grasp the natural system as a whole: one gradually works toward such a view by provisionally inhabiting the perspective of every possible subject position within it. Thus, as we would expect, Derham makes quick work of the objection that a good God would not make "noxious Creatures," pointing out that creatures we find noxious don't seem so to themselves, and indeed often prove useful to others (55, 59). But rather than denying the existence of this category, he relativizes it, for it resurfaces in his discussion of the long-tailed maggot, a "singular and remarkable Work of God" who unfortunately is tormented by earwigs. Thankfully, the wise and indulgent Creator has given her everything she requires to defend herself against these "noxious Animals." Something interesting happens here: although he rejects the term *noxious* as unphilosophical, Derham is obliged to retrieve it in order to do justice to the maggot's perspective (234, 249).

Derham does everything he can to facilitate our identification with the creatures as fellow subjects, "transact[ing] their Business" and "refresh[ing] and recruit[ing] themselves" (27). A tantalizing reference to the mole's "Pastimes and Pleasures" follows a sensitive reconstruction of the preferences that must condition the "Subterraneaous" mode of life in all its variety: "Some [burrowing animals] delighting in a lax and pervious Mold, admitting them an easy passage; and others delighting in a firmer and more Solid Earth, that will better secure them against Injuries from without."[33] Derham's descriptions of animal behavior always register a participatory zeal, whether his subject is the midge, who "lies quietly on top of the Water, now and then gently wagging it self this way, and that," or the insects that "catch, hold, and tear their Prey," "pierce and wound," or "gnaw, and scrape out their Food" (190n20, 190–191). These insects are literally bloodthirsty, but his rendering of their energetic pursuit of survival has little in common with the moralizing depictions that dominated the bestiary tradition. This is because the identificatory opportunity offered by each animal is finally the same: like us, the creatures simply want to live—"to enjoy their own Beings," as Ray puts it.[34]

Lynn Festa has observed that "It is not that we automatically have feelings for others because they are human; it is because we have feelings about others that they come to be seen as human."[35] This helps explain why perusing creation for evidence of God's protective care for his creatures led the physico-theologist to confer personhood on the objects of that care. To rationally reconstruct God's design choices when making the creatures, one had to invest such care oneself, considering every circumstance "conducing to their happy, comfortable living in this World."[36] For the responsible physico-theologist, the spider's lack of a neck was a problem to be solved; since spiders can't move their heads, "it is requisite that defect should be supplied by the multiplicity of Eyes."[37] He had to provide a good reason for why snails and moles lacked acute vision: because "their slowness allows them time to dwell longer on an Object."[38] Ray's discovery that only some animals, such as frogs, have nictitating membranes almost seems to generate a crisis of faith: if the purpose of the membrane is to protect the eye, why wouldn't every eye in creation be fitted with one? Attempting to come up with a "tolerable" explanation, he considers that frogs are regularly obliged to leap among branches, twigs, and leaves; without the membrane, they would have to leap with their eyes shut, or run the risk of having their corneas sliced. But this does not finally satisfy him, since men and apes could easily find themselves in similar circumstances. Later in the text, he revisits the problem: if the membrane's purpose is to clean the cornea, its absence in the eyes of apes and men is entirely understandable, since they have hands with which to rub their eyelids.[39] His relief is palpable; it's as if justice has actually been served.

Sometimes, of course, it was impossible to determine the purpose of a given feature, but even these failures could be turned to account. Henry More works himself up into a lather imagining a physico-theological skeptic who might look askance at the turkey's apparently useless snood:

> Nor are we to cavill at the red pugger'd attire of the *Turkey*, and the long *Excrescency* that hangs down over his Bill, when he swells with pride and anger; for it may be a Receptacle for his heated bloud, that has such free recourse to his head, or he may please himself in it as the rude *Indians*, whose Jewells hang dangling at their Noses. And if the bird be pleasur'd we are not to be displeased, being alwaies mindfull that Creatures are made to enjoy themselves, as well as to serve us, and it is a grosse piece of Ignorance and Rusticity to think otherwise.[40]

By the end of the passage, the excrescency's purpose hardly seems to matter. Whether it has a practical function or is simply an ornament in which the turkey takes a proud delight, the crucial lesson for the physico-theological novice to grasp is that the creatures were "made to enjoy themselves," not merely to serve man. (Thus, although More compares the turkey to a "rude" people, the rudeness ends up being imputed to the ignorant and rustic observer who finds them rude.) In Ray, this theme becomes an obsession; the once unexceptionable doctrine that the creatures were created with humanity's needs in mind comes to seem as scandalous as atheism itself.[41]

When it comes to nature's losers (which every creature finally is), the challenge is severe. More's effort to show that the rabbit is "fitted on purpose for her Condition" engages him in a series of difficult questions. Why, he asks, is the rabbit "ever rearing up and listning whiles she is feeding? and why is she so exceeding *swift* of *foot*, and has her *Eyes* so *prominent*, and placed so that she can see better behind her then before her? but that her flight is her onely safety, and it was needful for her perpetually to eye her pursuing enemy, against whom she durst never stand at the Bay, having nothing but her long soft limber *Eares* to defend her."[42]

After reading this passage, one is struck less by the brilliance of rabbit-design than by the success with which More has imagined what it's like to be constantly stressed out by fear of predators. He later speculates, with some desperation, that continual frights make rabbits "more sensibly relish their present safety," so that "they are more pleased with an Escape then if they had never met with any Danger."[43] Every major physico-theologist after More takes up the rabbit problem, Ray noting that her back legs are longer than her front legs, "by which means she climbs the Hill with much more ease than the Dogs . . . and many times escapes away clear, and saves her Life," and Nehemiah Grew pointing out that her ear contains a "natural *Otocoustick*," enabling her to hear sounds far behind her.[44] The attempt to sympathize with the rabbit comes to seem like an end in itself, since these detailed reconstructions of an existence spent on the run convey chronic terror and exhaustion above all, suggesting few opportunities to enjoy the creaturely pleasures afforded by a beneficent God.

While physico-theologists had little trouble demonstrating that the organs of each creature were created to serve its particular needs, their relentless emphasis on God's benign foresight and protective providence could not help but focus attention on the ultimate failure of such protection at the end of every life. Even more inconveniently, their serial assumption of divergent creaturely perspectives makes it impossible to forget that the thriving of one creature tends to come at the expense of another. Theodicy and natural history are not natural allies. It's not surprising, then, that many seventeenth- and eighteenth-century readers who immersed themselves in this literature, along with the authors themselves, were driven to imagine improvements on the natural order—afterlives for the animals, a future in which each creature would be raised at least a few degrees on the scale of being. In his last decade, Wesley became increasingly preoccupied with the possibility of a final salvation for animals, a truly general deliverance. He found support for his hope in the speculations of his friend the physician and medical writer George Cheyne: "It is utterly incredible that any creature . . . should come into this state of being and suffering for no other purpose than we see them attain here. . . . There must be some infinitely beautiful, wise, and good scene remaining."[45] The literature of physico-theology made these questions feel *more* urgent, not less. The descriptive demands of the mode, then, seem to have been flatly at odds with its ostensible purpose. But perhaps our sense of physico-theology's purpose needs revising. Such a revision would require us to entertain the not particularly radical possibility that what seems obvious to us was also obvious to contemporaries: a persuasive case for the benevolence of the Christian God can't be made on the grounds of nature alone. Evidently, what contemporaries valued about this literature was the occasions it provided for entertaining perspectives that were divorced from their own persons, and cultivating the emotions proper to an impartial—or rather, a consecutively partial—spectator. Each new demonstration of design, each new creature, offered the reader another occasion to surrender a perspective shaped by human ends and interests: to experiment with the idea that other creatures are also people.

"We are not a little pleased," Joseph Addison and Richard Steele's Mr. Spectator remarks, "to find every green Leaf swarm with Millions of Animals, that at their largest Growth are not visible to the naked Eye." While we may claim angels as "our brethren," we can now also say to the lowly worm, "thou art my mother and my sister." Riffing on Job 17:14 ("I have said to corruption, Thou art my father: to the worm, Thou art my mother, and my sister"), Addison converts Job's humiliation into something very like its opposite; man's kinship with the worm now testifies to an expanded field of sociability, a line of acquaintance that extends from celestial beings all the way to pond scum. James Granger would repurpose this passage for his sermon in 1772 on the rights of animals, describing the "righteous man" as one willing not only to acknowledge the angel as his brother but to "say to the worm, *Thou art my sister*."[46]

Despite the variety of creaturely phenomenologies, the identificatory opportunity offered by each creature is finally the same. All of these people are engaged in the same activity: trying to make an earthly living. Accordingly, Derham's momentary identification with the maggot is not based on very much aside from the physical circumstances of its being. And yet we tend to think of round characters—characters that seem worth identifying with—as characters that authors have invested with individuating detail: details that, taken together, make up a "personality." But it seems to be the absence of such details that makes it so easy for Derham, and for us, to take up residence in the physical facts of the maggot's existence. Had the maggot been invested with a "character," we might have met with an obstacle; its absence eases our way.

It seems paradoxical that divesting a creature of any particular qualities could qualify it to become a fully-fledged person. Here a comparison to the characters of the early English novel might be useful. Frederick Keener has addressed the question of how fully fictional characters need to be developed in order to seem real; to read *Robinson Crusoe* or *Moll Flanders* is to conclude: not very much.[47] What permits us to care about Defoe's characters as though they were real people is their own survival imperative, a drive with which we instantly identify, since the material conditions of their existence are represented so fully—much more fully than their "personalities." We don't require any initial individuating characteristic or deviant detail to lure us into identification with these people; having made the leap, however, we can accept these characteristics and even cherish them. Likewise, when Derham (writing around the same time as Defoe) wistfully describes how snails can "stick unto, and creep up Walls and Vegetables at their Pleasure," or when Grew, trying to imagine a life without eyes, observes that the worm "finds what it searches after, only by Feeling, as it crawls from one thing to another," we feel an intimacy with these animals that perhaps could never be generated by a discussion of their character traits.[48] And while the "animalcules" revealed by the microscope displayed little in the way of personality, their "*perpetual Commotion*" clearly demonstrated volition: the purposeful activity of "bare life," which was enough to elicit the interest and even the care of those who observed them.[49]

The Vicarious Representative

Reflecting on the effort engaged in "plenary attention, serious and protracted and repeated," Walter Ong concludes that "contemplation of this sort involves love, and the question is whether it can be carried on, or how far it can be carried on, without some suggestion of reciprocity."[50] The attention physico-theologists paid to every aspect of the creatures' well-being, as if they were actually responsible for securing it, entails a kind of love, and it was lavished on the creatures without any expectation that it would be reciprocated. To be sure, their solicitude for the creatures was expressed in an entirely virtual manner. While they dutifully reverse-engineered God's choices, it was only "as if" they were caring for the living products of those choices.

But there was one very practical way in which they could consider themselves to be helping the creation they described so carefully. Scripture calls upon all the creatures to praise God, which, as Ray observed, "they cannot do by themselves."[51] The creatures itemized in Psalm 148, from trees to cattle to birds, could offer "the sacrifices of praise and thanks" only through "the mouth of man." Understood as "a *Priest* in this magnificent *Temple* of the *Universe*," the physico-theologist was a "vicarious Representative, the common High Priest of the inanimate and irrational World." This, as experimentalists insisted, was the priesthood described in Genesis 2:19–20, which would have been "the only *Religion*, if men had continued innocent in *Paradise*, and had not wanted a *Redemption*."[52] Contemporary critics were only too happy to agree. The problem with this "innocent" form of devotion was that there was nothing particularly Christian about it. So much focus on God the Creator made his incarnation as the Redeemer seem quite ancillary. From the perspective of contemporaries worried about the spread of deism (often imagined as the starter drug to full-blown atheism), Derham's publication history was not exactly reassuring: *Physico-Theology* runs to five hundred pages; its sequel, *Astro-Theology* (1715) is over three hundred pages; last, and very much the least, in the series is *Christo-Theology* (1730), coming in at twenty-six pages.

And yet, as we have seen, even the arguments physico-theologists make on behalf of the Creator often do him little honor. The imaginative and affective practice *associated* with those arguments, however, reprises what Diane McColley describes as the Son's "descent down the scale of being for which he is exalted" quite successfully.[53] In physico-theological literature, and indeed in much experimentalist practice, the first and second Adams come together and join forces. In the physico-theologists' ambition to provisionally assume the phenomenological limitations of other natures, it is possible to glimpse a distinctly experimentalist version of the *imitatio Christi* (which, as Daniel Shore has shown, began to be freer, more approximate in the seventeenth century), one based on the kenosis rather than the crucifixion.[54] Physico-theology can be understood as a forerunner of modern ecotheology's attempt to imagine a lordship of the crucified, a stewardship modeled on an incarnate God who vicariously participates in the suffering of all his creatures.[55] According to this logic, the human capacity for identifying and sympathizing with the lower creatures associates us with the very pinnacle of the chain of being: one identifies with Christ by pursuing other identifications, elsewhere.

Charles Taylor (himself no admirer of physico-theology) has observed in another connection that "what has always been stressed in Christian agape is the way in which it can take us beyond the bounds of any already existing solidarity. The Good Samaritan was in no way bound to the man he helped. That was, indeed, part of the point of the story."[56] There is a reason why an emphasis on conventionally despised animals is a virtual requirement of the mode. But if physico-theology's experiments in cross-species identification extend fellowship beyond the borders of the human, expanding the category of Christian community, they in no way depend on the surrender of a belief in human exceptionalism. The capacity for identifying and sympathizing with other creatures affirms the very human superiority that it repeatedly unsettles. The ease with which the reader can plunge into these identifications, in other words, is the surest sign of his difference from its objects. This is not a problem but the point. To truly look *at* a creature necessitates seeing and feeling *as* that creature, but once one makes the attempt, one is no longer *like* that creature at all. As a "vicarious Representative, the common High Priest of the inanimate and irrational World," the physico-theologist (and through him the reader) feels *for* the creatures and, by this means, feels *like* the heavenly mediator he is imitating.

Here we might consider the blended consciousness that characterizes both sympathetic identification and the incarnate God. Even while affirming the doctrine that the Son emptied himself of his godhead in order to take on the nature of man, many seventeenth-century commentators believed that this process resulted in a fusion of the perspectives appropriate to each. If this blended consciousness suggests the quasi-supernatural consciousness that focalizes free indirect style in the novel, any practical human effort to approximate this state of mind suggests the sympathetic individual described by Adam Smith, who is forever consulting with the impartial spectator within, an ideal witness Smith describes in suggestively Christlike terms: "the great demigod within the breast . . . though partly of immortal, yet partly too of mortal extraction."[57] It is this quasi-divine perspective we must try to approximate if we are to sympathize adequately with others. Critiques of sentimental literature often emphasize distance and detachment as conditions of sympathetic identification, but the process also works in reverse: for devotees of physico-theology, objective insight depended on the cultivation, not the suppression, of sympathetic response.

I have suggested that, in his capacity as a mediator between the reader and the rest of creation, the physico-theological narrator achieves some of the effects we associate with free indirect style, which delivers what D. A. Miller has called "an impersonal intimacy," granting us "at one and the same time the experience of a character's inner life as she herself lives it, and an experience of the same inner life as she never could."[58] To feel the inability of the worm to be anything but a specialist in its own relation to the world is to experience something utterly beyond its ken—to feel for it in a way it can't feel for itself. We seem again to be in the world described by Smith's *Theory of Moral Sentiments*, in which how one feels when one imagines oneself in the position of another is inevitably quite different from what the other feels. Indeed, the continual iterations of "people" as agents negotiating their way through systems beyond their understanding place us on the threshold of the world described by political economy, in which, as Eleanor Courtemanche puts it, "to take part in society is to be partially blind to its larger workings." A meaningful order can thus only be imagined as the result of "overlapping interactions by different socially located points of view." Such a system will perhaps inevitably be represented through a blend of "impersonal discourse" and "the feeling-filled language of inside views."[59] In physico-theology, the effect, both ennobling and diminutive, can be something like mock epic delivered deadpan (as when Thomas Browne discovers the "house of the solitary maggot" in the "thicket on the head of the teasel" and finds there "the seraglio of Solomon . . . regularly palisadoed and stemm'd with flowers of the royall colour")—but the resulting disorientation is greater, since the standards of value that have been upset never recover their stability.[60]

I have suggested that physico-theology gave all creatures a shot at full personhood, but only a shot: all this discourse *guarantees* each creature is the reduced personhood of the stripped-down subject of political economy. As we have seen, despite the teeming variety of the natural world revealed in this literature, its people are all engaged in the same activity: trying to make a living. This may suggest that animals could become people only when people ceased to be persons—when people could be imagined first and foremost as producers and consumers, economic agents rather than traditional actors in the Aristotelian (and Arendtian) sense recently elucidated by Richard Halpern.[61] The dignity with which physico-theology invests the creatures, then, may simply be the dignity proper to vocational subjects, pursuing their "principal Business" between "their resting, leisure times."[62] Yet physico-theology enjoins its reader to take on a vocation beyond his or her business. The model of reading nature it promotes makes religious devotion, scientific insight, and aesthetic pleasure equally dependent on the imaginative and affective experience of feeling like someone or something other than one's self. The insight gained from the exercise of imagining and speaking "as if" one were something or someone else cannot, of course, be confused with intimate knowledge of the other's experience. Such exercise provides fitful (and, over time, easier) access to a blended consciousness, proper to a "vicarious representative."

The fact that physico-theology, political economy, and the novel rose up together and evolved along similar lines (even sharing some practitioners) does not suggest that any can be reduced to any other. But in their shared legacy there is still much to be cultivated. While they are not emphasized in our current system of education, the exercises these discourses sustained are not entirely foreign to us. What philosophers prefer to call "thought experiments" are also a form of affective training; John Rawls's veil of ignorance helps model legal and even economic justice because of the imaginative demand it makes on us actually to experience someone else's self-interest as our own. Deep ecology's ritual of the "council of beings"—in which the project of speaking for the creatures, from cows subjected to factory farming to lichen damaged by acid rain, gives voice to laments shaped by knowledge to which the creatures could have no access—turns free indirect discourse into a continually evolving liturgy.[63] To imagine experiencing familiar needs as they might be felt by alien subjects, or to virtually register alien sensitivities as if they were one's own, and then to testify to the "results," instills habits of attention and care that perhaps cannot be developed any other way. Such habits produce insight. Pierre Hadot has defended a strain of philosophical writing, extending from the ancients to modern practitioners like Thoreau, that understands philosophy as a project of forming rather than informing: an induction into a way of life.[64] Early modern experimentalists braided the projects of forming and informing together. If we have attended only to the propositions in the textual record they left rather than to the imaginative and affective processes these propositions sustain, it is our fault, not theirs.

Notes

1. Max Horkheimer and Theodor W. Adorno, *Dialectic of Enlightenment: Philosophical Fragments*, ed. Gunzelin Schmid Noerr, trans. Edmund Jephcott (Stanford: Stanford University Press, 2002), 2. The present essay draws from an argument made previously in "Milton and the People," in *Milton in the Long Restoration*, ed. Blair Hoxby and Ann Baynes Cairo (Oxford: Oxford University Press, 2016), 483–502.

2. Akeel Bilgrami, "What Is Enchantment?," in *Varieties of Secularism in a Secular Age*, ed. Michael Warner, Jonathan VanAntwerpen, and Craig Calhoun (Cambridge, MA: Harvard University Press, 2010), 145–165, 148–149. On desymbolization and exogenous meaning, see also Charles Taylor, *A Secular Age* (Cambridge, MA: Belknap Press of Harvard University Press, 2007), 35; and William Ashworth, "Natural History and the Emblematic World View," in *Reappraisals of the Scientific Revolution*, ed. David C. Lindberg and Robert S. Westman (Cambridge: Cambridge University Press, 1990), 303–332, at 318, as well as his "Emblematic Natural History of the Renaissance," in *Cultures of Natural History*, ed. Nicholas Jardine, James A. Secord, and Emma C. Spary (Cambridge: Cambridge University Press, 1996), 17–37. Brian W. Ogilvie nuances Ashworth's claims in *The Science of Describing: Natural History in Renaissance Europe* (Chicago: University of Chicago Press, 2006).

3. John Evelyn, *Sylva: A Discourse of Forest-Trees* (London, 1664); Evelyn, *Fumifugium: Or the Inconveniencie of the Aer and Smoak of London Dissipated: Together with Some Remedies Humbly Proposed* (London, 1661).

4. Catherine Gallagher, *The Body Economic: Life, Death, and Sensation in Political Economy and the Victorian Novel* (Princeton: Princeton University Press, 2006), 10.

5. Stephen Gaukroger, *The Emergence of a Scientific Culture: Science and the Shaping of Modernity, 1210–1685* (New York: Clarendon, 2006), 1–2.

6. This paragraph draws on chapters 1–3 of my *Labors of Innocence in Early Modern England* (Cambridge, MA: Harvard University Press, 2010).

7. The phrase comes from Steven Shapin and Simon Schaffer, *Leviathan and Their Air-Pump: Hobbes, Boyles, and the Experimental Life* (Princeton: Princeton University Press, 1985).

8. Harold Fisch, "The Scientist as Priest: A Note on Robert Boyle's Natural Theology," *Isis* 44, no. 3 (1953): 252–265.

9. For future refinements of this susceptibility, see Amanda Goldstein, *Sweet Science: Romantic Materialism and the New Sciences of Life* (Chicago: University of Chicago Press, 2017).

10. See most recently his book *The Territories of Science and Religion* (Chicago: University of Chicago Press, 2015).

11. Joseph Glanvill, "The Usefulness of Real Philosophy to Religion," in *Essays on Several Important Subjects in Philosophy and Religion* (London, 1676), sig. S2.

12. See Neal C. Gillespie, "Natural History, Natural Theology, and Social Order: John Ray and the 'Newtonian Ideology,'" *Journal of the History of Biology* 20, no. 1 (1987): 1–49, esp. 4, 28.

13. Charles E. Raven, *Natural Religion and Christian Theology* (Cambridge: Cambridge University Press, 1953), 110. See also John Gascoigne, "From Bentley to the Victorians: The Rise and Fall of British Newtonian Natural Theology," *Science in Context* 2, no. 2 (1988): 219–256.

14. Until recently: see, for example, Courtney Weiss-Smith, *Empiricist Devotions* (Charlottesville: University of Virginia Press, 2016); and Alexander Wragge-Morley, *Aesthetic Science: Representing Nature in the Royal Society of London, 1650–1720* (Chicago: University of Chicago Press, 2020).

15. See note 12. Hence the old joke that no one doubted the existence of God until the Boyle lecturers undertook to prove it, a witticism rehearsed and all but endorsed by John Brooke and Geoffrey Cantor, *Reconstructing Nature: The Engagement of Science and Religion* (Edinburgh: T and T Clark, 1998), 150–151. The argument is made at length in Michael J. Buckley, SJ, *At the Origins of Modern Atheism* (New Haven: Yale University Press, 1987), 346–363. See also John Dillenberger, *Protestant Thought and Natural Science: A Historical Interpretation* (New York: Doubleday, 1960), 152–153.

16. Derham, *Physico-Theology* (London, 1714), 174, 299. I quote from the second edition.

17. Ray, *The Wisdom of God Manifested in the Works of Creation* (1701), 220. I quote from the third edition.

18. Ray and Willughby, *The Ornithology of Francis Willughby* (London, 1678), sig. A4.

19. See Gordon Teskey, *Allegory and Violence* (Ithaca: Cornell University Press, 1996).

20. Just as the novel continued to include many "emblematic" elements, so residual and emergent approaches to reading nature overlapped.

21. *The Seasons*, ed. James Sambrook (Oxford: Clarendon, 1981): *Autumn*, line 1176; *Spring*, line 165; *Autumn*, line 838; *Winter*, line 87; *Summer*, line 311. Alexander Pope, *Essay on Man*, Epistle 1, line 210. For a recent consideration of this term in Thomson, see Heather Keenleyside, "Personification for the People: On James Thomson's *The Seasons*," *English Literary History* 76, no. 2 (2009).

22. *Spring*, lines 510, 594; *Summer*, lines 302, 344, 737; *Winter*, line 811.

23. Precedents include Virgil and Sylvester's Du Bartas. See the beginning of book 4 of *The Georgics*: "Admiranda tibi levium spectacula terum / magnanimosque duces totiusque ordine gentis / mores et studia et populos et proelia dicam" ("peoples" are translated in the Loeb edition as "tribes"; see also "parvos Quirites" at 201); and *Divine Weeks and Works*, "Fourth Day of the First Weeke," line 84 ("God made the people that frequent the Water"), ed. Susan Snyder (Oxford: Clarendon, 1979), 1:208.

24. Tobias Menely, "Zoöphilpsychosis: Why Animals Are What's Wrong with Sentimentality," *Symploke* 15, nos. 1–2 (2007): 244–267, 249.

25. A point made by Lorraine Daston and Gregg Mitman in their introduction to *Thinking with Animals: New Perspectives on Anthropomorphism*, ed. Lorraine Daston and Gregg Mitman (New York: Columbia University Press, 2005), 6.

26. See *OED online*, *people*, n. def. 1. and 2.a, and *person* n. 2a.

27. See Picciotto, *Labors of Innocence*, chap. 2.

28. Obviously women had a less straightforward claim to Adamic personhood than men.

29. Ashworth, "Natural History and the Emblematic World View," 318–319.

30. Anthony Pagden, *European Encounters with the New World: From Renaissance to Romanticism* (New Haven: Yale University Press, 1993), 56.

31. I discuss the ethos of "indifferency" among the Royal Society virtuosi as a precursor to modern objectivity in *Labors of Innocence*, but for a more sophisticated treatment of this ethos, informed by affect theory, see David Carroll Simon, *Light Without Heat: The Observational Mood from Bacon to Milton* (Ithaca: Cornell University Press, 2018).

32. Walter J. Ong, SJ, *The Presence of the Word: Some Prolegomena for Cultural and Religious History* (New York: Simon and Schuster, 1967), 202. The literature on the bestiary tradition is vast, but see Hanneke Wirtjes, ed., *The Middle English Physiologus* (Oxford: Oxford University Press, 1991); and Ron Baxter, *Bestiaries and Their Users in the Middle Ages* (Stroud: Sutton; London: Courtauld Institute, 1988), which emphasizes their use as source material for sermons, though as Debra Hassig observes, some nonmoralized creatures gained entrance (*Medieval Bestiaries: Text, Image, Ideology* [Cambridge: Cambridge University Press, 1995], xvii). For Milton's simile, see the first book of *Paradise Lost*, lines 192–210.

33. Derham, *Physico-Theology*, 94, 63. Derham draws heavily from More's description of the mole. See Henry More, *An Antidote Against Atheisme* (London, 1653), 88–89.

34. Ray, *The Wisdom of God*, 379.

35. Lynn Festa, "Humanity Without Feathers," *Humanity* 1, no. 1 (2010): 7.

36. Derham, *Physico-Theology*, 57.

37. Derham, 92n9.

38. George Cheyne, *Philosophical Principles of Natural Religion: Containing the Elements of Natural Philosophy, and the Proofs for Natural Religion, Arising from Them* (London, 1705), 281.

39. Ray, *The Wisdom of God*, 269–271, 346.

40. More, *Antidote Against Atheisme*, 92.

41. See, for example, Ray, *Wisdom of God*, 228–229, and Derham, *Physico-Theology*, 55n3, 57. For a useful treatment of this issue, see John Hedley Brooke, "'Wise Men Nowadays Think Otherwise': John Ray, Natural Theology and the Meanings of Anthropocentrism," *Notes and Records of the Royal Society of London* 54, no. 2 (2000): 199–213.

42. More, *Antidote Against Atheisme*, 90.

43. More, 99.

44. Ray, *The Wisdom of God*, 153; Nehemiah Grew, *Cosmologia Sacra* (London, 1701), 24.

45. Cheyne, *An Essay on Regimen* (London, 1740), 91, quoted and discussed with reference to Wesley's sermons "The General Deliverance" and "New Creation" in Randy L. Maddox, "John Wesley's Precedent for Theological Engagement with the Natural Sciences," in *Divine Grace and Emerging Creation: Wesleyan Forays in Science and Theology of Creation*, ed. Thomas Jay Oord (Eugene, OR: Wipf and Stock, 2009).

46. *Spectator* 519 in Donald F. Bond, ed., *The Spectator*, vols. 1–5 (Oxford: Clarendon, 1965); James Granger, *An Apology for the Brute Creation, or Abuse of Animals Censured* (London, 1772), 5–6. His text is Proverbs 12:10, "A righteous man regardeth the life of his beast." For a different perspective on this material, see Lynn Festa, "Person, Animal, Thing: The 1796 Dog Tax and the Right to Superfluous Things," *Eighteenth-Century Life* 33, no. 2 (2009): 30.

47. Frederick M. Keener, *The Chain of Becoming: The Philosophical Tale, the Novel, and the Neglected Realism of the Enlightenment: Montesquieu, Voltaire, Johnson, and Austen* (New York: Columbia University Press, 1983), 6.

48. Grew, *Cosmologia Sacra*, 83; Derham, *Physico-Theology*, 410n11.

49. Walter Charleton, *Physiologia-Epicuro-Gassendo-Charltoniana: Or a Fabrick of Science Natural, Upon the Hypothesis of Atoms* (London, 1654), 126; Giorgio Agamben, *Homo Sacer: Sovereign Power and Bare Life,* trans. Daniel Heller-Roazen (Stanford: Stanford University Press, 1998). However, lively particles of matter were ideally suited for homologous reflections on systems, in which "atoms" (the term was used quite loosely) could be considered functionally identical to citizens (here the homophone of Adam and atom proved useful).

50. Walter Ong, "The Jinee in the Well-Wrought Urn," in *An Ong Reader: Challenges for Future Inquiry,* ed. Thomas J. Farrell and Paul A. Soukrop (Cresskill: Hampton, 2002), 199–208 at 203–204.

51. Ray, *The Wisdom of God*, 201.

52. Matthew Hale, *The Primitive Origination of Mankind, Considered and Examined According to the Light of Nature* (London, 1677), 372, quoted in Gaukroger, *Emergence,* 152; Thomas Sprat, *The History of Royal Society* (London, 1667), 349–350. Psalm 148 was a touchstone of physico-theological verse as well; see, for example, Richard Blackmore's *Creation: In Seven Books* (London, 1712), 398.

53. Diane McColley, "'All in All': The Individuality of the Creatures in Paradise Lost," in *"All in All": Unity, Diversity, and the Miltonic Perspective,* ed. Charles W. Durham and Kristin A. Pruitt (Selinsgrove: Susquehanna University Press, 1999), 21–38, 26.

54. To paraphrase Regina Schwartz, the imitation of Christ gave way to identification with Christ. See Daniel Shore, "WWJD? The Genealogy of a Syntactic Form," *Critical Inquiry* 37, no. 1 (2010): 1–25; and Regina Schwartz, "The Toad at Eve's Ear: From Identity to Identification," in *Literary Milton: Text, Pretext, Context,* ed. Diana Trevino Benet and Michael Lieb (Pittsburgh: Duquesne University Press, 1994), 1–21, 3. Many thanks to Anne-Lise François and David Marno for helping me think about this.

55. See, for example, John F. Haught, *God After Darwin: A Theology of Evolution* (Boulder: Westview, 2000); Douglas John Hall, *Imaging God: Dominion as Stewardship* (Grand Rapids, MI: W. B. Eerdmans, 1986), 186; and Peter Manley Scott, *God's Bounty?: The Churches and the Natural World,* ed. Peter Clarke and Tony Claydon (Woodbridge, Suffolk, UK: Ecclesiastical History Society; Rochester, NY: Boydell and Brewer, 2010), 431–457.

56. Taylor, *A Secular Age*, 246.

57. Knud Haakonssen, ed., *The Theory of Moral Sentiments* (Cambridge: Cambridge University Press, 2002), 6.3.25, 3.2.32.

58. D. A. Miller, *Jane Austen: The Secret of Style* (Princeton: Princeton University Press, 2003), 60.

59. The *"Invisible Hand" and British Fiction, 1818–1860: Adam Smith, Political Economy, and the Genre of Realism* (Basingstoke, UK: Palgrave Macmillan, 2011), 21–22; Deidre Lynch, *The Economy of Character: Novels, Market Culture, and the Business of Inner Meaning* (Chicago: University of Chicago Press, 1998), 212. It will be clear that my account diverges sharply from Lynch's, but this is because I am focused on forms of identification that are deindividualizing.

60. Thomas Browne, "The Garden of Cyrus," in *Thomas Browne: Selected Writings,* ed. Kevin Killeen (Oxford: Oxford University Press, 2014), 568.

61. Richard Halpern, *Eclipse of Action: Tragedy and Political Economy* (Chicago: University of Chicago Press, 2017).

62. Derham, *Physico-Theology*, 94, 199.

63. John Rawls, *A Theory of Justice* (Cambridge, MA: Belknap Press of Harvard University Press, 1971, 1999); John Seed, Joanna Macy, Pat Fleming, and Arne Naess, *Thinking like a Mountain: Towards a Council of All Beings* (Gabriola Island: New Catalyst, 2007).

64. Pierre Hadot, *Philosophy as a Way of Life: Spiritual Exercises from Socrates to Foucault*, ed. Arnold I. Davidson, trans. Michael Chase (Oxford: Blackwell, 1995).

CHAPTER 10

THE GREEN GROWTH PATH
TO CLIMATE STABILIZATION

ROBERT POLLIN

When the twenty-first United Nations Climate Change Conference opened in Paris in November 2015, annual global carbon dioxide (CO_2) emissions were at about thirty-two billion metric tons. This figure is 10 percent higher than the twenty-nine billion tons that were emitted in 2009, when the last major UN Climate Change Conference took place in Copenhagen.

There is no escaping the conclusion that we are playing Russian roulette with the environment by allowing CO_2 emissions to continue rising. The Intergovernmental Panel on Climate Change (IPCC) provides what is widely regarded as conservative benchmarks as to what is required to stabilize the average global temperature at its current level of around 60.3° Fahrenheit, which is 3.6° F (2° Celsius) above the preindustrial average of 56.7°. According to the IPCC, we need to bring global CO_2 emissions down by 40 percent within twenty years, to twenty billion tons, and by 80 percent as of 2050, to seven billion tons.

In the run-up to the Paris conference, the United States, China, all of Europe, Russia, India, Japan, South Africa, Brazil, and 140 other countries have pledged to cut their CO_2 emissions. But taken together, these pledges are not nearly enough to bring global emissions down to twenty billion tons as of 2035. On top of that, even these inadequate pledges are not likely to be made legally binding in Paris. Political leaders throughout the world, in other words, continue to court ecological disaster rather than mount a viable climate stabilization program.

This situation is shameful under any circumstances, but especially so because a simple and viable path to climate stabilization is, today, right before us, ready to be seized. That is, the global economy can bring global CO_2 emissions down to the IPCC target of twenty billion tons within twenty years if most countries—especially those with either large GDPs or large populations—devote between 1.5 and 2 percent per year of GDP to investments in energy efficiency and clean, low-emissions renewable energy sources. The consumption of oil, coal, and natural gas will also need to fall by about 35 percent over this same twenty-year period—i.e., at an average 2.2 percent rate of decline per year. These investments that are aimed at dramatically raising energy efficiency standards and expanding the supply of clean renewable energy sources—what we can term a *global green growth program*—will also generate tens of millions of new jobs in all regions of the world. This is because building a green economy requires hiring workers at much greater rates than maintaining the world's currently dominant fossil fuel–based energy infrastructure. In its essentials, this is the *entire* global green growth program.

If it is all so simple, why then do global emissions keep rising when they desperately need to fall? Raw greed and corrupt politics loom large, of course. But there is more at play, including serious misunderstandings on the left as to what can and should be done.

Three Barriers to Climate Progress

There is one fundamental reason why most countries throughout the world, at all levels of development, are unwilling to cut their CO_2 emissions sufficiently, notwithstanding the ever-mounting ecological threat. It is because the only way countries can achieve serious CO_2 emissions cuts is to stop burning so much oil, coal, and natural gas to produce energy. Confronting this unavoidable reality in turn creates three problems that are distinct but interrelated, and all formidable in their own right.

The first is that workers and communities throughout the world whose livelihoods depend on people consuming fossil fuel energy will face major losses—layoffs, falling incomes, and declining public-sector budgets that support schools, health clinics, and public safety. The second is that profits will fall sharply and permanently for the colossal fossil fuel companies, such as Exxon-Mobil, Shell, and the range of energy-based businesses owned by the US meta-billionaires David and Charles Koch. The world's publicly owned energy companies—such as Saudi Aramco, Gazprom in Russia, and Petrobras in Brazil, which together control about 90 percent of the world's total oil reserves—will take still larger hits to their revenues. The third problem pushes us beyond the fossil fuel industry itself and into broader issues of jobs and prospects for economic growth: that is, according to most analysts, economies will face higher energy costs when they are forced to slash their fossil fuel supplies. It will therefore become more expensive for them to operate the full gamut of buildings, machines, and transportation equipment that drive all economies forward.

How do we successfully attack these three problems? The challenges are undeniably huge. But there is also very good news here: they are not nearly as daunting as most analysts claim, including most commentators on the left.

For example, in a widely cited *Nation* article in 2014, "The New Abolition-ism,"[1] the progressive journalist Chris Hayes argued that the losses of money that fossil fuel companies will have to bear to bring global CO_2 emissions down sufficiently will be equivalent to those experienced by US slave owners as a result of the Civil War and abolition. This is a huge overstatement, as we will see. Leftists are also increasingly embracing the view that the solution to climate change is to oppose economic growth in general and advance an alternative "degrowth" agenda. Versions of this approach are developed, for example, by the influential political economists Tim Jackson, in his book *Prosperity Without Growth,* and Juliet Schor, in her work *Plenitude,* published in 2010.[2] Ironically, their views converge with the arguments long advanced on the right that achieving climate stabilization will inevitably require fewer job opportunities and the end of economic growth. The main difference between these left and right degrowth perspectives is that the right denounces this prospect while the left embraces it.

In reality, one critical step to mounting a successful global climate stabilization movement is to recognize that these perspectives are themselves obstacles to progress. Defeating the oil companies will be grueling, but it can certainly be accomplished without having to fight something like another Civil War. Victories are already mounting. Even more basically, it is simply not true that there must be large, painful trade-offs between stabilizing the climate on the one hand and supporting jobs and economic growth on the other. Quite the contrary: mounting a global green growth project is rather the one climate stabilization strategy that can realistically succeed and is available to us right now.

How Green Growth Delivers Climate Stabilization

Of course, executing this green growth plan—with investments of between 1.5 and 2 percent of global GDP every year to dramatically raise energy efficiency standards and expand the clean renewable energy supply—is easier said than done. To begin with, energy efficiency investments in all regions of the world will need to span each country's stock of buildings, transportation systems, and industrial processes. Efficiency levels will need to rise, among other places, in office towers and homes, with residential lighting and cooking equipment, and in the performance of automobiles and provision of public transportation. Expanding the supply of clean renewable energy will require major investments in solar, wind, geothermal, and small-scale hydropower, as well as in low-emissions bioenergy sources, such as ethanol from switchgrass, agricultural wastes, and waste grease. By contrast, expanding the supply of high-emissions bioenergy sources, such as burning corn ethanol and wood, provides no benefit at all relative to fossil fuel sources. Dependency on these high-emissions bioenergy renewables needs to be slashed at the same rate as fossil fuels.

To increase global clean energy investments by 1.5 percent of global GDP will require the development of effective industrial policies for countries at all levels of development. This will mean large-scale public investments in clean energy—for example, raising energy efficiency standards in government-owned buildings, dramatically expanding good public transportation systems, and substituting clean renewable energy for oil, coal, and natural gas with the government's own energy purchases. Providing abundant and affordable financing for private businesses will also be critical. The situation with energy efficiency investments makes this clear. On average, energy efficiency investments pay for themselves within three years. But they still require financiers to put up the funds to cover upfront investment costs.

Countries that successfully advance policies to invest between 1.5 and 2 percent of GDP in efficiency and clean renewables should not face any more difficulties in maintaining healthy economic growth rates than if they kept burning oil, coal, and natural gas at their current rates. To begin with, energy efficiency investments make it cheaper to cook meals, heat and cool homes and offices, use cooking equipment, travel by cars, buses, or trains, or operate industrial machinery.

Similarly, the International Renewable Energy Agency (IRENA) reports that, in all regions of the world, average costs of generating energy with most clean renewable energy sources are now at rough parity with fossil fuels.[3] This means that households, businesses, and public enterprises that substitute clean renewables for fossil fuels will not need to pay more to meet their energy needs. This is without even factoring in the environmental costs borne by everyone from continuing to burn fossil fuels. It is true that not all clean renewable energy sources are at rough cost parity with fossil fuels. Solar energy is the most significant case in point. But solar energy costs are falling sharply as increased investment levels spur innovation, and will fall still faster as global solar production expands.

Green Growth and Jobs

The investments in energy efficiency and clean renewable energy between 1.5 and 2 percent of GDP will also produce major expansions in job opportunities for countries at all levels of development. Research that I have conducted with coauthors has found this relationship to hold in Brazil, China, Germany, India, Indonesia, South Africa, South Korea, Spain, and the United States.

For the United States, for example, we found that increasing investments by around $200 billion per year in raising energy efficiency standards and expanding clean renewable production—about 1.2 percent of current GDP—would drop US emissions by 40 percent within twenty years, while also creating a net increase of 2.7 million jobs. This is after taking full account of the jobs that would be lost as oil, coal, and natural gas production falls by 40 percent.

For India, we found that, through increasing energy efficiency and clean renewable investments by 1.5 percent of GDP every year, CO_2 emissions could be stabilized at their current low level, which is one-tenth that of the United States on a per capita basis. This would be while India's GDP would still be growing at an average of 6 percent per year—a growth rate that produces a near tripling of average incomes within twenty years. Over the twenty-year investment program, these energy efficiency and clean renewable investments would also create an average of about ten million more jobs per year than if India continued to rely on its existing fossil fuel–dominant energy infrastructure. India could also eliminate nuclear energy altogether through this green growth program.

For the case of Spain, in a study that we produced for the progressive anti-austerity party Podemos, we showed how green growth could be a cornerstone for a broader antiausterity agenda. For example, we found that, through increasing investments in energy efficiency and clean renewables by 1.5 percent of GDP, Spain could reduce its emissions by more than 60 percent within twenty years, while also generating an increase of about four hundred thousand jobs relative to maintaining its existing energy infrastructure. This program would also enable Spain to steadily curtail its current heavy dependency on imported oil. At present, Spain's oil import bill expands rapidly whenever the economy starts growing. This becomes a major barrier to busting out of its ongoing austerity trap.

How Green Growth and "Small Is Beautiful" Converge

Throughout the world, the energy sector has long operated under a variety of ownership structures, including public/municipal ownership and various forms of private cooperative ownership in addition to private corporate entities. Public ownership already predominates throughout the global fossil fuel industry. But this doesn't mean that Saudi Aramco, Russia's Gasprom, or Brazil's Petrobras are prepared to fight climate change simply because we face a global emergency. National development projects, lucrative careers, and political power all depend on continuing the flow of large fossil fuel revenues.

At the same time, green growth investments will create major new opportunities for alternative ownership forms, including various combinations of smaller-scale public, private, and cooperative ownership structures. For example, community-based wind farms have been highly successful for nearly two decades in Germany, Denmark, Sweden, and the United Kingdom. A major factor in the success of these projects is that they operate with lower profit requirements than big private corporations.

Falling costs for clean renewable energy, solar in particular, are also opening huge new opportunities for people to install and operate their own small-scale "distributed energy" systems that are off the grid altogether. In January 2015, the *Financial Times* reported that "across the U.S., about 45,300 businesses and 596,000 homes have solar panels. . . . Over the past four years, the numbers have risen threefold for businesses and fourfold for homes, as the costs of solar power have plunged."[4] The prospects for distributed energy are still greater in developing countries such as India, where over 40 percent of rural households do not have access to grid-based electricity. Distributed renewable energy will enable rural communities to leapfrog over grid-based delivery systems entirely, just as mobile phone technology has enabled them stop depending on corporate-controlled landline phone companies.

The Problems with Degrowth

As expressed by Giacomo D'Alisa, Frederico Demaria, and Giorgos Kallis, in their new book *Degrowth: A Vocabulary for a New Era*, "The foundational theses of de-growth are that growth is uneconomic and unjust, that it is ecologically unsustainable and that it will never be enough."[5] These concerns are real, and degrowth proponents are making important contributions by addressing them forcefully. But on the specific issue of climate change, they are far too sweeping in their indictment of economic growth. Consider some very simple arithmetic. Following the IPCC, we know that global CO_2 emissions need to fall from their current level of thirty-two billion tons to twenty billion tons within twenty years. Now assume that global GDP contracts by 10 percent over the next two decades, following a degrowth scenario. That would entail a reduction of global GDP four times larger than what we experienced over the 2007–2009 financial crisis and Great Recession. In terms of CO_2 emissions, the net effect of this 10 percent GDP contraction, considered on its own, would be to push emissions down by precisely 10 percent—that is, from thirty-two to twenty-nine billion tons—i.e., exactly the global emissions level when the Copenhagen Climate Summit convened in 2009. So the global economy would still not even come close to bringing emissions down to twenty billion tons within twenty years.

What this example makes clear is that, even under a degrowth scenario, the overwhelming factor pushing emissions levels down will not be a contraction of overall GDP but *massive growth* in energy efficiency and clean renewable energy investments (which, for accounting purposes, will contribute toward increasing GDP) along with similarly dramatic cuts in fossil fuel production and consumption (which will register as reducing GDP). Moreover, any global GDP contraction would result in huge job losses and declines in living standards for working people and the poor. Global unemployment rose by over 30 million during the Great Recession. I have not seen any degrowth proponent present a convincing argument as to how we could avoid a severe rise in mass unemployment if GDP were to fall twice as much as during 2007–2009.

It is true that, for nearly forty years now, the gains from economic growth in virtually all countries have persistently favored the rich. Yet it is also true that the prospects for reversing inequality in all countries will be far greater when the overall economy is growing than when the rich are fighting everyone else for shares of a shrinking pie. Even thinking in strategic terms alone, attempting to implement a degrowth agenda would have the effect of rendering the global clean energy project as utterly unrealistic politically. This is exactly the opposite of what we need now.

A Superfund for Fossil Fuel Workers

In order for the global green growth project to succeed, it must provide adequate transitional support for workers and communities throughout the world whose livelihoods currently depend on the fossil fuel industry. The late US labor leader and environmental visionary Tony Mazzocchi pioneered thinking on what is now termed a "just transition" framework for these workers and communities. Mazzocchi developed the idea of a "Superfund" for workers who lose their jobs as a result of necessary environmental transitions. Mazzocchi's use of the term refers to the US environmental program that was implemented in 1980 to clean up sites at which corporations had dumped hazardous wastes from petrochemical and nuclear energy production. As Mazzocchi wrote as early as 1993: "Paying people to make the transition from one kind of economy—from one kind of job—to another is not welfare. Those who work with toxic materials on a daily basis . . . in order to provide the world with the energy and the materials it needs deserve a helping hand to make a new start in life. . . . There is a Superfund for dirt. There ought to be one for workers."[6]

The critical point in Mazzocchi's idea is that providing high-quality adjustment assistance to today's fossil fuel industry workers will represent a major contribution toward making a global climate stabilization project viable. It is a matter of simple justice, but it is also a matter of strategic politics. Without such adjustment assistance programs operating at a major scale, the workers and communities facing retrenchment will, predictably and understandably, fight to defend their livelihoods. This, in turn, will create unacceptable delays in proceeding with effective climate stabilization policies.

Well-funded worker Superfund policies therefore need to be incorporated into each country's green growth spending program. For the US case, I estimate that a generous Superfund program would be in the range of $1 billion per year, i.e., barely making a dent in an overall public and private spending level at around $200 billion. In addition, the impact on workers and communities from retrenchments in the fossil fuel sectors will not depend only on the support provided through an explicit Superfund budget. The broader set of opportunities available to workers will also be critical. The fact that clean energy investments will themselves generate a net expansion in employment in all regions of the globe means that there will be new opportunities for displaced fossil fuel sector workers within the energy industry itself. But further than this, the single best form of protection for displaced workers is an economy that operates at full employment. In a full-employment economy, the troubles faced by displaced workers—regardless of the reasons for their having become displaced—are greatly diminished simply because they should be able to find another decent job without excessive difficulties.

Divestment and Reinvestment to Defeat Big Oil

The giant fossil fuel corporations will obviously take a big hit under green growth. But overstating what is at stake doesn't help matters. In fact, it is critical to have a clear grasp of the scale of the losses that fossil fuel companies are likely to face.

A study published in 2013 jointly by Carbon Tracker and the Grantham Institute on Climate Change and the Environment at the London School of Economics examined the current holdings of the largest two hundred private-sector fossil fuel companies, as listed in the world's various stock exchanges. This study estimated that "60–80 percent of coal, oil, and gas reserves of [these] firms are unburnable."[7] From these figures, we can roughly estimate that these companies are looking at around $3 trillion in lost value over the next twenty years, and the certainty of further declines subsequently.

Of course, $3 trillion is real money. The fossil fuel companies are not about to relinquish it without a brutal fight. At the same time, $3 trillion is also equal to only about 1.3 percent of the $225 trillion in total worldwide private financial assets as of 2012. Still more, the $3 trillion in losses that the fossil fuel corporations will face will not happen in one fell swoop. The declines will rather be incremental over a twenty-year period. On average, this amounts to asset losses of $150 billion per year. By contrast, as a result of the US housing bubble and subsequent financial collapse in 2007–2009, US homeowners lost $16 trillion in asset values in 2008 alone—about one hundred times the annual losses fossil fuel companies would face.

The fact that the decline in fossil fuel asset values will occur incrementally over decades also means that investors will have ample opportunity to diversify their holdings. Many are already doing so. As one important example, in June 2014 Warren Buffet, the best-known corporate investor and third richest person in the world, announced that his company, Berkshire Hathaway, was doubling its holdings in solar and wind energy companies to $15 billion. This is even while Berkshire continues to carry large positions in conventional utility companies.

More significant still has been the rapidly growing global fossil fuel divestment movement. The divestment movement includes foundations, universities, religious organizations, and municipalities located throughout the United States and Western Europe as well as in Canada, South Africa, Australia, and New Zealand. These organizations have been pushed by grassroots activists to take principled stands with their investment portfolios and they have embraced the challenge. As of this writing, nearly two hundred organizations have sold off roughly $50 billion in fossil fuel assets. Big Oil can buy politicians and scream all it wants, but it can't stop investors from walking on them.

The divestment movement does still need to broaden its goals, to become just as actively committed to *reinvestment* in energy efficiency and clean renewables as it is presently to dumping fossil fuel stocks. This is how we can force the global economy onto a workable path toward both stabilizing the climate and generating tens of millions of jobs for working people and the poor throughout the world.

Notes

1. Chris Hayes, "The New Abolitionism," *Nation*, May 12, 2014, www.thenation.com/article/new-abolitionism/.
2. Tim Jackson, *Prosperity Without Growth: Foundations for the Economy of Tomorrow*, 2nd ed. (London: Routledge, 2017); Juliet Schor, *Plenitude: The New Economics of True Wealth* (New York: Penguin, 2010).
3. International Renewable Energy Agency, *Renewable Power Generation Costs in 2017* (IRENA: Abu Dhabi, 2018).

4. Ed Crooks, "US Energy: Off the Grid," *Financial Times*, January 13, 2015, www.ft.com /intl/cms/s/0/b411852e-9b05-11e4-882d-00144feabdco.html#axzz3gJfvRdyM.

5. Giacomo D'Alisa, Frederico Demaria, and Giorgos Kallis, eds., *Degrowth: A Vocabulary for a New Era* (New York: Routledge, 2015).

6. Tony Mazzocchi, "A Superfund for Workers," *Earth Island Journal* 9, no. 1 (1993): 40–41.

7. Carbon Tracker and the Grantham Research Institute, LSE, *Unburnable Carbon 2013: Wasted Capital and Stranded Assets* (2013), http://carbontracker.live.kiln.digital /Unburnable-Carbon-2-Web-Version.pdf.

CHAPTER 11

ALL TOO HUMAN

Orienting Environmental Law in a Remade World

JEDEDIAH BRITTON-PURDY

The scope and depth of the human transformation of the world are staggering to consider. If "Pave the planet!" remains an ironic slogan of college-age environmentalists, it is also a distillation of our collective impact. Earth is a world remade through, and as, human infrastructure. We are, in turn, an infrastructure species, whose activity and survival depend on our thoroughgoing remaking of the planet and its systems. Perforce, the other forms of life with which we share it are also infrastructure species, or they are species without futures—as is true of many of them. What might a full appreciation of these facts mean for the law, particularly for environmental law, the special legal redoubt of the nonhuman world?

The "great acceleration" of human activity and impact has transformed the planet with unprecedented intensity. Real global GDP has risen approximately twelvefold since 1950, primary energy use more than fourfold, fertilizer consumption by a factor of more than fifteen, water use fourfold, and paper production eightfold.[1] In the same period, the quantity of nitrogen synthesized annually for fertilizers rose from under four millions tons to over eighty-five million tons, and plastics production increased in weight from under one million tons to over three hundred million tons.[2] A study in 2017 estimated the mass of the global "technosphere," the material habitat that humans have created for themselves in the form of roads, cities, rural housing, the active soil in cropland, and so forth, at thirty trillion tons, five orders of magnitude greater than the weight of the human beings that it sustains.[3] (Restricting the estimate to buildings, pavement, and the like, in categories such as cropland and trawled sea floor that seem "too natural," cuts the estimate of the techno-sphere's weight by only a little more than one-third.)[4] That is approximately four thousand tons of transformed earth per human being, or twenty-seven tons of technosphere for each pound of a 150-pound person.

The transformation of planetary systems is less jaw-dropping on its face, but more profound. Famously, atmospheric concentration of carbon dioxide has increased from preindustrial levels of about 270 parts per million (ppm) to a little over three hundred ppm in 1950 and over four hundred ppm today.[5] Concentrations of methane and nitrous oxide, the other leading greenhouse gases, have increased by about 50 percent and 10 percent, respectively, since 1950.[6] Nitrogen runoff to coastal zones from farming and sewage increased from under twenty metric tons in 1950 to nearly one hundred tons today, marine fish capture from about ten million tons to about seventy million (with a recent decline driven by falling fish populations and a surge in polluting aquaculture), and the loss of tropical forest area is now nearly 30 percent from its historical baseline, doubling the loss since 1950 and increasing nearly six times since 1900.[7] More than a third of the Earth's land area is now classified as "domesticated" intensive human use, a proportion that may seem less mod-est in light of the fact that less than a third of land area is arable; much of what there is to occupy has been occupied.[8]

For a sense of the cumulative effect of all of this on cognate species, con-sider the following: Human biomass in 2000 was eleven times that of all other wild terrestrial mammals combined; in 1900, the ratio was only 1.3:1.[9] Already in 1900, domesticated animals outweighed wild terrestrial mammals by a fac-tor of 3.5; but in 2000, the total weight of the domesticated species was twenty-four times that of the wild ones.[10] In other words, there is little out there besides us and the beasts that we have bred, housed, and developed grain industries to feed.

Phenomena of this kind have led many commentators to propose that Earth has entered a new geological era, frequently called *the Anthropocene*, in which humans have become a force, arguably the dominant force, in shaping the planet.[11] Confronting the Anthropocene, the argument usually goes, implies discarding any fixed boundary between the natural and the artificial in favor of recognition that human activity is perennially making and remak-ing the world, from atmospheric chemistry to biodiversity to the shapes of continents (as sea levels rise).[12] The law cannot rely upon natural baselines, which are irretrievable in practice and, in any event, themselves the joint product of human and nonhuman causes. Whether the law's goal is a certain level of atmospheric carbon concentration or the management of a tract of public land as "wilderness," it is not seeking to preserve a nature that is simply "out there," but rather participating in deliberate world-making by selecting and pursuing a certain condition for a part of Earth.[13] In legal scholarship and lawmaking, these conclusions have led in several directions: a shift from "pre-serving" to actively managing natural phenomena under statutes such as the Endangered Species Act;[14] a burgeoning interest in areas of law governing sectors not historically regarded as "environmental," such as agriculture and energy, that contribute massively to the technosphere;[15] and attention to the distributional effects of environmental policy, in recognition that massive changes in (among others) weather, oceans, and seasons bring disparate burdens to various regions and populations.[16]

My goal here is to build on these developments by proposing a way of understanding environmental law as constitutively engaged in world-making. Such world-making is inseparable from the governance of "nonenvironmental" areas such as energy, agriculture, finance, and trade, and all these forms of world-making pervasively implicate distributional questions. The habits of thought in environmental law and policy remain significantly shaped by an older and simpler set of presuppositions: that the category of "environmental" issues can be delineated fairly clearly, or is even self-evident; that environmental topics are in important ways distinguished by their natural rather than artificial character; and that the values that should govern them are relatively self-evident by reason of their natural status. These values have tended to be of quite different kinds. In areas such as the preservation of wilderness and charismatic species, advocates, lawmakers, and statutory regimes have often assumed some self-evident and intrinsic value to be honored.[17] By contrast, in the management of instrumental goods such as raw materials, energy sources, soil fertility, and water for irrigation, it has often seemed equally self-evident that the nonhuman world is simply a stock of useful materials, to be managed for some version of maximum social benefit.[18] My central question here is how to think of environmental law after abandoning these assumptions and acknowledging the new conditions that I have sketched.

My proposal, spelled out further in section 2, is that environmental law, and the resources and phenomena it governs, should be approached as kinds of *infrastructure*. The concept conventionally embraces "the underlying framework of a system," or "resources shared to many ends," such as roads or railways, telephone networks, postal services, schools, court systems, and sewers.[19] Much infrastructure is at once foundational and essentially instrumental, in the sense that it may be used equally well for many specific purposes or projects; another way of making this point is by observing that much infrastructure directly serves human capabilities at a high level of generality, such as mobility, communication, and literacy, which potentially serve many goals.[20] Infrastructure has deeply to do with facilitating and shaping cooperation, both materially and through more abstract means such as enforceable contracts.[21] It is the way that people gain access to one another and enter into common projects.[22] It is deeply political, in that lawmaking and more direct forms of state action generate and shape infrastructure, determining in turn the kinds of cooperation and other activity that it facilitates, thus involving it in choices of value as well as questions of efficiency. The state is so central to infrastructure that institutions of political decision-making should be regarded as "*political* infrastructure," the "coordination regime [that] can enable the production of new regimes of coordination."[23]

To approach environmental law by way of infrastructure is to understand the nonhuman world as pervasively created by human activity that is facilitated and shaped by the "regimes of coordination" and systemic frameworks that infrastructure comprises, so that to make our infrastructure is to make our world. It is also to understand the nonhuman world as itself a species of infrastructure, both in the massive "technosphere" of built and managed systems that form much of the human environment and in the other systems, such as the carbon, nitrogen, and water cycles; ocean acidity; and patterns of regional and global climate that we rely on for survival. Infrastructure is both what sustains us as organisms and what enables us to amplify our capabilities and radicalize our interconnection. In both respects, it comprises systems that are deeply engineered, that serve as preconditions for many basic capabilities with multifarious applications, and in which those capabilities are enmeshed in profound interdependence, so that, although their extension or denial to some need not imply universal extension or denial, they are available only in widely shared regimes of coordination and management.

The shaping of environmental infrastructure has important distributive consequences. It deals out benefits and burdens and, in doing so, concretely expresses judgments about distributive justice: which needs and interests should be treated as the basis of what one might call "baseline claims," which every member of society can take for granted. As members of a species that depends on infrastructure for mobility, communication, sustenance, shelter that suits our climates (that heats or cools us, for instance), and, by a modest estimate, the bare survival of some 95 percent of us, people lead lives that are not separable from the decisions of their polities, concretely expressed in the infrastructure they inhabit.[24]

Section 1 asks why environmental law did not come sooner to these themes, and argues that the reasons are rooted in the peculiar historical conditions of the 1960s and 1970s, notably the consequences of an anomalous period of relative economic equality. There were many historical resources for an infrastructure-themed environmental law, but they were in eclipse as modern environmental law took statutory and institutional form. Section 2 sets out the questions that an infrastructure approach brings into focus: the interplay of natural and artificial phenomena in world-making and the consequences of world-making legal and political decisions for both distribution and the qualitative shape of social life.

I. The Infrastructure Movement That Wasn't:
A Partial Genealogy of Environmental Law

An infrastructural view of environmental law is not so much new as recovered from a period of forgetting. It was present in the conservationist policy-making and institution-building of the late nineteenth and early twentieth centuries, when reformers built up US state capacity to manage a range of systems for the utilitarian conception of the public interest: forests, soil systems, and watersheds, but also municipal public-health systems, labor markets, and market concentration. For those who crafted these policies, managing natural resources was (like the Homestead Act and other property-creating laws of an earlier generation) a matter of institution-building, landscape-forming statecraft, distributional judgments, and the political economy of citizenship. These organizing themes did not recede as twentieth-century environmental politics moved from utilitarian resource management into the distinctive twin agendas of modern environmental law: (1) pollution control and (2) the preservation of wild nature for aesthetic reasons and on account of its intrinsic value. Instead, these agendas arose in movements that were deeply concerned with the continuity between the human and nonhuman environments, the distributional aspects of environmental policy, and the role of the state (and politics more generally) as shaper of a common world. Behind Rachel Carson's advocacy and the research that helped to inspire the Clean Air Act and Clean Water Act stands a history of industrial toxicology addressed to industrial workplaces and urban health, and aligned with the labor movements of the early twentieth century. Behind the Wilderness Act, which critics cite as the apogee of environmentalism's distorting fixation on pristine, nonhuman nature, stands a movement populated by self-described "socialists," regional planners, and students of the postnatural interweaving of artificial and given elements in both landscape and political economy. Yet many of these themes fell into eclipse in the legislative and institutional creation of modern environmental law between 1970 and roughly 1977.

More recently, the infrastructure approach to environmental law has fallen mainly to a set of movements and theoretical perspectives grouped together under the label of environmental justice. Environmental-justice movements and scholars have worked to integrate considerations of distributive justice into areas of environmental law that are otherwise treated in aggregative cost-benefit terms or governed by other considerations—notably, but not only, the setting and enforcement of pollution limits and the siting of hazardous facilities. They have insisted that the scope of "environmental" questions should include the built, residential, and industrial environments. And they have proposed that, for these reasons, at least some of the decisions now entrusted to objectivity-seeking expertise should incorporate stronger democratic elements.[25] One purpose of this genealogy is to understand how environmental law lost its emphasis on infrastructure and had to be called back to it by the environmental-justice movement. In that respect, one could understand the reorientation of environmental law toward infrastructure as a way of asking what it would mean for the field to understand the themes of environmental justice as foundational.

A. The "Environment" in Environmental Law

Environmental law carries forward conceptions of "the environment" and law's role in managing it that formed in the 1960s and 1970s in the United States, when a set of problems were newly grouped together under the rubric *environmental*. Modern environmental law is very substantially the product of a burst of legislation and institution-building that took place between the end of the 1960s and the beginning of the 1980s. President Richard M. Nixon signed the National Environmental Policy Act on January 1, 1970, created the Environmental Protection Agency in December of that year, and signed the Clean Air Act on New Year's Eve.[26] In the ten years that followed, Congress passed, among other statutes, the Clean Water Act, the Endangered Species Act, laws governing the use and disposal of pesticides and other toxic substances, comprehensive reform of federal public-lands management, and the Surface Mining Control and Reclamation Act, to give an incomplete list.[27] This period of lawmaking closed out with the adoption in 1980 of the toxic-waste cleanup law commonly known as Superfund.[28] Congress passed the 1970s statutes by margins that today suggest typographic errors: only one vote against the Clean Air Act in 1970 in all of Congress; average votes of 76–5 in the Senate and 331–30 in the House for major environmental statutes.[29]

The same years saw the rise of advocacy organizations and centers of professional expertise that still do much to shape environmental law. The Environmental Law Institute, the key professional clearinghouse for the field, was founded in 1969, and the Natural Resources Defense Council (NRDC), arguably the touchstone modern environmental group, was created in 1970. The Environmental Defense Fund and the think tank Resources for the Future already existed but were substantially built up in these years, when they moved their previous science-and-policy missions toward the emerging field of environmental law, as defined by the post-1970 statutes. Long-standing organizations such as the Sierra Club became mass-membership groups and built up litigation arms that followed the NRDC's example.[30]

These same years saw the retreat of environmental politics from engagement with issues of power and justice in the human environment, a "great forgetting" of strands of environmentalism concerned centrally with just these questions.[31] The forgetting came, ironically, when these themes were most needed, but seemed most dispensable. To understand this, the first step is to grasp what was forgotten, and why. The rest of this section locates a genealogy of modern environmental law within the unusual period of American history in which it arose, one of (actual and perceived) relative economic equality, ill-founded confidence in the capacity of nonenvironmental areas of law to enforce principles of equity that environmental law left unaddressed, and a certain elite-institutional tilt toward a relatively narrow conception of which problems counted as "environmental."

B. The Forgotten Long History of Infrastructure Environmentalism

Environmental law developed in the last fifty years around two organizing concerns: the control of pollution and the preservation of biodiversity, natural beauty, and wild places. Both have struck critics as too narrow in scope, attached to Romantic ideas of pristine nature, and focused on protecting a suburban idyll from poisonous intrusion. This narrow focus comes at the cost of inattention to the full scope of the human environment in a half-artificial world, including multifarious threats to health in the workplace and in poor and vulnerable communities. These themes, however, were richly represented in the immediate predecessors of modern environmentalism. What must be understood is not simply their absence, but their disappearance.

1. The Wilderness Society, Regional Planning, and "Socialism"

"Environment," wrote Benton MacKaye in 1928, "is the influence upon each inner mind of the thing shared by every mind . . . the filament which binds our separate lives . . . the total life which every life must share."[32] More concretely, MacKaye meant by "environment" the world formed by the braiding of human and nonhuman systems. His paradigm of an environmental way of thinking was a description of New York City as a nexus of many kinds of "flows": the Hudson River, the Atlantic tides, the weather-bearing wind out of the West; and steel from the Great Lakes, commodities from Europe and South America, and people pulsing daily through the veins and arteries of highways, railways, and ferries.[33] In an "environment" that was a skein of many systems, humanity now faced what MacKaye called "the wilderness of civilization."[34] He meant by "wilderness" something newly encountered and not yet fully understood, something alien, despite being a human creation, and wondrous. To call civilization "wilderness" was to invite exploration of it.

In contemporaneous work, MacKaye contributed to making "wilderness" a key word in the mainstream environmental lexicon and a key feature of the environmentalist agenda. He was a founder in 1935 of the Wilderness Society, which very substantially shaped the Wilderness Act of 1964, a statute that has preserved about 110 million acres of federal land from development, mechanical transportation, and commercial activity.[35] He is also commonly credited with the creation of the Appalachian Trail, which he substantially designed and long championed. The Wilderness Society's priorities have been the target of a prominent criticism of environmental politics, which holds that a fixation on the wild and pristine encourages devaluing the places where people mostly live.[36]

The forgotten history of the wilderness movement cuts against this criticism. MacKaye rejected pastoral idylls, antiurban politics, and the binary between the natural and the artificial. Opposing his view to the antiurban aesthetics and politics of Sierra Club founder John Muir, he argued that one could equally well admire cities, workplaces, and "wild" places.[37] The Appalachian Trail that he advocated was to be not a mere walk in the woods, but a link among rural settlements of artists, artisans, and farmers, which would in turn be linked by roads to larger towns and cities.[38] The wild portions of the trail, like designated wilderness areas on other public lands, were parts of a graduated series of engineered environments.

A New Dealer, MacKaye regarded workers' struggles as part of the larger campaign for a better human environment. Writing of strikes in steel mills near Pittsburgh, he reflected, "The workers dwelling in these steel towns are, as is well known, in profound rebellion against their condition in life."[39] Although all observers understood that strikers "are fighting . . . for higher pay and for longer hours of leisure," MacKaye argued that the strikers should be seen as aiming at not just "time," but "space" in which to develop and explore their own capacities.[40] Reshaping space to enrich human life for members of all classes was MacKaye's environmental agenda.

MacKaye's Wilderness Society cofounder Robert Marshall agreed that wilderness was only an element in a much broader program for reshaping the landscape.[41] As president of the Society, Marshall formulated the administrative classification of "wilderness" areas for federal forest management when he was also chief of forestry for the Bureau of Indian Affairs during the partial restoration of tribal ownership and self-government that is often termed the "Indian New Deal." An enthusiast for the managerial state, he was in his own terms a "socialist" who looked forward to the replacement of "the profit system" by administration and cooperation.[42] Marshall argued for nationalizing most of the country's commercial timberland, both to manage a boom-and-bust sector and to break what he called the "whip hand" of the timber industry over both workers and regulators.[43] He envisioned public forestry management as part of a larger program of "rural reorganization."[44]

The wilderness movement, then, was conceived as part of a much larger reconstruction of American landscapes, residential patterns, resource use, and the boundaries between state and market. Advocates such as MacKaye and Marshall were keenly interested in the broad shape of American political economy, the distributive contests over economic and political power that they saw in the fields of both politics and labor, and the qualitative shape that lawmaking gave to the places where people live, work, play, and learn.

2. Silent Spring *and the Industrial Environment*

In *Occupational Tumors and Allied Diseases*, published in 1942, Wilhelm Hueper argued that "the new artificial environment" of industrial chemicals created an imperative for government to secure "the fundamental requirements for a healthful living, not merely for a small, select, and socially privileged class, but for the entirety of its citizens ... by suitable laws adequately enforced."[45] Hueper was working in a decades-long tradition of reformist public-health research, but his work also confirmed that tradition's turn from fieldwork to lab work and from narrative to risk measurement. The previous generation had been defined by the pioneering industrial-health researcher (and the first woman on the Harvard faculty) Alice Hamilton, who studied the health of factory workers intensively between 1908 (her appointment to the Illinois Commission on Occupational Diseases) and 1935 (her retirement from Harvard), including major investigations into the health effects of lead and phosphorous.[46] Hamilton had previously studied the epidemiology of typhoid in neighborhoods surrounding Chicago's Progressive enclave Hull House, where she was a resident, and her industrial work was an application of that style of fieldwork to the factory.

A sympathetic observer of the Lawrence, Massachusetts, textile strike of 1931–1932, Hamilton regarded the political power of manufacturers as a key impediment to reforming industrial conditions, and unions as playing an essential role in bringing about such reform.[47] In her support for organized labor as a necessary part of the governance of the industrial environment, Hamilton reflected not just the general fact that conflicts over union organizing figured prominently in the politics of the early twentieth century, but also the more specific engagement of unions in workplace health. Between 1921 and 1928, the Workers' Health Bureau, which began as a project of the labor and public-health reformers Grace Burnham and Harriet Silverman, established a beachhead as a labor-based institution for research and advocacy on industrial health.[48] With an independent relationship with union locals, the Bureau by 1927 received dues from 190 member locals in twenty-four states and collaborated with leading public-health researchers.[49] Burnham insisted on setting research agendas "FROM THE STANDPOINT OF THE WORKER," the "individual . . . caught and bound fast in the great web of machine industry as the fly is caught in the thousand-strand web of the spider."[50] The roots of industrial toxicology were thus thoroughly enmeshed with movements for reform and efforts to build both workers' power and systems of industrial governance in the early part of the twentieth century. Their concern was the democratic and humanitarian management of the "artificial environment" that the industrial economy produced.

Rachel Carson's *Silent Spring*, an eloquent brief against reckless pesticide use, builds on Hueper's work in its treatment of contaminated drinking water, DDT, cancer in wild fish, and the imperative of eliminating carcinogenic agents from the environment as a public-health prophylactic.[51] Carson, however, put the tradition of industrial toxicology to a new use. Like the larger environmental movement of the 1960s and 1970s, which she contributed importantly to shaping, Carson at once expanded the scope of environmental questions by thematizing the human relationship to the natural world as a whole, rather than specific places and resources, and narrowed environmental discourse by treating the social world as populated by an undifferentiated humanity, whose emblematic middle-class members lived almost exclusively in small towns and suburbs. The rhetorical switch uprooted the tradition of industrial toxicology from its original engagement in social and economic reform.

3. A Narrowing Agenda

The political program of wilderness preservation began within a New Deal vision of public ownership, planning for quality of life, and, above all, an integration among different kinds of human environment, with wilderness just one type of managed environment. By the middle of the 1950s, this vision had narrowed. The Wilderness Society formed a strategic alliance with the Sierra Club, whose longtime head, David Brower, had recently won attention-getting battles and was looking for a new flagship issue. The Sierra Club had long been apolitical outside its advocacy for preserving public lands, and the Wilderness Society now perfected its own version of single-issue advocacy, in which wilderness was the singular goal, its values readily translatable to a middle-class vacation itinerary.

Meanwhile, Rachel Carson's *Silent Spring* was published in 1962. Carson's book, which gets deserved credit for formulating the sense of threat that informed so much of environmental politics in the ensuing fifteen years, shared much with the public-minded studies of industrial toxins that Alice Hamilton and others had pioneered. Carson, however, crafted an environmental rhetoric centered on threats to small-town and suburban domesticity, on the one hand, and, on the other, threats to a natural world that she portrayed as full of perennial harmonies among species and their settings. Carson opened *Silent Spring* with an image of "a town in the heart of America where all life seemed to live in harmony with its surroundings," which was mysteriously visited by "a strange blight" and "a shadow of death." She described an undisturbed world in which "life reached a state of adjustment and balance with its surroundings," and contrasted it with the world after the application of pesticides, where "the whole closely knit fabric of life has been ripped apart."[52]

These two developments, in the wilderness movement and public-oriented toxicology, are emblematic of a change in which environmental themes, once closely tied with working conditions, economic power, and the larger question of how to shape American life, were adjusted to fit the constraints of mid-twentieth-century consensus politics.[53] Within those constraints, the basic questions about how Americans were to live seemed settled: they would live in suburbs, modeled on idyllic small towns, separated from their workplaces spatially but also by the distance between an ethics of commerce and an ethics of domesticity.[54] Carson did not portray these spheres as involved in disputes either over the rules that should prevail within them or over their boundaries and relations with one another—the kinds of disputes that feminism, labor politics, and the civil rights movement had launched and would soon amplify. The break in the harmony came, on this account, from failing to respect the perennial balance of nature.

These activists gave the following cast to the ecological crisis: On the one hand, everything must change, in the form of collective self-restraint to respect and restore natural harmonies; on the other hand, nothing in particular must change, in that the environmental crisis was not linked to calls for other changes in the legal or social order. This environmentalism was a defense of society, imagined as a whole, against an exogenous ecological crisis. That formula led *Time* and President Nixon, among others, to identify the environment as a unifying issue for the 1970s, in contrast to conflicts over race and war.[55] It was substantially this environmentalism that shaped the statutes and institutions of modern environmental law.

B. Drivers of the Great Forgetting

Why the great forgetting of the older, more encompassing agendas? At least three conditions contributed to the shape that environmental law and politics took in the formative years of the 1970s. First is the anomalous period of relatively equitable distribution of wealth and income that followed World War II and made distributive questions appear relatively inessential to environmental law. Second was the expectation that, where structural inequality made the consequences of facially neutral pollution-control statutes inequitable in practice, constitutional remedies should be available in the form of equal-protection claims based on theories of disparate impact that were not rejected until the latter part of the 1970s. Third was the distinctive culture of legal advocacy in which environmental law took institutional shape, in good part by way of the role of the Ford Foundation in the formative years of the field.

1. The Postwar Anomaly and the Forgetting of Economic Inequality

In the period of roughly from 1946 to 1973, high levels of economic growth coincided with a relatively egalitarian distribution of income and wealth, producing the widespread impression that economic inequality was a problem substantially solved.[56] In this generation-long experience (and longer-lasting perception) of inclusive growth as a "new normal," political elites characterized the remaining challenges of economic disparity as problems of exclusion, not inequality: Certain populations had been shut out, signally African Americans and Appalachian whites; but once dealt into the system, they too could expect to share in its benefits.[57] Even in a time of great reformist energy, lawmaking could set aside direct concern with inequality.

As environmental-justice critics have pointed out, the major environmental statutes do not address the distribution of benefits and burdens. They lack measures to avert disparate impact, whether from siting decisions, market-mediated migration, or the interplay of multiple and historically accumulated hazards, all of which tend to burden poor and nonwhite populations. What do these omissions reveal about the statutes and the attitudes of those who wrote them?

In spring 1970, Senator Edmund Muskie, a major architect of the antipollution statutes, offered an expansive picture of environmental law's social purpose.[58] On the first Earth Day, he argued that "Man's environment includes more than natural resources. It includes the shape of the communities in which he lives: his home, his schools, his places of work."[59] Arguing for protection of the "total environment," he insisted that "the only kind of society that has a chance" is "a society that will not tolerate slums for some and decent houses for others, rats for some and playgrounds for others, clean air for some and filth for others."[60] Muskie linked the environmental crisis rhetorically to the War on Poverty, to the Great Society idea that the country's challenge had changed from achieving prosperity to building widely shared flourishing within affluence, and to the civil rights movement.[61]

Does the design of the major environmental statutes give the lie to these sentiments, or do the statutes represent a window into what their sponsors thought necessary to implement these ideals in their time? Muskie's statements express a worldview in which the Clean Air Act, as written, was *already* an environmental-justice statute, part of a legal and economic infrastructure of security and fair opportunity. Antipollution laws, in this view, formed part of a comprehensive approach to regulating what John Kenneth Galbraith had influentially called "the affluent society."[62] (Muskie used the term in setting out the program of the Clean Air Act on the Senate floor.)[63] In this view, the antipollution statutes were part of a comprehensive renovation of the human "total environment," alongside other programs of the Great Society and the War on Poverty.

The common goal of this agenda was to overcome social and economic isolation, especially clusters of persistent poverty, and build up institutions that promoted learning and development.[64] Establishing national standards for healthful air and water played a key part in this program, as did controlling toxins. But when it came to distributing hazards, those measures were crafted in reliance on other initiatives in the larger program, and on background conditions of shared prosperity. The danger that, absent specific protections, environmental burdens will be distributed along familiar lines of race and poverty is weighty in proportion to the intensity of those other dimensions of inequality. Many reformers in 1970 expected poverty generally, and its racialized versions, to give way to a combination of egalitarian macroeconomic tendencies and inclusive and redistributive policies. Although it may seem clear in hindsight that Nixon-era retrenchment had already begun even as the major environmental laws were passed, Muskie endorsed the view of Joseph Califano, President Johnson's principal aide for domestic policy, who argued in 1968 that the Johnson years had "cleared the liberal agenda," that conservatives, too, were now interventionists and redistributionists in matters ranging from affirmative action and day care to proposals for a basic income, and that the important question was how to carry forward not just piecemeal legislation but what Muskie called "a reshaping of our basic political institutions, and . . . a reshaping of our thinking about them."[65] In this view, the challenge of the time was to follow the lead of events, both macroeconomic and political, in the direction of greater economic and social equality.

Soon the political turn against the regulatory and redistributionist state would become more apparent. Moreover, at roughly the time the major environmental laws were passing through Congress with huge majorities, economic inequality began its forty-year increase.[66] In their time, though, the major environmental statutes incorporated their architects' expectation that the laws would operate within a general movement toward greater equality.

2. Disparate Impact and Constitutional Remedies
The passage of the major environmental statutes from 1970 to 1977 preceded or coincided with the Supreme Court's rejection of disparate-impact claims

under the Equal Protection Clause and adoption of the requirement that equal-protection plaintiffs demonstrate "discriminatory purpose" by an official actor.[67] The authors of most of the environmental laws of 1970–1977 had reason to believe that, where the statutes failed to achieve a reasonably equitable distribution of burdens, litigants could call on courts to review policies with racially disparate effects. It was because of this doctrinal development that environmental-justice plaintiffs were later thrown onto the environmental statutes as their primary source of remedies, and so confronted those statutes' facial indifference to the economic and racial distribution of hazards. The standard of formal equality is more plausible as a principle of legal justice when participation in economic life tends to produce substantive equitable results.[68] The rejection of constitutional disparate-impact claims thus came at a historically ironic time as macroeconomics took a freshly inegalitarian turn.

3. Legal Liberalism and the Institutionalization of Environmental Law

The architects of mainstream environmentalism's flagship organizations knew, or so they believed, which problems were "environmental" and what interest the public had in those issues. This self-confidence expressed the dominance of a network of elite reformers who substantially shared an outlook on these matters. These elites, in turn, were significantly selected and cultivated by funders in the Ford Foundation, which made many of its largest cumulative grants in the 1970s to the Natural Resources Defense Council, the Environmental Defense Fund, and the Sierra Club Legal Defense Fund (later Earthjustice) and, at the same time, shaped their development.[69] Ford expressed confidence that its grantees' priorities aligned with an uncontroversial "public interest," citing "the environmental cases" as "the best examples" of the Foundation's thesis that working with "established and well-informed groups or organizations" would ensure there were not "substantial interests in the community that do not get represented adequately."[70]

Ford, which had already been at the center of building up clinical programs in law schools and developing pro bono expectations for the bar, was shaping environmental law in line with what Steven Teles has termed "legal liberalism."[71] A program of reform linked with a view of the law's role in a democratic society, legal liberalism was defined by its emphasis on the use of litigation and adjudication-like procedures to protect individuals against arbitrary discrimination with respect to their basic interests—that is, to ensure the formal preconditions of their full participation in political, economic, and social institutions. Ford's own account of its reasons for supporting "the environmental law movement" exemplified this view: "in a pluralistic society the views and interests of all segments of opinion should have their day in court."[72] The ideal was to bring all "views and interests" before an impartial decision-maker, not to engage in political contests to form views and challenge or reshape interests. Legal liberalism took plausibility from the distributional optimism of the mid-twentieth century: its procedural emphases made sense on the view that formally equal and open economic participation overcame rather than reinforced embedded inequality.[73]

***

The mainstream environmentalism that emerged in the 1970s was shaped by the premises of the time. With economic inequality seemingly in decline and other egalitarian policies in place and expected to grow, it was easy for the environmental statutes' drafters to imagine that they need not guard against compounding inequality. Legal liberalism helped to channel the enforcement of these statutes into organizations that defined environmental issues in relatively narrow fashion. Ironically, just at this watershed moment of environmental legislation, 1970–1977, the anomalous period of widely shared growth was ending. There followed four decades of increasingly unequal distribution of wealth and income, even as egalitarian political and doctrinal trends receded. Thus, precisely at the time when environmental law's character as infrastructure and political economy might have come to the fore, the field was instead domesticated, with a limited topical remit and scant distributive significance.

I turn next to what a very different environmental law might be.

II. An Infrastructure Approach to Law and Environment

My reasons for working with the idea of infrastructure are pragmatic and experimental. Although infrastructure is somewhat conceptually shaggy, lacking precise delineation and entailments, it emphasizes themes that are often neglected: the ways that law makes a world, and the consequences of that world-making for both distribution and the forms of life and cooperation that flourish or perish there. It makes politics central by highlighting that, if an infrastructure species is to make its own history rather than become an epiphenomenon of its artificial world, it must have a site of agency over its own world-making. Of particular interest to environmental law, an infrastructure approach sets aside a persistent, field-defining emphasis on "natural" environments to attend instead to the inseparability of the built and the given in many Earth systems that are essential to life.

A. What We Talk About When We Talk About Infrastructure

1. *Infrastructure as Resources, Cooperation, and Politics*

A. INFRASTRUCTURE AS RESOURCES

Scholars have given various inflections to the concept of infrastructure, some characterizing it as an economic category of resources, others as something nearer a social- and legal-theoretic concept enabling us to understand how the material and nonmaterial systems we create both empower and constrain our activity. On the economistic end of the spectrum, Brett Frischmann characterizes it as a set of resources for which consumption is nonrivalrous, at least for a considerable range of demand, which have many potential "downstream" uses, and for which demand is driven by those downstream uses.[74] Key examples are transport networks, renewable environmental resources, technologies and ideas, and communication networks such as the Internet.[75] Such resources frequently exhibit increasing returns to scale and positive network externalities, so that their value tends to increase with the number of users of a given infrastructure system, whether because of the low marginal cost of additional provision or because of the increased benefit on account of access to a larger number of other participants.[76]

B. INFRASTRUCTURE AS COORDINATION REGIME

An alternative and complementary approach emphasizes that infrastructure has deeply to do with facilitating and shaping cooperation, and always represents a choice among multiple possible modes of cooperation. David Grewal emphasizes that infrastructure comprises "regimes of coordination," the technologies (such as communications networks), bodies of law (such as the law of contract), and even intangible standards such as language that guide and facilitate cooperation.[77] Infrastructure is, in this respect, the way that people gain access to one another and enter into common projects. Imagine a population and add or subtract roads or public transit, telephones, the Internet, laissez-faire labor contracts or contracts governed in their detail by the French Code du Travail, and a series of cooperative endeavors becomes possible or recedes into impossibility—both in terms of whether they can happen at all and in terms of the distribution of the resulting benefits.

C. INFRASTRUCTURE AS POLITICAL CHOICE

The emphasis on infrastructure as regimes of coordination points to a third (compatible) emphasis, addressed to the ways that infrastructure expresses political choices. A coordination regime may favor any of a variety of values and goals.[78] If infrastructure substantially determines the kinds of activity open to us, then in a real sense it determines us—what sort of species we are, what activities are characteristic of us, what alternatives are open to us—except inasmuch as we conversely determine it, whether through technological innovation or through political decision. A basic role of the modern state, via both the direct creation of physical infrastructure and the provision and definition of legal regimes, has been to shape the terms of cooperation. For this reason, Grewal puts special emphasis on the state as "*political* infrastructure": the "coordination regime [that] can enable the production of new regimes of coordination," a collective means of shaping collective means, and thus of making intentional the terms of cooperation.[79]

2. Infrastructure and/as Environment

To think of the environment as infrastructure central to law's political economy is one way of taking on board the intellectual consequences of what I called in *After Nature* "the Anthropocene condition." That is the situation in which human activity has become a geological force, a shaping influence on the planet, from remaking the chemistry of the upper atmosphere (and with it the planet's climates and seasons) to determining which species survive and which go extinct to thoroughly remaking the surface of every inhabited continent. The Earth's ecology is now in significant part the product of human economy, or economies, the webs of activity in which we use its resources and return our emissions to its systems, making them ever more *our* systems.[80]

A. THE TECHNOSPHERE AS A SPECIES OF INFRASTRUCTURE

Vast shares of the Earth's surface and its systems have been reworked as human habitat. A study in 2017 estimated the mass of the global "technosphere," the material habitat that humans have created for themselves in the form of roads, cities, rural housing, the active soil in cropland, and so forth, at thirty trillion tons, five orders of magnitude greater than the weight of the human beings that it sustains.[81] (Restricting the estimate to buildings, pavement, and the like, in categories such as cropland and trawled sea floor that seem "too natural," cuts the estimate of the technosphere's weight by only a little more than one-third.)[82] An early treatment of the "technosphere" estimates that without it, "the human population would quickly decline toward its Stone Age base of no more than ten million."[83] If this seems dire, as it does to me, it might still be reasonable to expect a global population closer to estimates for the beginning of the Common Era, about two hundred million: that is, of today's global population of 7.5 billion, somewhere between 7.3 billion and 7.49 billion survive only because of this global habitat. We are a species of infrastructure, thoroughly unsuited to live in Stone Age or pastoral idylls, no matter the pseudo-Paleolithic diets or rustic farming vignettes that inform our consumption patterns.

B. ECOLOGY AS INCIPIENT INFRASTRUCTURE

The massive increase in human influence on the natural world brings into the category of infrastructure what was long regarded as a merely "given" backdrop, such life-sustaining features of the natural world as the global atmosphere, the chemistry of the oceans, and the fresh-water cycle. If these ecological substrates of life are too far from active engineering to count as the technosphere, they are nonetheless deeply and pervasively destabilized, reshaped, and in some measure maintained by human activity.[84] The study of infrastructure often means bringing into consciousness those systemic substrates that people have made and then forgotten and treated as given—the sewers and power grids of a city, the standards, wires, and settings of an information network;[85] in this case, what needs to come into consciousness is something that people never made, but that has been invisibly or obscurely remade until the line between natural and artificial substrate ceases to have much use.

C. ECOLOGY AS A PRODUCT OF POLITICAL DECISION

I have been discussing natural systems as *being* infrastructure: partly artificial, widely shared for many kinds of ends, and dependent in important ways on public preservation (if not, as with many forms of traditional infrastructure, public creation or provision in the first instance). What we might call environmental infrastructure, like other forms of infrastructure, is also the product of political decisions. The signal fact of the Anthropocene is that economic activity has become a sort of collective, planetary-scale landscape architecture: energy extraction, production, and transmission, agricultural production and global transport networks for food, and transport and shelter generally all contribute to the remaking of the global atmosphere, the chemistry of the seas, the nitrogen and fresh-water cycles, and so forth. These economic activities are nested within regulated and state-owned utilities, an agricultural sector deeply shaped by legislation and the path-dependent patterns of existing transport networks and utility grids. Even to the degree that they merely take place "in the market," they are the products of pricing schemes, or the absence of such schemes, for greenhouse gases and other drivers of environmental change. Political decisions thus form the center point in a nested sequence of world-making, shaping legal and material order that, in turn, shapes natural systems. The natural world itself is at stake in law and politics, not merely in the binary sense of whether it is saved or lost, but in the sense of how it is made and remade.

B. How Infrastructure Matters

We can distinguish between *material* infrastructure, such as roads, canals, and cables, which law helps to produce and govern, and legally *constitutive* infrastructure, in which the law is itself the means of cooperation, such as the law of contract. I shall emphasize two dimensions in which both kinds of infrastructure matter: a distributive dimension, concerning how benefits and burdens are shared, and a qualitative dimension, concerning the kinds of activities, relationships, and projects that a given infrastructure fosters or inhibits.

1. Material and Constitutive Infrastructure

The traditional paradigm of material infrastructure is the roadway, canal, sidewalk, or utility grid. Such infrastructure has usually exhibited characteristics, if not the pure form, of public goods, from which it is effectively infeasible to exclude anyone; returns to scale, in which the marginal cost of providing the good to additional users diminishes compared to the marginal benefit; network effects, in which increased numbers of users increase the potential value of the infrastructure to any user; and "natural monopoly," with capital requirements that inhibit new market entrants and make redundant investments exceedingly wasteful.[86]

Much of the conservationist agenda of public ownership and administrative state-building in the late nineteenth and early twentieth centuries concerned natural resources that, when they did not precisely exhibit these characteristics, were nonetheless prone under private decisions to over-harvesting (forests), wasteful extraction (oil and natural gas), exhaustion (soil), or under-investment of essential capital (irrigation). For these reasons, conservationists argued that these resources should be under intensive state management that made them, in effect if not in their origin, objects of public provision. This long-established dimension of environmental infrastructure, consisting in intensively transformed and managed portions of the natural world that are essentially annexed to human activity, now includes the larger domains of the "technosphere" and the still more inclusive domain of the many natural systems now deeply changed by human activity.

A second dimension of legal infrastructure is the law's role as itself constituting the terms of interpersonal cooperation. Employment law or contract law does not just set boundaries on how people may cooperate (analogous to speed limits) but is instead the condition of possibility for these relationships, analogous to such material infrastructure as the roads themselves.[87] Similarly, the law of property offers a series of pathways or formats, so to speak, that people may adopt in setting the terms of marriage, business partnership, cohousing, and so forth.[88] What is often termed private law is, seen in this light, the publicly provided means by which various "private" cooperative endeavors simply would not take place. This role of law is central to Friedrich Hayek's influential theory of markets and to the closely related information-costs approach to property law, in which a simple set of rights provides an efficient platform for the coordination of multiple, diverse projects with relatively low informational transaction costs.[89]

2. Distributive and Qualitative Importance

In considering the *distributive* dimension of law's infrastructure, certain material-infrastructure examples are straightforward: the construction of an energy grid, or the provision or prohibition of medical facilities necessary to exercise the abortion right.[90] The constitutive dimension of legal infrastructure also has an important distributive aspect: consider the legal realists' analysis of what they called the nonneutrality of the marketplace of property and contract. For example, antitrust law and labor law facilitate, block, or shape the division of benefits in a range of transactions between producers and consumers, suppliers and processors, employers and employees, and so forth.

These disparate effects of infrastructure, which one might call *descriptively* distributive, give only a partial picture of the issue. A fuller picture of the distributive significance of infrastructure includes the recognition that the public provision and shaping of infrastructure concretely embody collective judgments about distributive justice: which human needs and interests form the basis of what one might call "baseline claims," those resource-dependent capacities whose exercise every member of society can take for granted. How frequently do we treat roads as if they were as natural as the ground, electricity and Internet access as if they were air, and air as if its life-sustaining properties were a natural given rather than the product of extensive legal and technological engineering (to be contrasted with the air's role in life in, say, Beijing or Delhi)? To be an able participant in social, economic, and political life just is to be massively resourced via public provision (often via regulation or regulated utilities rather than direct provision). The extent of this material dependence reveals itself in breakdown, such as power outages (or even Internet failures!), when orderly life fragments into some array of festival, riot, and isolation, or in contrast with those resources that are withheld: those who take medical infrastructure for granted, for instance, will be struck by the disruption in the life of a woman in South Carolina who seeks an abortion (constitutionally protected but quite probably not available nearer than Chapel Hill, North Carolina), or, for that matter, of an uninsured person diagnosed with a chronic illness.

For much of the early to mid-twentieth century, legal scholars and institutional economists grappled with the role of infrastructure in distributive justice, in the extremely concrete contexts of rate-setting for public utilities.[91] The "art of fixing prices" was a task closely related to the art of making markets in antitrust law. Each asked, of an industry and a class of capital, what the public interest implied for the ways in which they should be regarded as quasi-public resources, subject to the affirmative legal shaping of the field on which they operated.[92] Robert Hale's work on property law was also an attempt to think of this consummately "private" and common-law topic in public-law and regulatory terms, to ask what was the public interest that should guide and bound the definition of private powers in this domain.[93] The decline of this approach to economic regulation is an integral part of the erasure of distributional political economy from the leading discourses of law and policy-making during and after the "great compression" of inequality.[94] If the modern era of environmental law and politics had not arisen during the late stages and early (frequently invisible) retreat of that era of presumed relative equality, questions such as the nature and extent of the public interest in the atmosphere might have been rather more central to thinking around climate change, where instead questions about distribution have been regarded as inconvenient political barriers to efficient policy-setting.[95]

The distributive dimension of environmental infrastructure has often been explicit, such as when the federal government allocates resources to create, or contribute to, the building blocks of certain economic sectors: the privatization or reservation of federal lands, the allocation of water (ultimately owned by the governments of most Western states and frequently distributed through federal infrastructure), and the issue or withholding of rights to timber, mine, drill, graze, and hunt have been flashpoints of dispute from the nineteenth century to today. This history of open distributive conflict makes it all the more striking that the distributive significance of pollution control and other post-1970 regulation, as discussed earlier, was at first obscure to environmental regulators and established advocacy groups, and came to prominence in the 1980s and thereafter mainly through the efforts of environmental-justice activists from outside those circles. But even in that period, as we have seen, legislators such as Senator Muskie understood the antipollution statutes as creating, in effect, a "right" to clean air and water: that is, they understood regulation as embodying a distributive baseline claim to environmental infrastructure. Indeed, the earliest protoenvironmental regulations, such as the doctrine of public access to waterways, harbors, beaches, and certain of their resources, were also quite explicitly infrastructure regulations of the distributive-justice kind, preventing monopolies in these potential chokepoints of travel and commerce so as to maintain a shared baseline of access and participation in social and economic life. Moreover, the resolution of pitched distributive conflict, such as that over federal public lands, has often taken the form of a judgment about baseline claims to public resources: for instance, universal recreational access to most public lands (in contrast of course to private recreational lands), and a shared public claim, expressed in utilitarian terms, to the long-run material benefit of other resources, such as forests and managed watersheds.

Today, the front line of environmental reform is constitutive and fraught with distributive questions. Any pricing scheme (whether in the form of taxes or tradable permits) for carbon dioxide and other greenhouse gases will be as profound a market-making instrument as property or antitrust. The intense distributive stakes of the allocation of costs for such a scheme were evident in the design of the failed Waxman-Markey legislation, which included significant funds for redevelopment of coalfield regions, subsidies for rural and low-income energy users, and so forth.[96] Even if it might be possible in principle for a domestic carbon-pricing scheme to be revenue-neutral and minimize distributive consequences (itself a distributive choice, of course), any international coordination in carbon pricing quickly implicates the most basic questions of international distributive justice in a setting of historical global inequality.[97] Many of the countries most at risk from climate change, such as Bangladesh, have contributed the least to the historical burden of atmospheric carbon, and for the same reason—because they were substantially excluded from the Industrial Revolution. Parsing the respective duties of the United States, Bangladesh, and intermediate cases such as China (long a minor contributor to greenhouse-gas concentration, now the world's largest) implicates theoretical disputes between cosmopolitanism and more partial views of obligation and hard-to-answer empirical questions about the sources of uneven development.

In appreciating the *qualitative* dimension of infrastructure, the antitrust example is again instructive: As recent scholarship has reminded nonhistorians, the field originated in part as a political economy of citizenship, with the goal of preserving an economy of small- and mid-scale proprietors and producers in an era of national and international markets.[98] In this manner, the distributive aspect of legal infrastructure has been intimately linked with qualitative goals: establishing certain pathways of economic collaboration with an eye to preserving a certain kind of political culture.[99] Cognate instances are conceptions of collective bargaining as linked with "industrial democracy" and arguments that effective citizenship requires a measure of social provision and security.[100]

The qualitative dimension of environmental infrastructure is multifarious. Public lands provide a relatively straightforward example: forms of recreation, forms of community around the sharing of that recreation, and identity connected with those forms of sociability all depend on access to wilderness, national parks, and other public acreage, and many of these competing uses, such as wilderness solitude and snowmobiling, are mutually exclusive. Other public-land uses that might appear "simply economic," such as timbering or grazing, also involve forms of community and identity—one reason that disputes over these questions are especially fierce. In other areas, climate change threatens nomadic, fishing, hunting, and agricultural ways of life embedded in historically stable ecosystems that may soon pass, or are already passing. The legal shape of the agricultural economy shapes not just a landscape of crops, but also a landscape of labor, either creating a substantial, often intensely vulnerable agricultural proletariat of the kind that does most of the nonmechanized work in American food production or creating (larger or smaller) potential for the smaller-scale and more agriculturally diverse production that forms the ideal of the current agrarian revival in American culture.

The preceding discussion suggests that distributive and qualitative judgments about infrastructure are very closely aligned, though their connection is sometimes obscured by the impression of neutrality that certain accounts of wealth-maximizing policy imply. This is true in a pair of ways. First, choices in infrastructure design inevitably facilitate some forms of activity and cooperation and inhibit others. Second, infrastructure design embodies a normative view of what it means to be a member of the society in which it operates, making access to certain foundational resources "a given," while making the lack of others a danger that organizes market activity (e.g., keeping health care, being able to afford to live in a safe neighborhood with good public transport) and specifies the stakes of inequality (i.e., to be disadvantaged means to risk lacking those things that are not provided universally and on equitable terms). Infrastructure is, as it were, the material constitution of a legal order.

C. The Politics of Infrastructure

1. Constraints on Politics and Conscience

It has been a favorite observation of students of infrastructure that "infrastructure has politics"—that is, it embeds distributive decisions and qualitative priorities into the order of things. It also has implications for the kind of politics that we infrastructure-dwelling creatures need to have if we are to revise our half-made, half-given environment, and the barriers we face in doing so.

David Grewal has recently argued that the ambition to create an open-source biotechnology akin to open-source software code is likely futile, despite the intelligence, energy, and ambition of those pursuing it. The reason is the absence of both (1) public investment in the basic research necessary to produce essential genetic information and (2) a legal regime that facilitates something like a share-and-share-alike obligation among users of the technology. The reformers' goal is to create a new infrastructure for collaborative research, composed of both genetic material and a cooperative regime for sharing that material toward diverse projects. The problem is that this sought-after infrastructure can only be built upon prior infrastructure, both material and legal. Although addressed specifically to the example of open-source biotechnology, Grewal's argument, like Amy Kapczynski's earlier case for restoring an emphasis on the state to an intellectual-property discourse enchanted by its own anarchist conceits, makes a more general case that neglect of infrastructure, and particularly of the state's role in providing and shaping it, leaves reformers purblind to the requirements of their own aims.[101]

Consider in this light the prospects for containing climate change through a reshaping of the energy economy, which is generally agreed to be a necessary if not sufficient condition for avoiding catastrophe. Individuals, communities, and movements have scant choice about whether to participate in these dimensions of infrastructure, or about the environmental consequences of doing so. Yet the very inexorability of these systems, once in place and given direction, is the reason that so much hope is placed in their reengineering, through some combination of technological innovation and greenhouse-gas pricing, to save us from ourselves. Achieving a switch to renewable energy sources would mean that humanity had re-created itself as a carbon-neutral species. For the vast majority of people, however, this remaking might well be a passive, even invisible process. Sustaining the planet would be a metabolic side effect of our species-existence, as catastrophic climate change now looks to be. The drivers of the change would be a mix of technological determination and market engineering—again, material and constitutive infrastructure.[102]

I argued in *After Nature* and earlier work that such changes frequently need political impetus and legitimation to take place in the first place and to stabilize themselves, and this still seems true. This formulation, however, now strikes me as needing to be tempered by an appreciation of just how far the infrastructural dimension of environmental problems inhibits familiar forms of political agency. On what level does anyone, or any collective, make a difference? On that of the technosphere, which is to say, through basic legal reform or radical technological innovation, the latter (for those without engineering genius) most likely by means of either venture capital or industrial policy. It is this level of systemic steering that substantially determines the survival or disappearance of species, ecosystems, regions, and everything else that depends on the interplay of the extractive economy and global climate change.

The tendency in some environmental discourse (including legal scholarship) to look to technological or technocratic solutions reflects a failure to appreciate the importance of political movements that establish and revise basic public commitments. The pathway to this elision of politics, however, is more tragic than hubristic or obtuse. Several considerations inhibit the development of a politics of environmental infrastructure.

2. Scales of Crisis and Response

First is the disproportion between the scale of challenges in the infrastructure environment and the scale of the capacity for response. The former is substantially global, the latter mostly national. This difficulty has been well canvassed, of course, in analyses of the collective-action problems confronting responses to climate change.[103] But the problems may be even greater than they have often appeared. To be effective, change in the infrastructure environment needs to be broad, indeed global, which is the usual basis of the observation that it is hard to achieve in a world of multiple states; but it must also be deep, going to the shape of the energy economy, transport, communication, and food production, to name a few of its central topics. This is all the harder to accomplish in a world in which all of these systems are integrated into global networks of commodity production and exchange and capital movement. Technologies must be compatible, and returns on investment competitive, for deep changes to be viable— meaning they must take hold at a global level where no political agent exists to enforce deep change at the level of industrial policy, pricing structure, or the basic format of, say, land use. It may happen that, in the decades ahead, China's industrial policy, the technical features of solar technology, and the movement of states and regions such as California and the European Union to incentivize clean-energy production through market-modeled carbon pricing will hurry an eventual global transition to non-carbon-based energy. If so, that will be a great achievement, but not cause for future confidence. The availability in time of a viable technological alternative, under all relevant political, economic, and material constraints, would be more deus ex machina than stress-test result. The accelerating ecological footprint of the technosphere all but ensures further systemic crises—those of extinction and the collapse of marine and terrestrial ecosystem diversity are already underway—and the same structural barriers will attend responses to those.

3. One World, Many Humanities: The Perplexity of Global Crisis

Second is the fact that the infrastructure environment makes humanity a unified global phenomenon in a new way, even as the same environment emerges historically from profoundly unequal and systemic differentiation among human populations. This point is often registered in the conventional observation that the regions of the world most vulnerable to harms from climate change are typically the poorest, which is to say, the ones that have benefited least from, and whose historical populations have contributed least to, the centuries of industrial development that produced the present crisis. The observation often appears dressed as one about the collective historical responsibility of nations and their populations; but it can also be made from a different angle: the world's current unprecedented wealth and its current environmental crises have been produced and distributed within global regimes of political economy, which mixed varieties of colonial exploitation with uneven development among formally sovereign countries. The attractive and sensible proposal that fighting climate change should involve "common but differentiated responsibilities" among countries reflects the deeper fact that the problem is a dimension of overwhelming unity for an intensely differentiated humanity—and, to repeat, vulnerability to the problem is of course refracted through the prior and continuing differentiation of humanity.

There has been much criticism of the idea of the Anthropocene on the grounds that it tends to obscure these patterns of inequality by invoking a unified *Anthropos* as the source of climate change and of potential responsibility for it. The criticism is misplaced, not because the observation about historical inequality is wrong, or because there is no risk of occluding that inequality by invoking a common humanity (there is plenty), but because the human unity of the Anthropocene is as real as the historical differentiations of imperialism and subsequent uneven development. The medium of unity is the global infrastructure of de facto engineered (though hardly controlled) atmosphere and climate, a thing that humanity inhabits together in a concrete and consequential way. We really are in this together, are in some fashion a "we" by virtue of being in it together, but are at the same time profoundly divided in historical and contemporary relation to the problem. There is no agency of collective action with adequate potential to change the terms on which we are brought, unequally, together.

4. The Invisibility of the Ubiquitous

A third and final consideration is the deep tendency of infrastructure to be invisible in daily life, so that, as one scholar puts it, the salient thing about infrastructure is that it is not salient.[104] Environmental politics generated, over a long struggle, an awareness of "the environment" not as a given and stable backdrop but as an object of urgently needed and ongoing political attention. It did so with significant use of long-standing cultural habits of attachment to, even reverence toward, various synecdochal parts and places of the natural world, which, if it had been "given," also came into salience in aesthetic, theological, and artistic formulation and experience. There hardly seem to be comparable resources for infrastructure, let alone ones to rely on in a time of global emergency.

✳ ✳ ✳

Thinking through infrastructure has a paradoxical set of consequences. On the one hand, it highlights the scope and depth of the human transformation of the world, not in the naïve sense of damaging a natural realm "out there," but in the sense of perennially reworking our own half-artificial habitat. Humanity emerges in this image in a position of world-making long reserved cosmologically to deities. Hobbes famously compared the political construction of the "artificial man" of sovereignty to "that *Fiat*, or *Let us make man*, pronounced by God in the Creation."[105] *Let us make a world* is the analogue here.

But how? A recurring theme of this chapter has been the disparity between our world-making effects and our capacity to shape them within an intentional form of law rooted in institutions of political will-formation. Attention to infrastructure at least helps to identify and articulate the character of this problem. There is, as yet, scarcely a *we* commensurate to that "Let us make a world."

Humanity has become an infrastructure species and, willy-nilly, brought the rest of terrestrial life into the same situation. If such a species is to act deliberately in the shaping of its collective life, it must recognize its law as an inevitably world-making enterprise. The work of this chapter has been to take hold of environmental law in that light.

Notes

1. Will Steffen, Wendy Broadgate, Lisa Deutsch, Owen Gaffney, and Cornelia Ludwig, "The Trajectory of the Anthropocene: The Great Acceleration," *Anthropocene Review* 2 (2015): 81–98 (summarizing these changes).
2. John R. McNeill and Peter Engelke, *The Great Acceleration: An Environmental History of the Anthropocene Since 1945* (Cambridge, MA: Harvard University Press, 2016), 4, 11, 1–288.
3. Jan Zalasiewicz, Mark Williams, Colin N. Waters, Anthony D. Barnosky et al., "Scale and Diversity of the Physical Technosphere: A Geological Perpsective," *Anthropocene Review* 4, no. 1 (2017): 9–22.
4. Zalasiewicz et al., 12.
5. Steffen et al., "Trajectory of the Anthropocene," 7.
6. Steffen et al., 7.
7. Steffen et al., 7.
8. Steffen et al., 7.
9. Vaclav Smil, "Harvesting the Biosphere: The Human Impact," *Population and Development Review* 37, no. 4 (2011): 613, 619.
10. Smil, 613, 619.
11. See, e.g., Steffen et al., "Trajectory of the Anthropocene," 7; Dipesh Chakrabarty, "The Climate of History: Four Theses," *Critical Inquiry* 35, no. 2 (2008): 197 (arguing that the Anthropocene joins geological and social scales of time and confronts historians with a new kind of "species history"); Jedediah Purdy, *After Nature: A Politics for the Anthropocene* (Cambridge, MA: Harvard University Press, 2015) (using the concept for both historical narration and present typology of environmental law and

politics); Bruno Latour, *Facing Gaia: Eight Lectures on the New Climatic Regime*, trans. Catherine Porter (Cambridge: Polity, 2017) (on forms of mythic and theoretical imagination appropriate to a new geological era); Amitav Ghosh, *The Great Derangement: Climate Change and the Unthinkable* (Chicago: University of Chicago Press, 2016) (on the forms of narrative and imagination characteristic of modernity and their difficulty getting hold of Anthropocene phenomena); Jason W. Moore, *Capitalism in the Web of Life* (London: Verso, 2015) (proposing a picture of political economy and its history in which the natural and the social are neither separate nor separable, but mutually constitutive); cf. Andreas Malm, *Fossil Capital: The Rise of Steam-Power and the Roots of Global Warming* (London: Verso, 2015) (similarly braiding natural and social processes in economic history, but preferring "capitalocene" as a term designating the centrality of capitalism); Donna J. Haraway, *Staying with the Trouble: Making Kin in the Chthulucene* (Durham: Duke University Press, 2016) (proposing the eponymous term as a way of displacing the centrality of "human" to the account of global ecological disruption).

12. A note on terminology: In these materials, I use "environmental," especially but not only in "environmental law," to refer to a broad category that regulates the human relationship to and use of the nonhuman world. In this use, the term includes questions sometimes classified as "animal law," "public lands law," and "agricultural law," as well as the core areas often taught as environmental and natural resources law. It also refers to the dimensions of, for instance, property law that touch on the human relationship to and shaping of the nonhuman world. I take this usage to be fairly conventional and intuitive. I also frequently use "the natural world" as a synonym for the nonhuman world. Although I have argued (elsewhere and at length) that the idea of "nature" as apart from humans is in key ways both obsolete and inherently flawed, and am partly persuaded by others' arguments that even the category of humanity is more vexed than is often allowed, I find self-aware conventional usage far more serving than neologism.

13. See, e.g., Wilderness Act, Pub. L. 88–577, 78 Stat. 890, codified at 16 U.S.C. sec. 1131 et seq. (1964) (codifying definition of wilderness developed with Wilderness Society advocacy).

14. See, e.g., Alejandro E. Camacho and Robert L. Glickman, "Legal Adaptive Capacity: How Program Goals and Processes Shape Federal Land Adaptation to Climate Change," *University of Colorado Law Review* 87, no. 3 (2016): 711; Camacho, "Assisted Migration: Redefining Nature and Natural Resources Law Under Climate Change," *Yale Journal on Regulation* 27 (2010): 171; J. B. Ruhl, "Climate Change and the Endangered Species Act—Building Bridges to the No-Analog Future," *Boston University Law Review* 88, no. 1 (2008); Douglas A. Kysar, *Regulating from Nowhere: Environmental Law and the Search for Objectivity* (New Haven: Yale University Press, 2010).

15. See, e.g., William Boyd, "Public Utility and the Low-Carbon Future," *UCLA Law Review* 61, no. 1614 (2014); Mary Jane Angelo and Joanna Reilly-Brown, "Whole-System Agricultural Certification: Using Lessons Learned from Leed to Build a Resilient Agricultural System to Adapt to Climate Change," *University of Colorado Law Review* 85, no. 689 (2014).

16. See, e.g., Jedediah Purdy, "The Long Environmental Justice Movement," *Ecology Law Quarterly* 44 (2018): 809–864 (surveying environmental-justice literature); Dale Jamieson, "Climate Change and Global Environmental Justice," in *Changing the Atmosphere: Expert Knowledge and Environmental Governance*, ed. Clark A. Miller and Paul N. Edwards (Cambridge, MA: MIT Press, 2001), 287–308.

17. See Endangered Species Act, Pub. L. 93–205, 87 Stat. 884 (codified at 16 U.S.C. sec. 1531 et seq.); Wilderness Act, Pub. L. 88–577.

18. See, e.g., Gifford Pinchot, *The Fight for Conservation* (New York: Doubleday, 1910) (presenting a utilitarian account of natural resources and natural-resources policy); National Forest Service Organic Act (1891), 16 U.S.C. sec. 471, et seq. (establishing exclusive goals of timber production and watershed management); U.S. v. New Mexico, 438 U.S. 696 (1978) (confirming the primary of these goals in federal forest management).

19. Brett Frischmann, *Infrastructure: The Social Value of Shared Resources* (Oxford: Oxford University Press, 2012), 4.

20. Amartya K. Sen, "Goods and People," in *Resources, Values, and Development* (Cambridge, MA: Harvard University Press, 1984) 509, 509–10 ("The process of economic development is best seen as an expansion of people's 'capabilities'"); Sen, "Opportunities and Freedoms," in *Rationality and Freedom* (Cambridge, MA: Harvard University Press, 2003), 583, 585 ("More freedom gives us more *opportunity* to achieve those things that we value, and have reason to value. This aspect of freedom is concerned primarily with our *ability to achieve*").

21. As David Grewal expresses it, a "platform," a species of technological infrastructure available for many kinds of "applications," is "not essentially material, but rather a materially embedded social practice, . . . a coordinating convention that enables derivative forms of cooperative action." Grewal, "Before Peer Production: Infrastructure Gaps and the Architecture of Openness in Synthetic Biology," *Stanford Technology Law Review* 20, no. 143 (2017): 209.

22. Grewal, "Before Peer Production," 209.

23. Grewal, 209.

24. Peter K. Haff, "Technology as a Geological Phenomenon: Implications for Human Well-Being," in *A Stratigraphical Basis for the Anthropocene?*, ed. Colin N. Waters, Jan A. Zalasiewicz, Mark Williams, Michael A. Ellis, and Andrea M. Snelling, *Geological Society, London, Special Publications* 395 (2014): 396.

25. See Purdy, "Long Environmental Justice" (surveying this literature).

26. National Environmental Policy Act, Pub. L. 91–190, 83 Stat. 852 (codified as amended at 42 U.S.C. sec. 4321 et seq.); Reorganization Plan No. 3, 35 Fed. Reg. 15623 (July 9, 1970) (establishing Environmental Protection Agency); Clean Air Act, Pub. L. 88–206, 77 Stat. 392 (codified as amended at 42 U.S.C. sec. 7401 et seq.).

27. Federal Water Pollution Control Act, Pub. L. 92–500, 86 Stat. 816 (codified at 33 U.S.C. sec. 1251 et seq.) (colloquially, "Clean Water Act"); Endangered Species Act, Pub. L. 93–205, 87 Stat. 884 (codified at 16 U.S.C. sec. 1531 et seq.); Federal Insecticide, Fungicide, and Rodenticide Act, Pub. L. 61–152, 36 Stat. 331 (codified at 7 U.S.C. sec. 136 et seq.); Toxic Substances Control Act, Pub. L. 94–469, 90 Stat. 2003 (codified at 15 U.S.C. sec. 2601 et seq.); Resource Conservation and Recovery Act, Pub. L. 94–580, 90 Stat. 2795 (codified at 42 U.S.C. sec. 6901 et seq.); Federal Land Policy and Management Act, Pub. L. 94–579, 90 Stat. 2743 (codified at 43 U.S.C. sec. 1701 et seq.); National Forest Management Act, Pub. L. 94–588, 90 Stat. 2958 (codified at 16 U.S.C. sec. 1600 et seq.); Surface Mining Control and Reclamation Act, Pub. L. 95–87, 91 Stat. 445 (codified at 30 U.S.C. sec. 1201 et seq.).

28. Comprehensive Environmental Response, Compensation, and Liability Act, Pub. L. 96–510, 94 Stat. 2767 (codified at 42 U.S.C. sec. 9601 et seq.).

29. Richard J. Lazarus, *The Making of Environmental Law* (Chicago: University of Chicago Press, 2006), 69.

30. These developments are discussed in parts 2.C and 4.C.

31. With this term, I am indicating the resonance between this argument and the historical interpretation that Joseph Fishkin and William Forbath have advanced in their manuscript on the "great forgetting" of egalitarian constitutional political economy during the period after the New Deal. See, e.g., Joseph Fishkin and William Forbath, "Reclaiming Constitutional Political Economy: An Introduction to the Symposium on the Constitution and Economic Inequality," *Texas Law Review* 94, no. 1287 (2016) (Introduction to *Texas Law Review* symposium, "The Law and Economic Inequality").

32. Benton MacKaye, "Environment as a Natural Resource," in *The New Exploration* (New York: Harcourt, Brace, 1928), 134. It was, he continued, "the last common denominator of our inner selves."

33. MacKaye, 5–25.

34. MacKaye, 15.

35. Wilderness Act, Pub. L. 88–577, 78 Stat. 890, codified at 16 U.S.C. sec. 1131 et seq. (1964).

36. William Cronon, "The Trouble with Wilderness; or, Getting Back to the Wrong Nature," in *Uncommon Ground: Rethinking the Human Place in Nature*, ed. William Cronon (New York: Norton, 1995), 69–90.

37. MacKaye, "Environment as a Natural Resource," 215–218.

38. His treatment of this, in the Wilderness Society bulletin and elsewhere, and secondary sources.

39. MacKaye, "Environment as a Natural Resource," 143.

40. MacKaye, 143.

41. See, e.g., Robert Marshall, "The Problem of the Wilderness," *Scientific Monthly* 30, no. 2 (1930): 141 (arguing, rather eccentrically but with great intensity, that the pleasure wilderness devotees took in the unspoiled outdoors was so qualitatively distinct from other satisfactions that it swamped the utilitarian calculus that was otherwise appropriate in management decisions for public lands).

42. James M. Glover, *A Wilderness Original: The Life of Bob Marshall* (Seattle: Mountaineers Books, 1986), 149 (quoting letters that Marshall wrote in 1932–1933, saying, "the only eventual solution will be Socialism" and "I wish very sincerely that Socialism could be put into effect right away and profit system eliminated"). For Marshall's leadership in the ACLU, see, e.g., Glover, *A Wilderness Original*, 185; for Marshall service under Collier, see Glover, *A Wilderness Original*, 157–166.

43. Robert Marshall, *The People's Forests* (New York: H. Smith and R. Haas, 1933), 89–97 (attacking private ownership and management of commercial forests), 123–140 (advocating public ownership for both conservation and worker-welfare reasons), 141–158 (advocating public acquisition of forests by transfer, as "the fact is inescapable that with the country functioning on a capitalistic basis it is out of the question to consider confiscation as a feasible means of acquiring public forests").

44. Marshall, *The People's Forests*, 165–170, 166. Marshall wrote in the same vein, "Many towns and even counties should be abandoned to the forests."

45. Wilhelm C. Hueper, *Occupational Tumors and Allied Diseases* (Springfield: Charles C. Thomas, 1942), 3–5, 848, quoted in William C. Boyd, "Genealogies of Risk: Searching for Safety, 1930s–1970s," *Ecology Law Quarterly* 39, no. 4 (2012): 895, 924.

46. Alice Hamilton, *Exploring the Dangerous Trades* (Boston: Little, Brown, 1943), 114–126 (recounting work on lead and phosphorous exposure), 138–160 (lead, silica, and solvents).

47. Hamilton, *Exploring the Dangerous Trades*, 357–358 (on the "industrial feudalism" of the Lawrence mills, in which low wages combined with denial of "self-respect and a sense of human dignity" to inspire conflict); 12 ("the National Association of Manufacturers has fought the passage of occupational-disease compensation as it has fought laws against child labor, laws establishing a minimum wage for women and a maximum working day"); 6, 13 (contrasting the "hot, dirty, and dangerous work [and] contempt from more fortunate Americans" that plagued the unorganized workers whom she observed at the beginning of her career with structure negotiation among trade unions, industry representatives, and experts). Boyd claims that Hamilton saw responsibility for factory conditions as residing mostly with supervisors, and argues that this represents a pre–New Deal view of the limits of the regulatory state. See Boyd, "Genealogies of Risk," 924n111. This claim strikes me as exaggerating the contrast between Hamilton and later reformers, although it is true that in (one version of) classic Progressive style, she had many warm words for individual managers who took responsibility for factory conditions.

48. David Rosner and Gerald Markowitz, "Safety and Health as a Class Issue: The Workers' Health Bureau of America During the 1920s," in *Dying for Work: Workers' Safety and Health in Twentieth-Century America*, ed. David Rosner and Gerald Markowitz (Bloomington: Indiana University Press, 1989) 53, 53–64.

49. Rosner and Markowitz, 55–57.

50. Grace Burnham, *A Health Program for Organized Labor* (October 1922).

51. Rachel Carson, *Silent Spring* (New York: Houghton Mifflin, 1962), 18, 50, 221–225, 235, 239, 240–243.

52. Carson, 1–2, 6, 67.

53. See, e.g., Louis Hartz, *The Liberal Tradition in America: An Interpretation of American Political Thought Since the Revolution* (New York: Harcourt Brace, 1955). Although Hartz is often invoked as a consensus-school thinker, he is somewhat peculiar in that he took consensus as historical fact and sought to understand it, not uncritically. This puts him in a rather different light from, say, Daniel Boorstin, whose work is more emblematic of the political cast of the consensus school. See, e.g., Daniel J. Boorstin, *The Genius of American Politics* (Chicago: University of Chicago Press, 1953).

54. *Haven in a Heartless World, Feminine Mystique*, and the like.

55. See, e.g., Richard M. Nixon, "1970 State of the Union Address" (Washington, DC, January 22, 1970) (arguing that answering the environmental crisis could unite Americans otherwise divided over war and race); "Issue of the Year: The Environment," *Time*, January 4, 1971 (environmental crisis a "problem which American skills . . . might actually solve, unlike the immensely more elusive problems of race prejudice or the war in Vietnam").

56. Thomas Piketty, *Capital in the Twenty-First Century*, trans. Arthur Goldhammer (Cambridge, MA: Harvard University Press, 2013), 304–376 (advancing this finding).

57. See, e.g., John Kenneth Galbraith, *The Affluent Society* (New York: Houghton Mifflin, 1958), 429 ("as an economic and social concern, inequality has been declining in urgency [because of] increased production [which is] an alternative to redistribution or even to the reduction of inequality. The oldest and most agitated of social issues, if not resolved, is at least largely in abeyance"); 577–580 (nonetheless, "poverty does survive," especially in the form of "insular poverty" characterized by an "island [where] everyone or nearly everyone is poor," exemplified by black and white Southerners, "slums," and Appalachia) in Galbraith, *The Affluent Society, and Other Writings*, ed. James K. Galbraith (New York: Library of America, 2010). As a piece of

evidence for the persistence of this impression, Galbraith in the original edition of *The Affluent Society* (published in 1958) flatly asserted that economic inequality was declining, and hung a great deal of his analysis on that fact. In subsequent revisions through 1984, he acknowledged that his analysis could no longer be attributed to falling inequality, as it was in fact growing, and responded by amplifying the role of ancillary cultural and psychological considerations in his argument. See also Lyndon Baines Johnson, "The Great Society" (speech), Ann Arbor, MI, May 22, 1964 (arguing that, after fifty years of "unbounded invention and untiring industry to create an order of plenty for all our people," the challenge was now "to use that wealth to enrich and advance our national life, and to advance the quality of our American civilization"); Johnson, "State of the Union," Washington, DC, January 8, 1964 (calling for "war on poverty" "in city slums and small towns, in sharecropper shacks or in migrant worker camps, on Indian Reservations, among whites as well as Negroes" and with special attention to "the chronically distressed areas of Appalachia").

58. News conference of Senator Edmund Muskie, May 13, 1970, entered in 116 Cong. Rec. 15607 (May 14, 1970) (entered by Rep. Boland).

59. Edmund Muskie, Earth Day speech, Bates College, April 22, 1970, http://abacus.bates .edu/muskie-archives/ajcr/1974/Earth%20Day.shtml.

60. Muskie, Earth Day speech.

61. Muskie, Earth Day speech. Muskie had earlier "war" on "poverty" and "hunger" with "another war" on "the pollution of our environment," and then insisted that only creating "a whole society," in King's sense, which he defined as a "healthy total environment," could count as victory. He also argued, in terms closely akin to those of Galbraith's *Affluent Society*, that "Our technology has reached a point where it is producing more kinds of things than we really want, more kinds of things than we really need, and more kinds of things than we can really live with," and that this condition represented "a moral frontier" where material increase must be exchanged for a "society in which all men live in brotherhood . . . where each member of it knows that he has an opportunity to fulfill his greatest potential." Muskie, Earth Day speech.

62. See discussion and citations, Muskie, Earth Day speech.

63. 116 Cong. Rec. 32900 (referring to "wasteful practices of an affluent society"), September 21, 1970 (statement of Senator Muskie).

64. Muskie, Earth Day speech (discussing Great Society and War on Poverty programs).

65. 1968 Cong. Rec. 11158, May 1, 2008 (statement of Sen. Muskie) (inserting into record a speech of Joseph A. Califano before the Nieman Fellows of Harvard University, April 23, 1968). Muskie repeatedly sounded this note about the environmental statutes, describing them as paradigm-shifting changes in the duties as well as the rights of citizenship. See, e.g., 118 Cong. Rec. 36874, 1972 (statement of Sen. Muskie) ("The whole intent of [the Clean Water Act] is to make a national commitment. . . . Can we afford clean water? . . . Can we afford life itself? . . . Those questions were never asked as we destroyed the waters of our Nation, and they deserve no answers as we finally move to restore and renew them. These questions answer themselves. And those who say that raising the amounts of money called for in this legislation may require higher taxes, or . . . contribute to inflation simply do not understand the language of this crisis"); 116 Cong. Rec. 42392, 1970 (statement of Sen. Muskie) ("There has to be a commitment to [clean air] by every citizen, not only with respect to the activities of others, but with respect to each citizen himself").

66. Piketty, *Capital*, 291 (showing income inequality in the US, 1910–2010, with sharp increase beginning in mid- to late 1970s); 349 (same for inequality in wealth, in this case for a timeline spanning 1810–2010).

67. *Washington v. Davis*, 426 U.S. 229, 1976 (finding no equal protection claim from a showing of disparate racial impact by an exam given to prospective employees of the Washington, DC, police department in the absence of evidence of purposeful discrimination by an official actor); *Personnel Administrator of Massachusetts v. Feeney*, 442 U.S. 266 (1979) (holding that where a law has a foreseeable disparate impact, it must have been passed because of, not despite, that impact, i.e., must have been motivated by or aimed at the impact).

68. See, e.g., Jedediah Purdy, "Neoliberal Constitutionalism: Lochnerism for a New Economy," *Law and Contemporary Problems* 77, no. 2 (2014).

69. Ford made some of its largest cumulative environmental and natural resources grants of the 1970s to these groups: $2,995,000 to NRDC, $1,114,500 to EDF, and, in addition, $1,584,000 to the Center for Law in the Public Interest (based in Los Angeles) and $688,000 to the Sierra Club Legal Defense Fund, which later became EarthJustice. See *Ford Foundation Grants in Resources and the Environment* (1978) (publication of the Ford Foundation). Ford also made a cumulative grant $15.5 million to Resources for the Future, a research-oriented organization that remains a key resource for informed policy-making. See *Ford Foundation Grants*. In the same years, Ford developed a proposal to fund Earth Action, the advocacy and organizing group proposed by the creators of the original Earth Day in 1970, but abandoned it in late stages. Robert Gottlieb, *Forcing the Spring: The Transformation of the Environmental Movement* (Washington, DC: Island, 1993). The pattern of Ford's field-shaping investments was decisively in favor of expert knowledge and advocacy, not grassroots political organizing.

70. Sanford M. Jaffe, "Public Interest Law—Five Years Later," *ABA Journal* 62 (1976): 985. (Ford's transitions: 1962 is technocratic and democratic, sounds conservationist, and is oriented to the problems that would be the crucible of mainstream environmentalism; 1970–1975 is taking the new skepticism, radicalism, and resistance into the body of these ideas, and seeing public-interest law as tending to become governance; by 1978, there is somewhat more awareness of deep pluralism.)

71. Steven M. Teles, *The Rise of the Conservative Legal Movement* (Princeton: Princeton University Press, 2008), 22–57 (describing "the rise of the liberal legal network"). The term *liberalism* is also associated with Laura Kalman, *The Strange Career of Legal Liberalism* (New Haven: Yale University Press, 1998). Kalman uses the term to refer to a series of scholarly and institutional developments in which legal scholars sought to justify, preserve, and expand the reformist jurisprudence of the Warren Court (and to some extent the early Burger Court, in cases such as *Roe v. Wade*). Teles refers to a different phenomenon, the central place that legal institutions, practice, and concepts achieved in the institutional and intellectual life of center-left reform movements between the late 1950s and the late 1970s, which he calls "the legalization of reform" (52, internal quotation marks omitted).

72. *Ford Foundation Grants*, 14 (emphasis added).

73. Teles doesn't make this argument, but David Grewal and I have, drawing on his version of "legal liberalism," in David Singh Grewal and Jedediah Purdy, "Inequality Rediscovered," *Theoretical Inquiries in Law* 18 (2017): 61.

74. Frischmann, *Infrastructure*, 61–90.

75. Frischmann, passim.

76. How much more efficient to use "the railroad," rather than "my railroad" or "your railroad"; how much better to be able to email everyone in your professional community than everyone in your hall! Frischmann argues that because resources with these characteristics often produce the greatest social benefit when governed as

commons, that is, free and open without distinction to members of a (relatively large) relevant community, a free and open Internet fits a larger theory of infrastructure. See Frischmann, *Infrastructure*, 317–357.

77. Grewal, "Before Peer Production" (treating infrastructure as a "platform," a species of technological infrastructure available for many kinds of "applications," is "not essentially material, but rather a materially embedded social practice, . . . a coordinating convention that enables derivative forms of cooperative action").

78. Susan Leigh Star, "The Ethnography of Infrastructure," *American Behavioral Scientist* 43, no. 3 (1999): 377, 388 (on "how values are inscribed into technical systems"); Langdon Winner, "Do Artifacts Have Politics?," in *The Social Shaping of Technology*, ed. Donald MacKenzie and Judy Wacjman (Buckingham: Open University Press, 1986), 26–37. Grewal elaborates on the social-theoretic significance of values and interests being embedded in coordination regimes in Grewal and Singh, *Network Power: The Social Dynamics of Globalization* (New Haven: Yale University Press, 2008).

79. Grewal, *Network Power*.

80. Jedediah Purdy, *After Nature: A Politics for the Anthropocene* (Cambridge, MA: Harvard University Press, 2015). I survey some of the more arresting instances of these developments in the Introduction.

81. Zalasiewicz et al., "Scale and Diversity," 9–22.

82. Zalasiewicz et al., 12. Some of this thoroughly remade material habitat has affinities with traditionally environmental "nature," such as areas of intensive cropland and industrially trawled sea floor; some of it, such as private homes and buildings, may seem at first blush to be simply "private property," although its suitability as habitat depends on its intimate nesting in a web of infrastructure: electricity, water, roads, communication, and, arguably, the law-constituted market. Just consider the prospects of our suburban, exurban, and urban exoskeletons as habitat in a world where they were stripped of their infrastructural circulatory and nervous system. See, e.g., Alan Weisman, *The World Without Us* (New York: Picador, 2007) (detailing the likely timeline of the decay of human infrastructure were humans to disappear).

83. Haff, "Technology as a Geological Phenomenon."

84. I canvass some of the signal instances of this transformation in the introduction. Haff, "Technology as a Geological Phenomenon."

85. See Haff, "Technology as a Geological Phenomenon" (giving these examples).

86. See, e.g., Frischmann, *Infrastructure*. These characteristics overlap, of course, and run into countervailing considerations in some cases, such as the problem of congestion, which may diminish network effects. Public provision and oversight of these dimensions of infrastructure were long a major concern of public-utilities law, a field that sought, among other things, to delineate the proper spheres of public and private ownership (as well as the hybrid "regulated monopolies") and to establish criteria for rate-setting to avoid creating rents in the owners or beneficiaries of regulated infrastructure. See, e.g., public utilities casebooks.

87. Hobbes (discussion of contract and property in *Leviathan*); Grewal, *Network Power* (on standards of cooperation). There are deep debates in legal and social theory over the degree to which these legal forms, and that of property, arise spontaneously in the manner of grammar or, alternatively, depend on state definition and enforcement. See, e.g., Istvan Hont, *Jealousy of Trade: International Competition and the Nation-State in Historical Perspective* (Cambridge, MA: Belknap Press of Harvard University Press, 2010) (on this debate among early-modern jurists). It is in any case true that in

large and complex societies today, states assume considerable responsibility and potential latitude in the definition and enforcement of these regimes.

88. See, e.g., Thomas W. Merrill and Henry E. Smith, "What Happened to Property in Law and Economics?," *Yale Law Journal* 111, no. 2 (2001): 357 (on the information-theoretic advantages of relatively simple and uniform property rights); Jedediah Purdy, "People as Resources: Recruitment and Reciprocity in the Freedom-Promoting Approach to Property," *Duke Law Journal* 56 (2007): 1047 (arguing for an analytics of private-law cooperation emphasizing the motives to which bargainers may appeal, such as survival, material gain, and flourishing); Hanoch Dagan, *Property: Values and Institutions* (Oxford: Oxford University Press, 2011) (classifying a variety of property regimes according to the kinds of relationships they foster, such as arm's-length cooperation or familial intimacy).

89. See generally Friedrich Hayek, *Law, Legislation, and Liberty* (Chicago: University of Chicago Press, 1973) (portraying economic order as a legally constructed information-processing system).

90. See, e.g., Cary Franklin, *Infrastructures of Provision* (draft, on file with author).

91. See, e.g., William Boyd, *The Art of Fixing Prices* (draft, on file with author); Robert Lee Hale, *Freedom Through Law: Public Control of Private Governing Power* (New York: Columbia University Press, 1952); John R. Commons, *Institutional Economics* (New York: Macmillan, 1934).

92. See, e.g., K. Sabeel Rahman, *Democracy Against Domination* (Oxford: Oxford University Press, 2016).

93. Barbara Fried, *The Progressive Assault on Laissez-Faire: Robert Hale and the First Law and Economics Movement* (Cambridge, MA: Harvard University Press, 1998) (examining the thought of Hale and other realists); Morris R. Cohen, "Property and Sovereignty," *Cornell Law Review* 13, no. 8 (1927) (arguing for a conception of property rights as a delegation of sovereign power).

94. See, e.g., Robert H. Bork, *The Antitrust Paradox* (New York: Free Press, 1978); Rahman, *Democracy Against Domination*.

95. See, e.g., Stephen M. Gardiner, *A Perfect Moral Storm: The Ethical Tragedy of Climate Change* (Oxford: Oxford University Press, 2011) (on the failure to develop such a view); cf. Peter Singer, *One World: The Ethics of Globalization* (New Haven: Yale University Press, 2002) (seeking to develop such a view); Jamieson, "Climate Change."

96. "Waxman-Markey Bill," American Clean Energy and Security Act of 2009, H.R. 2454.

97. See, e.g., Singer, *One World*; Gardiner, *Perfect Moral Storm*.

98. Rahman, *Democracy Against Domination*; Barry Lynn; Louis D. Brandeis, *Curse of Bigness* (1967); Sitaraman (*Crisis of the Middle-Class Constitution*); Fishkin and Forbath (forthcoming); Sandel (*Democracy's Discontent*); cf. Bork (setting in motion the move from a political-economy to a pure consumer-welfare maximization view of the field).

99. Charles Reich, "The New Property," *Yale Law Journal* 73, no. 5 (1964).

100. Such ideas also figured in eighteenth- and nineteenth-century liberal arguments for "free labor": advocates of this program argued that the relationships and personalities that emerged in contractual, consent-based economic orders were better than those of status-based duty or prerogative—more aware of and sensitive to the interests and perspectives of others, as well as more disciplined and self-responsible. I once argued for a typology of "regimes of recruitment," networks of legal pathways to cooperation, based on relative prominence in them of three kinds of appeal: to fear

for survival, to material self-interest above the level of need, and to a sense of purpose, vocation, or self-expression. Purdy, "People as Resources," *Duke Law Journal* 56
(2007).

101. Amy Kapczynski, "Intellectual Property's Leviathan," *Law and Contemporary Problems* 77 (2014). Cf. Cary Franklin [current version of paper on infrastructure in the
development of the abortion right and litigation].

102. I am guided in these reflections by Professor Michael Warner's forthcoming Tanner
Lectures on "the ethics of infrastructure and the infrastructure of ethics" in climate
change, which he has kindly shared with me.

103. See, e.g., Jonathan Wiener, Eric Posner; Gardiner, *A Perfect Moral Storm*; cf. Richard
Tuck, *Free Riding* (Cambridge, MA: Harvard University Press, 2008), on the contestable assumptions of collective-action analysis, and Gardiner on reasons to suspect
that some of the scholarly fixation on the collective-action barriers to addressing the
problem is simply a convenient distraction.

104. Paul N. Edwards, "Infrastructure and Modernity: Force, Time, and Social Organiza-
tion in the History of Sociotechnical Systems," in *Modernity and Technology: The
Empirical Turn*, ed. Philip Brey, Arie Rip, and Andrew Feenberg (Cambridge, MA:
MIT Press, 2003), 386.

105. Hobbes, *Leviathan*, ed. Richard Tuck (Cambridge: Cambridge University Press,
1991).

CHAPTER 12

LIFE SUSTAINS LIFE 1

Value, Social and Ecological

JAMES TULLY

I would like to address the question of social and ecological value by bringing two approaches to this question into conversation with each other and show their connections. The two approaches are those of Jonathan Schell and Akeel Bilgrami. The connection between the two approaches is their shared interest in the "conditions that sustain life" on Earth. The answer to the question of what the conditions are that sustain life is, in my opinion, "life sustains life": that is, living ecological systems sustain themselves and the living systems with which they interact (symbiosis).

Jonathan Schell's Project

I see Jonathan Schell's current project as trying to provide a perspicuous representation of the ecological crisis of the present.[1] He begins with the concept of the Anthropocene yet goes beyond it. He is searching for a perspicuous representation of the Anthropocene that performs two roles.

The first role of his perspicuous representation of the Anthropocene is to bring out *deep and crucial aspects* of the present crisis that are concealed, obscured, or misrepresented in other formulations of the Anthropocene (by Crutzen, Hansen, Lovelock, Stern, IPCC, and so on):[2] that is, his representation of the Anthropocene is *world-disclosing*.

He does this in part by describing both the global effects of the Anthropocene and the processes that bring them about. The processes that bring about the destructive effects of the Anthropocene are such things as the link between modern science and its application, the development of capitalism in Europe and its spread around the world, the domination and exploitation of nature, the rapid increase in population, militarism, and so on, and the complex ways in which the effects of these process on the living Earth and atmosphere generate positive feedback loops that tend to amplify their destructive effects: climate change, polar ice melt, deforestation, desertification, the acidification of the oceans, and so on. These complex effects in turn affect population growth, inequality, life chances, starvation, mass migrations, agricultural, resource and water wars, neocolonial appropriation of land and water, failed states, increased militarization, and the positive feedback effects these complex processes have on global warming and the destruction of the conditions of life on Earth—in increasingly vicious circles.[3]

I will call these processes over the last three hundred years, insofar as they are destructive, *processes of modernization* for shorthand (they are called globalization, commodification, growth, imperialism, industrialization, neoliberalism, and so on in various schools of thought).

The second role of his perspicuous representation is to bring out aspects of the Anthropocene and the processes that bring it about that help us to see how to respond to the crisis, and even, I hope, to be moved to respond as citizens: that is, the perspicuous representation is also *action-guiding* as well as *world-disclosing*.

The key feature of the Anthropocene for Jonathan Schell is, I believe, the following. The human activities and form of life embodied in the processes of modernization are not only destroying life on the planet (biodiversity) but also, and more fundamentally, destroying the very conditions of life on Earth (for many species). They are destroying the ecosystems that sustain life. This trend is destroying the conditions of life on Earth for thousands of species, including *Homo sapiens*, and it is bringing about a sixth mass extinction of species and ecosystems.[4]

The question at the heart of this way of disclosing the present is, therefore, What are the "conditions of life" and how can we "sustain rather than destroy" them?

Akeel Bilgrami's Project

It is my opinion that Akeel Bilgrami is engaged in a somewhat similar project. He seeks to give a perspicuous representation of the form of life that developed in Europe, beginning in the scientific revolution of the seventeenth century and then spread around the world by European imperial and economic expansion:

he is seeking to give a more historical and genealogical account of central features of the same dominant form of life as Schell—that is, the processes of modernization that are bringing about the Anthropocene crisis.

Bilgrami seeks to disclose central features of this modern way of life in terms of the concept of an *alienated form of life*.[5] Three of the main features of this way of life as alienated are

1. A disengaged or disembedded stance of humans vis-à-vis nature;
2. A working relationship of control, mastery, and domination of nature embedded in our working relationship to nature; and
3. The presupposition that nature is devoid of intrinsic value and norms. Values and norms are assumed to derive from the autonomous human mind and are imposed by humans on a nonnormative world.
4. I would like to amend (3) in the following way: when moderns do see values and norms in nature, they tend to see values and norms that naturalize or reinforce the patterns of organization of modern institutions: a war of all against all (Hobbes), asocial sociability (antagonism) (Kant), and struggle for existence (Malthus, Darwin) all serve to legitimate the institutionalized forms of competition among individuals, groups, companies, and states that, in their view, lead to the development of the human species.[6]

As we can see, these features overlap with the features in Schell's approach.

The second dimension of Bilgrami's project is to give a perspicuous representation of an unalienated way of life in contrast. Three of the main and contrasting features of an unalienated way of life are

1. Humans see themselves as participants in nature, in the ecosystems in which they live.
2. From this participatory perspective, when humans act, they engage with nature: they interact in ecological relationships. They do not stand above and control.
3. When humans act and experience the world in accordance with (1) and (2) the world is disclosed to them as alive (comprising living systems) and of value (there is no nonevaluative language of description of being in the world from this perspective). Nature is seen and experienced to be suffused with values and norms that can be seen to involve responsibilities, that is, norms that are *action-guiding*.

The distinction between alienated and unalienated ways of life maps onto Anthony Laden's distinction between two types of reasoning: reasoning-over and reasoning-with.[7]

Life Sustains Life Is Their Common Ground

I think it is easy to see the overlap and connections between the two complementary projects.

The dominant way of life that is described as "alienated" is the same way of life that is bringing about the destructive effects of the Anthropocene: that is, the destructive side of processes of modernization. It complements Schell's account (and many others).[8]

The unalienated way of life—as a response to the ills of this alienated way of life—is also a possible response to the crisis of the Anthropocene that this modernizing way of life brings about and reproduces.

Recall that the key feature of the way of life that brings about the Anthropocene is that it destroys the natural or ecological conditions of life. Now, if humans move around to the unalienated way of life and see themselves as participating in the ecological conditions of life, engaging with them, and if they discover norms that move them not to destroy these conditions of life, as in the current predicament, but rather to not harm them and sustain them, then the unalienated way of life is indeed a response to the Anthropocene.

I would like to argue that this is the case. When humans participate in and engage with the ecological conditions of life on Earth (ecosystems), they discover intrinsic, living, and action-guiding normative relationships of cooperation and competition that sustain rather than destroy these conditions of life. Learning from and getting in tune with these life-sustaining normative relationships of interdependency is the way to respond to the life-destroying features of the Anthropocene, that is, to see humans as members and citizens of a living commonwealth of all forms of life with civic responsibilities to sustain the conditions of mutual dependency and coevolution.

This is of course the deep ecology view. But it is also the view of the most advanced life sciences and Earth sciences over the last fifty years.

1.   The basic unit of life is not the individual or the species but ecosystems or living networks.[9]

2.   Moreover, the patterns of self-organization and self-reproduction (autopoiesis) of interdependent living systems are not struggles for existence, but, as Lynn Margulis has shown, predominately symbiosis and symbiogenesis.[10] Living systems tend more over time to reproduce and transform themselves and the living systems with which they interact than to destroy themselves and their neighbors, or life on Earth would not have evolved as it did.

3.   Further still, an emergent property of the webs of life as a whole is the self-regulation of the Earth and its atmosphere so as to sustain life on Earth: the Gaia theory of Sir James Lovelock.[11] The Gaia theory is symbiosis at the planetary level.

4. Finally, humans *can* learn from the symbiotic patterns of organization of these living systems how to organize their own living systems of communities of practice that sustain life (discussed later).

The mantra for this fourfold hypothesis is the saying "life sustains life." For example, Stephen Harding: "The key insight of the Gaia theory is wonderfully holistic and non-hierarchical. It suggests that it is the Gaian system as a whole that does the regulating, that the sum of all the complex feedbacks between life, atmosphere, rocks and water gives rise to Gaia, the evolving, self-regulating planetary entity that has maintained habitable conditions on the surface of the planet over vast stretches of geological time."[12] And Lynn Margulis: "Gaia is not an 'organism' but an emergent property of interaction among organisms. Gaia is the series of interacting ecosystems that compose a single huge ecosystem at the earth's surface. Gaia is symbiosis on a planetary scale."[13]

If this is correct, then the very norms that could guide humans from a way of life that is destroying the conditions of life toward a way of life that sustains the conditions of life can be found in the self-organizing and self-sustaining patterns of interaction of the living conditions of life themselves: that is, this understanding of the "conditions of life" connects Schell's and Bilgrami's projects.[14]

Misrepresenting and Destroying the Conditions of Life

Once we see this account of the living world and our engaged place within it, we can see how Bilgrami's three features of the alienated life of modernization misrepresent and occlude the "conditions of life." The disengaged and disembedded stance, the relation to the world of control and mastery, and the view that all value and norms come from the human mind alone literally alienate us from the living world. Human autonomy is purchased at the price of earthly alienation.

This dominant human form of life and its mode of representation make it difficult to see the connected ways in which our human activities are destroying the living relationships that sustain life.

For example, as Bilgrami puts it, through this mode of world disclosure nature is seen as a repository of "opportunities" to satisfy the "states of minds" of humans (values, interests, utilities, projects, and the like) through its use and exploitation by corporations and states. The role of the social sciences is to link the opportunities or resources of nature with the utilities of humans.

Another example is Karl Polanyi's analysis of the rise of the unique capitalist economy in the nineteenth century. In *The Great Transformation* Polanyi showed, for example, how the commodification of the natural world causes us to overlook its living systems and to treat the effects of commodification as externalities.[15] The commodification of living systems as "resources" disembeds them from the symbiotic relationships in which they exist and reembeds them in the abstract economic, legal, political, and institutional relationships of the global capitalist economy. The effects this radical transformation has on the underlying living ecosystems in which these processes of extraction, production, and consumption take place cannot be seen from within this alienated form of representation of them. When the damage and destruction are seen, they are treated as externalities and indirect costs, and thus not taken into economic account. This is the fundamental flaw at the heart of our system of economics and the major cause of the ecological crisis.[16]

Polanyi also argued that the commodification of individual and collective labor power of humans was another "fictitious commodity" that amplified this alienation. By turning labor power into a commodity, capitalism disembeds its exercise from the living human beings and the social relationships in which they live and on which they are interdependent. It then reembeds labor power and laborers into the abstract, competitive, and alienating relationships and institutions of the global economy. As a result, humans under capitalism internalize the form of representation and way of acting on nature and one another encapsulated in Akeel Bilgrami's three features of an alienated life (as Marx, Adorno and Horkheimer, and Gandhi also argued). The result, Polanyi predicted in 1944, would be the destruction of the social and natural world.

Discounting Ecological Destruction and Climate Change

Now, let's turn and look briefly at the responses to the Anthropocene and the scientific evidence that climate change, global warming, and so on are destroying the living conditions of life.

First, if it is correct to say that the alienated form of self-consciousness characteristic of modern subjectivity and brought about by the three features Bilgrami foregrounds causes us the overlook the living networks in which we exist and on which we codepend and coevolve, then this helps us to understand why people deny or discount the Anthropocene, the damage it brings about, and the threat it poses.

Aldo Leopold diagnosed this ailment of not seeing the damage one is doing to the conditions of life by the way we are living, because we inhabit our way of life and do not reflect on it as a way and its effects. We see the world through its frame (the three features Akeel lays out). Leopold: "One of the penalties of an ecological education is that one lives alone in a world of wounds. Much of the damage inflicted on land is quite invisible to laymen. An ecologist must either harden his shell and make believe that the consequences of science are none of his business, or he must be the doctor who sees the marks of death in a community that believes itself well and does not want to be told otherwise."[17] (Note the combination of false belief and weakness of the will in the last sentence.)

In *Our Ecological Footprint*, Mathias Wackernagel and William Rees list three main strategies of denial and discounting the overshooting of the carrying capacity of the conditions of life on Earth and its destructive consequences:[18]

1. The boiled frog syndrome: the brain functions so that slow changes, long-term implications, and multiple connections cannot be easily seen.
2. Mental apartheid: the psychological barrier between modern humans and the rest of reality: perceptual dualism since Descartes.
3. The idea of the tragedy of the ungoverned commons, from Hobbes's state of nature to Hardin's analysis, makes it appear that the only alternative is privatization (and this becomes our present tragedy). So, there appears to be no other possibility.

I think all three of these strategies can be seen as consequences of Bilgrami's alienated form of life.

But, I would add a fourth. I think the really basic one is that our alienated way of life causes us to overlook the living Earth and to represent it as resources for production and consumption, whether capitalist or Marxist. We do this because of the background picture of humans rising, through stages of historical development, to a position of independence of the world and standing in a stance of command and control over it. Thus, the only alternative is to be under the control of natural forces that we do not understand, in a position of heteronomy. This is the position attributed to "primitive" and "less-developed" societies in this modernist worldview, as Franz Boas classically argued in 1911:[19]

> Proud of his wonderful achievements, civilized man looks down upon the humbler members of mankind. He has conquered the forces of nature and compelled them to serve him. He has transformed inhospitable forests into fertile fields. The mountain fastnesses are yielding their treasures to his demands. The fierce animals which are obstructing his progress are being exterminated, while others which are useful to him are made to increase a thousand fold. The waves of the ocean carry him from land to land, and towering mountain-ranges set him no bounds. His genius has moulded inert matter into powerful machines which await a touch of his hand to serve his manifold demand.
>
> With pity he looks down upon those members of the human race who have not succeeded in subduing nature; who labour to eke a meagre existence out

of the products of the wilderness; who hear with trembling the roar of the wild animals, and see the products of their toils destroyed by them; who remain restricted by oceans, rivers or mountains; who strive to obtain the necessities of life with the help of few and simple instruments.

Such is the contrast that presents itself to the observer. What wonder if civilized man considers himself a being of higher order as compared to primitive man, if he claims that the white race represents a type higher than all the others!

The third possibility of humans being in relationships with the living Earth drops out of the picture.

In *The Nature Principle*, Richard Louv argues that the denial and discount are the result of a "nature deficit disorder." That is to say, if we grew up and lived in an unalienated form of life in which we participated in and engaged with the living networks in which we in fact live, and if this was also the representation and self-understanding of the human condition, then these blind spots, barriers, and tragedies would be overcome. That is, long-term implications and multiple connections in (1) would be easily seen; the psychological barrier between modern humans and the rest of reality would be overcome in (2), and the binary of either the ungoverned commons or global privatization of (3) would be seen to overlook a third possibility: sustainable modes of "cooperative commons" modeled on and interacting with nonhuman living systems of symbiosis and symbiogenesis.

The "Alienated" Response to the Crisis 1

Second, there is response to the Anthropocene that takes it seriously. This response sees the Earth as comprising living systems that form the conditions of life and these conditions are under imminent threat. These authors write *Our Dying Planet* (Peter Sale), *Collapse* (Jared Diamond), *The Revenge of Gaia* (James Lovelock), *The Social Conquest of the Earth* (E. O. Wilson), and "Does Success Spell Doom for Homo Sapiens?" (Charles Mann).[20] They also present a response they believe can save *Homo sapiens* from destruction or a sixth mass extinction.

This is perhaps the dominant response to the Anthropocene. It is advanced by neo-Malthusians and neo-Darwinians. Let's call this response the Medea Hypothesis, as Peter Ward does in his influential book of 2009.[21]

On this view, humans, like Medea, are destined to destroy their own children: that is, *Homo sapiens*, since the beginning of their migration out of Africa one hundred thousand to fifty thousand years ago, is programmed to destroy not only other humans and nonhuman species, as Darwin argued. But also, *beyond Darwin*, the consequence of success in these exterminating struggles for existence is that they also destroy the ecosystems and, eventually, Gaia: the conditions that sustain life on Earth.

These hyper-Darwinians argue that the only response is a massive and global project to bring human activities and population under the command and control of some kind of global authority that can act as "the mind of Gaia" and save humanity from itself. They see *Homo sapiens* as evolving in such a way that it comes to know and understand how Gaia works and how to save it at exactly the last moment before it is destroyed (a repetition of the nineteenth-century dogma that the greatest danger gives rise to the greatest insight and salvation).

This response is likened to the command and control of the war effort in World War II and other such projects. Even James Lovelock, the founder of the Gaia theory, is a proponent of this hyper-Darwinian analysis and Faustian response.

I think there are two dubious features of the Medea Hypothesis and its Faustian response. First, the Medea Hypothesis sees the natural world as alive and infused with normative relationships, but it mistakes norms of conquest and extermination for the dominant norms of living systems, whereas biologists, paleontologists, and archeologists have known since Kropotkin that genocidal competition and extinctions have always been enveloped within larger and deeper living relationships of cooperation and nonexterminating competition (such as cosustaining predator-prey relationships).

This hypothesis blows out of proportion the trend of the last two to three hundred years of overshooting and destroying the carrying capacity of the living Earth and its consequences in the age of industrialization, cheap carbon fuels, economic globalization, population explosion, and their consequences. It then projects these trends back fifty thousand to one hundred thousand years, or to the beginning of the agricultural age, eleven thousand years ago. It highlights and generalizes isolated incidents of struggles for existence that lead to the extermination of opponents and the population increase of the victors into "super organisms," and then this leads to the destruction of the carrying capacity of the regional biosphere—such as Easter Island.

It makes it appear that the coevolution of life on Earth is *predominantly* a series of conquests followed by the destruction of the underlying conditions that sustain the conquerors (a new "gravediggers' dialectic," so to speak).

If many biologists, archeologists, anthropologists, and ecologists are correct, this conquer-conquest-destruction representation of human and biological history is false. It overemphasizes destructive and exterminating struggles and radically underemphasizes the broader living networks of symbiosis on which these struggles are parasitic and which, for the most part, contain

competition in forms that do not lead to extermination but to sustaining ecosystems as a whole.

Gandhi diagnosed this kind of Darwinian view of history in *Hind Swaraj*
in 1909.[22] He argued that moderns are taught history and biology and economic development as a series of battles and conquests. However, if life were a
war of all against all or a struggle for existence, life on Earth would have ended
ages ago. Rather, humans and nonhumans learn ways to live together and set-
tle their disputes nonviolently or with types of violence—such as predator-
prey relations between deer and coyote—that keep species and the ecosystems
on which they interdepend in rough equilibrium, punctuated with occasional
rapid and drastic change.

That is, this response overlooks or underplays the living networks of symbiosis and mutual aid that have sustained the coevolution and complexification of
forms of life for millions of years. Life in general and *Homo sapiens* in particular
have evolved through the development of living networks, not through relations
of conquest and control. Donald Worster, America's leading ecological historian,
argues that it is doubtful that anyone from any other civilization than modern
Western civilization would even understand this view of nature, inherited and
inflated from its nineteenth-century origin in Townsend, Malthus, Lyell, Darwin,
Huxley, and the social Darwinians such as Spenser, Nietzsche, and Freud.[23]

So, I would agree with Schell, Bilgrami, Timothy Mitchell,[24] and many others that the situation we are confronted with in the Anthropocene age of the
last two hundred to three hundred years is unique, and it is unique for the
reasons Bilgrami gives: namely, the way in which, since the seventeenth century, it overlooks and overrides the participation in and engagement with living networks of complex forms of life.

There were of course exterminating struggles for existence throughout
history, including five mass extinctions, and *Homo sapiens* has been an
aggressive disturber since its spread from Africa to around the world over
150,000 years. But extermination, extinctions, biocide, and ecocide were not
the dominant factors; symbiosis and coevolution were larger factors, as the
Gaia Theory has rediscovered, or life would not have continued to grow and
become more complex. It is only in the last three to four hundred years—the
centuries of European imperialism and rapid economic development—that
biocidal or "imperial" factors have become predominate over symbiotic or
"Arcadian" factors, and thus ushered in the Anthropocene.[25]

The "Alienated" Response to the Crisis 2

The second feature of the Medea Hypothesis that places it firmly within the
alienation framework is its recourse to global organizations of command and

control as the means to save humanity from destruction. This is a double mistake.

First, it fails to see that the way life is sustained within living networks is not through relationships of conquest or command and control, but through relations of cooperation and nonexterminating competition within and among species and their ecosystems. The multiplicity of living networks that comprise the Earth System exhibits countless varieties and variations of cooperative and competitive interaction—of participation in and engagement with other forms of life.

Second, if life has survived and coevolved by these means, then humans can learn from them how to participate in and engage with these living networks in a nondestructive and mutually sustaining way. Humans can learn something about their own forms of living organizations from nature.[26]

This is heretical from the alienated perspective, according to which nature can teach humans nothing about sustainable forms of human organization. It is also heretical from the Medean perspective, according to which nature teaches us to employ command and control over nature to rescue humanity from the destruction that conquest, command, and control have brought about.

The great question is how we can learn from nature in time. It seems to me that there are three distinct types of learning processes.

The first of course is to understand how living systems bring living systems into being through symbiosis (autopoiesis) and sustain and complexify them, from microbes and bacteria to mammals, complex ecosystems, and Gaia. This is the great work of the life sciences since World War II.

The second is for humans then to learn how to live and interact with and in them in such a way that they harm them as little as possible, on the one hand, and care for and help to sustain them on the other (stewards of mother earth). And this includes of course how to repair the damage that we have done.

The third and youngest science is to learn from sustainable living systems how to organize and operate self-sustaining and mutually supportive living organizations (networks) of human beings in their various activities: forms of organizations that create zero emissions; setting up communities and networks of organizations that use one another's waste so everything is recycled and reused (the "cradle to cradle" approach); in short the plethora of initiatives that apply self-sustaining life rules to human organizations or "communities of practice."[27]

This is how Fritjof Capra sees these three types of education in *Hidden Connections*:

> The key to an operational definition of ecological sustainability is the realization that we do not need to invent sustainable human communities from

scratch but can model them after nature's ecosystems, which are sustainable communities of plants, animals and microorganisms. Since the outstanding characteristic of the Earth household is its inherent ability to sustain life, a sustainable human community is one designed in such a manner that its ways of life, businesses, economies, physical structures, and technologies do not interfere with nature's inherent ability to sustain life. Sustainable communities evolve their patterns of living over time in continual interaction with other living systems, both human and non-human. Sustainability does not mean that things do not change; it is a dynamic process of co-evolution rather than a static state.[28]

By "modeling" sustainable human networks after nature's ecosystems, Capra does not mean "imitate" nature. He gives full weight to the differences between *Homo sapiens* and other species (reflective consciousness, meaning, language, power, and so on) and summarizes the ecological norms that can be used to guide humans in building sustainable communities.[29]

Life Sustains Life Is Not Only the End but the Means, the Way

If this analysis is partially correct, then the right response to the Anthropocene is not more attempts to command and control human and nonhuman forms of life on the planet. It is not to treat forms of life as things or means to be moved by imperatives and coercion, but to interact with them as they interact: as interdependent living beings or "interbeings."

This would involve coming to see and to act in accord with the values or norms of "sustaining life by means of sustaining life" that, as we have seen, "suffuse" nature, to use Bilgrami's phrase (and also to see the subordinate role conquest and extermination have played until now). Learning from it would be the first step toward an unalienated life in Bilgrami's sense, and, at the same time, the first step in a response to the Anthropocene in Schell's sense.

On this view means and ends are one and the same. The way to an unalienated life and to transform the Anthropocene into a self-sustaining Gaia complex is by means of unalienated participation in and engagement with other living networks, that is, the three lessons mentioned earlier.

Now, this Gaia ethic is not something that we learn only from nature. Many if not all the great ethical and spiritual traditions teach the same lesson. They share the basic ethical norm of *ahimsa* in its negative and positive sense. This is the ethical precept that we should avoid harming any living being as much as possible on the one hand and should also help to care for and sustain living beings on the other. We should do this in everything we say and do in everyday life—in every breath we take.[30]

We only need to extend this predominantly human-centered ethic to the biotic communities in which we live and breathe. As Aldo Leopold put it in *The Sand County Almanac*:

> All ethics so far evolved rest upon a single premise: that the individual is a member of a community of interdependent parts. His instincts prompt him to compete for his place in the community, but his ethics prompts him also to cooperate (perhaps in order that there may be a place to compete for).
>
> The land ethic simply enlarges the boundaries of the community to include soils, waters, plants, and animals, or, collectively: the land.
>
> In short, a land ethic changes the role of Homo sapiens from conqueror of the land-community to plain member and citizen of it. It implies respect for his fellow-members, and also respect for the community as such.
>
> In human history, we have learned (I hope) that the conqueror role is eventually self-defeating. Why? Because it is implicit in such a role that the conqueror knows, ex cathedra, just what makes the community clock tick, and just what and who is valuable, and what and who is worthless, in community life. It always turns out that he knows neither, and this is why his conquests eventually defeat themselves.[31]

An ethic presupposes some "mental image of land" as a "biotic system." It is not only a mental image that is required, but also "we can be ethical only in relation to something we can feel, understand, love, or otherwise have faith in."[32]

Millions of ordinary human beings, individually and collectively in communities and networks, have begun to learn this lesson since 1949, and especially since Rachel Carson, *The Silent Spring*,[33] and then the effects of global warming and climate change mentioned by Schell and catalogued by Lester R. Brown, the Stern Report, and others. They have begun the difficult task of transforming themselves into unalienated plain members and engaged citizens of the ecosystems that sustain them.

There is no other way to bring about a sustainable world than by participating in and engaging with the living systems in which we exist. The end is the way, as Gandhi put it, or the way is constitutive of the end. The activity of disengaging from the alienated and destructive way of life and gradually acting in accord with the unalienated way of life involves difficult and transformative individual and collective practices of everyday life. As we begin to take the first two steps in Akeel Bilgrami's unalienated way of life—of participating in and engaging with the ecosystems in which we live and breathe and have our individual and collective being—the "world" begins to "show up" for us as not only valuable but the condition of all value: as a living system that sustains all life. The world is disclosed to us as Gaia. But it only "shows up" for us in this way if we engage in the daily sustaining practices of participation and engagement. The insight is based not on reasoned argument alone, but also on beginning to try to live and experiment with the mutually sustainable ways of life. In Wittgenstein's wonderful phrase, as we begin to engage in this "way of life," "the light dawns slowly on the whole."

The Relation Between Knowing and Being: The Fourth Feature of the Alienated and Unalienated Ways of Life

Once we see this relationship between practice and wisdom concerning our place in the world we overcome what I will call the fourth feature of the alienated way of life. This is the presumption that humans can know the good life without becoming good themselves. This presumption—the so-called Cartesian Moment—is the great dividing line between ancient and modern philosophy.[34] For the ancients there were always practices of the self that one had to undergo as the condition of knowledge, of wisdom, and these were of course ethical practices that, once engaged in, gradually disclosed the meaning and value of the world for the novitiate. Philosophy was literally a "way of life."[35]

Modern philosophy since Descartes has been based on the premise that we can know the truth and the good without engaging in the practices of becoming truthful and good ourselves. And this alienation of knowing from the means of knowing is paralleled in social sciences in the separation of means and ends in modernization theories: wars can lead to peace and authoritarian rule can lead to democracy are the two widely held examples, as Arendt argues in *On Violence*. She argued that the presumption of a "contingent relation between means and ends" legitimated rapid development and the arms race to spread and protect it, and that, as a result, these "processes of modernization" are out of control, can no longer be called "progress," and will lead to the destruction of life on Earth.[36]

This modern view of the contingent relation between knowing and being was challenged by William James and his "ancient" argument that "knowing is dependent on and conditioned by being" (or ways of knowing or ways of being). This internal relation between being and knowing is the foundation of pragmatism (and of a certain strand of Marxism). It has been articulated by Aldous Huxley, Maurice Merleau-Ponty, Jonathan Schell, Michel Foucault, Ludwig Wittgenstein, and others.[37] In the terms of the Santiago Theory of Cognition, "a mode of being (or living) brings forth a world."[38] This is what I will call the fourth feature of the unalienated life: means are constitutive of ends.

In many nonmodern civilizations, especially indigenous civilizations, when youth participate in and engage with the world in mutually sustainable ways the Earth is disclosed to them as "mother earth" and this in turn as a natural "gift economy" of which they are always already participants: that is, they see themselves as members of ecological relationships of gifts (the goods and services nature supplies), of the attitude of gratitude for the gifts, and of duties of reciprocity to the ecosystems that provide the sustenance; that is, they are inducted into the living cycles of gift-gratitude-reciprocity that sustain life on Earth and they are thereby moved to engage accordingly in these life-sustaining cycles. As they say, "we take care of mother earth and she takes care of us."[39]

This practice-based view of knowing-how and knowing-that has not been completely lost even in the alienated modern West. Steady-state economists like Herman Daly, Polanyi, and Charles Eisenstein; deep ecologists like Arne Naess, Richard Louv, and David Abram; ethicists like Stephanie Kaza and Patrick Curry all present arguments for this gift-economy orientation and model sustainable human economies on the natural-gift economy (as does Capra).[40] The idea that there is a "subsistence gift economy" in all premodern societies and it continues to exist within capitalist societies (the neighborhood and volunteer sectors, for example), and that capitalist economies are parasitic upon it, was first brought to prominence by Marcel Mauss in his book *The Gift*.[41] After writing *The Great Transformation* Polanyi turned to economic anthropology and rediscovered the gift economy in non-Western societies. In the 1980s in France Gilbert Rist published his famous critical history of the capitalist "developmental" or "growth" economy as based on blind "faith" rather than economic rationality, and argued for a move forward to renewed gift economies.[42] During the same period, Fritz Schumacher, in *Small Is Beautiful*, showed how gift economies could replace rapid development capitalist and communist economies and be more efficient.[43] This was carried forward by Charles Eisenstein in *Sacred Economies* and Vandana Shiva in *Earth Democracy*.[44] In these latter works, the human gift economy is modeled explicitly on the ecological gift economy and its symbiotic and cyclical relationships of interdependency. These works, especially Schumacher's, have been influential in the spread of local and networked gift economies (economic cooperatives and community-based economies) throughout the world.

Of equal importance, Aldo Leopold laid out steps in a practical education system that would bring students to see, appreciate, and revere nature and all forms of life in this way, and these are now steps in education systems throughout the world—against the grain of the dominant way of life.

Modern European languages retain traces of this view as well, especially in the term we use to describe the world as we experience it: that is, "the given." The world is a "gift" and what is entailed by a gift is gratitude and reciprocity. And, as we have seen, what is given to us as a gift is not given to us by some giver standing apart from the world and controlling it. Rather, the gift of life is given to us by the conditions of life—by life itself. Life sustains life. This is the miracle that induces a response not only of gratitude but also of wonder and perhaps awe, as Schell puts it.

However, to experience the given in this world-disclosing and action-guiding way we have to begin to take the steps of the unalienated life in our daily practice, and that is up to each one of us, not just to talk about the change, but to be the change. Whether or not this is too little too late is a question for another time. But, in thinking about the factor of time, it is important to remember that on the unalienated view, time is not abstracted and independent of life but interdependent, so it discloses a completely different way of thinking about temporality.

Notes

1. Jonathan Schell, "On the Anthropocene" and "The Human Shadow" (unpublished manuscripts).
2. For example: Paul Crutzen and Hans Günter Brauch, eds., *Paul J. Crutzen: A Pioneer on Atmospheric Chemistry and Climate Change in the Anthropocene* (New York: Springer, 2016); James Hansen, *Storms of My Grandchildren: The Truth About the Coming Climate Catastrophe and Our Last Chance to Save Humanity* (New York: Bloomsbury, 2009); James Lovelock, *Gaia: A New Look at Life on Earth*, 3rd ed. (1979; Oxford: Oxford University Press, 2000); Nicholas Stern, *The Economics of Climate Change: The Stern Review* (Cambridge: Cambridge University Press, 2007); *Climate Change 2014: Synthesis Report: Contribution of Working Groups I, II and III to the Fifth Assessment Report of the Intergovernmental Panel on Climate Change* (Geneva: Intergovernmental Panel on Climate Change, 2015).
3. Lester R. Brown, *World on the Edge: How to Prevent Environmental and Economic Collapse* (New York: Norton, 2011).
4. Elizabeth Kolbert, *The Sixth Mass Extinction: An Unnatural History* (New York: Henry Holt, 2014).
5. Akeel Bilgrami, *Secularism, Identity and Enchantment* (Cambridge, MA: Harvard University Press, 2014), 101–216.
6. We see this in Charles Mann, E. O. Wilson, and others, cited later.
7. Anthony Simon Laden, *Reasoning: A Social Picture* (Oxford: Oxford University Press, 2012).
8. Laden, *Reasoning*.
9. Harold Morowitz, *The Emergence of Everything* (Oxford: Oxford University Press, 2002).
10. Lynn Margulis, *Symbiotic Planet: A New Look at Evolution* (New York: Basic, 1998).
11. Lovelock, *Gaia*.
12. Stephen Harding, *Animate Earth: Science, Intuition and Gaia* (White River Junction: Chelsea Green, 2006), 64.
13. Margulis, *Symbiotic Planet*, 120.
14. One of the first people to try to sketch out a very general account of how this might work is Fritjof Capra, *The Hidden Connections: A Science for Sustainable Living* (New

York: Anchor, 2002). He tried to bring together the life sciences, the Santiago theory of cognition, complexity theory, systems theory and feedback loops, Gaia theory, social theory, and social networks theory.

15. Karl Polanyi, *The Great Transformation: The Political and Economic Origins of Our Time* (Boston: Beacon, 2004).

16. Brown, *World on Edge.*

17. Aldo Leopold, *The Sand County Almanac* (Oxford: Oxford University Press, 1949), 197.

18. Mathias Wackernagel and William Rees, *Our Ecological Footprint* (Gabriola Island: New Society, 1996).

19. Franz Boas, *The Mind of Primitive Man* (1911; New York: Macmillan, 1923), 4–5.

20. Peter Sale, *Our Dying Planet: An Ecologist's View of the Crisis We Face* (Berkeley: University of California Press, 2011); Jared Diamond, *Collapse: How Societies Choose to Fail or Succeed* (New York: Penguin, 2005); James Lovelock, *The Revenge of Gaia: Earth's Climate Crisis and the Fate of Humanity* (New York: Basic, 2007); Edward O. Wilson, *The Social Conquest of Earth* (New York: Norton, 2012); Charles Mann, "State of the Species: Does Success Spell Doom for Homo Sapiens?," *Orion Magazine* (November/December 2012).

21. Peter Ward, *The Medea Hypothesis: Is Life on Earth Ultimately Self-Destructive?* (Princeton: Princeton University Press, 2009).

22. Mahatma K. Gandhi, *Gandhi: Hind Swaraj, and Other Writings*, ed. Anthony J. Parel (Cambridge: Cambridge University Press, 2007), chap. 17.

23. Donald Worster, *Nature's Economy: A History of Ecological Ideas* (New York: Cambridge University Press, 1994).

24. Timothy Mitchell, *Carbon Democracy: Political Power in the Age of Oil* (London: Verso, 2011).

25. Worster, *Nature's Economy.*

26. Ellen LaConte, *Life Rules: Nature's Blueprint for Surviving Economic and Environmental Collapse* (Gabriola Island: New Society, 2012).

27. William McDonough and Michael Braungart, *Cradle to Cradle: Remaking the Way We Make Things* (New York: North Point, 2002).

28. Capra, *Hidden Connections*, 230–231.

29. Capra, 73, 231.

30. Stephanie Kaza, *Mindfully Green: A Personal and Spiritual Guide to Whole Earth Thinking* (Boston: Shambhala, 2008).

31. Leopold, *Sand County*, 239–240.

32. Leopold, 251.

33. Rachel Carson, *Silent Spring* (Boston: Houghton Mifflin, 1962).

34. Michel Foucault, *The Hermeneutics of the Subject: Lectures at the Collège de France 1981–1982*, ed. Frédéric Gros (New York: Picador, 2005).

35. Pierre Hadot, *Philosophy as a Way of Life: Spiritual Exercises from Socrates to Foucault*, ed. Arnold Davidson (Malden, MA: Blackwell, 1995).

36. Hannah Arendt, *On Violence* (Orlando: Harcourt, 1969).

37. Aldous Huxley, *The Perennial Philosophy* (London: Chatto and Windus, 1946); Hadot, *Philosophy*; Maurice Merleau-Ponty, *The Phenomenology of Perception* (London: Routledge, 2012); Jonathan Schell, *The Unconquerable World: Power, Nonviolence and the Will of the People* (New York: Henry Holt, 2002); Foucault, *The Hermeneutics of the Subject.*

38. Fritjof Capra, "The Santiago Theory of Life and Cognition," *La Revista Be-Vision* 9, no. 1 (1986).

39. James Tully, "Reconciliation Here on Earth," in *Resurgence and Reconciliation: Indigenous-Settler Relations and Earth Teachings*, ed. Michael Asch, John Borrows, and James Tully (Toronto: University of Toronto Press, 2018).

40. Herman Daly, *Beyond Growth: The Economics of Sustainable Development* (Boston: Beacon, 1987); Polanyi, *Great Transformation*; Charles Eisenstein, *Sacred Economics: Money, Gift, and Society in the Age of Transition* (Berkeley: North Atlantic, 2011); Arne Naess, *Ecology, Community, and Lifestyle* (Cambridge: Cambridge University Press, 1989); Richard Louv, *The Nature Principle: Reconnecting with Life in a Virtual Age* (Chapel Hill, NC: Algonquin, 2011); David Abram, *Becoming Animal: An Earthly Cosmology* (New York: Random House, 2010); Kaza, *Mindfully Green*; Patrick Curry, *Ecological Ethics: An Introduction* (Malden, MA: Polity, 2011).

41. Marcel Mauss, *The Gift: Forms and Functions of Exchange in Archaic Societies* (London: Routledge, 1990).

42. Gilbert Rist, *The History of Development: From Western Origins to Global Faith*, 4th ed. (London: Zed, 2014).

43. Fritz Schumacher, *Small Is Beautiful: Economics as If People Mattered* (New York: Harper, 2010).

44. Eisenstein, *Sacred Economics*; Vandana Shiva, *Earth Democracy: Justice, Sustainability, and Peace* (Berkeley: North Atlantic, 2015).

CHAPTER 13

LIFE SUSTAINS LIFE 2

The Ways of Reengagement with the Living Earth

JAMES TULLY

Over the last five hundred years the West has developed a social system that is socially and ecologically unsustainable and destructive. It overreaches and undermines the social and ecological conditions that sustain life on Earth for *Homo sapiens* and many other species and ecosystems.

This social system has been spread around the world by Western imperialism in the colonial and postcolonial periods and is now the dominant social system on the planet.

It is a "vicious" system in the technical sense that the regular feedback loops within the social system and between the social system and the ecosystems in which it is embedded reproduce and intensify the destructive effects of the system on the social sphere and ecosphere. These effects include such things as global warming, climate change, pollution and the diseases it causes, acidification of the oceans, desertification of once arable land and the recolonization of Africa, melting of the polar icecaps and the release of methane, the depletion of nonrenewable recourses and the military conflicts over what's left, the use of renewable resources and aquifers beyond their cyclical rate of renewal, millions of climate change refugees, global inequalities in life chances, petro-states, a planet of slums and gated enclaves, the domination of democracy by concentrations of private, media, and military power, the concentration of the means of production in a handful of multinational companies, counterviolences to the system's overt and structural violence that trigger counterinsurgencies and feed an ever-increasing arms race and arms trade, and so on.

We have known that the dominant social system (a historical assemblage of many social systems) is unsustainable and self-destructive socially and ecologically since the 1970s: Rachel Carson, Barry Commoner, the Club of Rome, and *The Limits to Growth*.[1] It has been reaffirmed ever since by experts from a wide variety of different fields: *The Limits to Growth Revisited* (2012), IPCC, Lester Brown, and Craig Dilworth.[2]

Yet it continues despite efforts to address it in various ways since the 1970s and in the face of increasing evidence of its destructiveness, such as the sixth mass extinction, global warming and its effects, and the naming of the predicament *as* "peak everything" and the "Anthropocene."[3]

The phenomenon of a life system becoming a vicious system rather than a virtuous and self-sustaining system, and thus overshooting and destroying the social and ecological conditions on which it depends for its sustainability and collapsing, is not unusual in the history of human systems or nonhuman living systems.

Furthermore, as the resiliency literature argues, it is also not unheard-of for a vicious life system tending toward self-destruction to transform itself into a virtuous one and avoid collapse, albeit usually at a qualitatively different dynamic equilibrium.

Thus, while life systems are the most complex systems in the world, they are not predetermined, and it is not impossible for humans to save themselves from self-destruction, for, after all, we are the subjects and agents who reproduce the unsustainable social system by participating in it in our everyday activities, and we have the freedom to act otherwise.

Thus, there are three central questions with regard to sustainability I wish to address:

1. How do we act in such a way as to transform the unsustainable system in which we find ourselves into a sustainable system?
2. How do we become motivated to act in a sustainable way once we know what that is?
3. And, before these two questions can be addressed, a third, prior question needs to be addressed: What are the salient features of sustainable and unsustainable life systems?

I want to address the question of the main features of unsustainable and sustainable life systems through the distinction Akeel Bilgrami makes between alienated and unalienated ways of life.[4] For his concept of an alienated way of life describes life from within the unsustainable social system we inhabit as moderns and his concept of an unalienated way of life describes life within a sustainable social system.

An alienated or unsustainable way of human life has the following features:

1. A disengaged or disembedded stance of humans vis-à-vis nature;
2. A working relationship of control, mastery, and domination of nature embedded in our working relationship to nature; and
3. The presupposition that nature is devoid of intrinsic value and norms. Values and norms are assumed to derive from the autonomous human mind and are imposed by humans on a nonnormative world.
4. If there is a normative dimension to nature then it is a basic, amoral antagonism or struggle for existence among independent living beings (humans and nonhumans) in which the fitter gain control over or exterminate the less fit and establish a new order or system that is a higher stage of development than the previous one in an unlimited set of stages of development and progress.

An unalienated or sustainable human way of life has the following features:

1. Humans see themselves *participants in nature*, in the ecosystems in which they live.
2. From this participatory perspective, when humans act, they *engage with* nature: they interact symbiotically in ecological relationships. They do not stand above and control.
3. When humans act and experience the world in accordance with steps (1) and (2) the world is disclosed to them as interdependent webs of life (comprising living systems) and of value. Nature is seen and experienced to be suffused with values and norms that can be seen to involve responsibilities, that is, norms that are *action-guiding*.

Bilgrami argues that an alienated way of life became dominant over the last five hundred years, from the seventeenth century onward. I agree and I believe that our task today is to begin the long project of reengaging or reconnecting with and within the symbiotic living ecosystems on which our social system is an interdependent subsystem.

We have brought forth and spread around the world a social system within which we see ourselves as disengaged from and independent or autonomous of the interdependent social relationships on which we are interdependent, and we also see ourselves and our entire social systems as independent of the ecosystems in which we and our social systems are embedded and interdependent.

From this independence perspective, the natural, social, and policy sciences see themselves as exercising control and mastery over the living Earth.

But the life sciences and Earth sciences over the last sixty years have shown this modernist picture to be a false representation or illusion. Our social system is a subsystem deeply embedded in and interdependent upon the ecosystems that compose the living Earth for every breath we take. This is how Barry Commoner famously made this fundamental point in 1971:

> To survive on earth, human beings require the stable, continuing existence of a suitable environment. Yet the evidence is overwhelming that the way in which we now live on earth is driving its thin, life-supporting skin, and ourselves with it, to destruction. To understand this calamity, we need to begin with a close look at the nature of the environment itself. Most of us find this a difficult thing to do, for there is a kind of ambiguity in our relation to the environment. Biologically, human beings *participate in* the environmental system as subsidiary parts of the whole. Yet, human society is designed *to exploit* the environment as a whole, to produce wealth. The paradoxical role we play in the natural environment—at once participant and exploiter—distorts our perception of it. . . .
>
> [That is,] all of modern technology leads us to believe that we have made our own environment and no longer depend on the one provided by nature. We have become enticed into a nearly fatal illusion: that through our machines we have at last escaped from dependence on the natural environment. . . .
>
> [Yet,] every human activity depends on the integrity and proper functioning of the ecosphere. Without the photosynthetic activity of green plants, there would be no oxygen for our engines, smelters, and furnaces, let alone support for human and animal life. Without the action of plants, animals and microorganisms that live in them, we could not have pure water in our lakes and rivers. Without the biological processes that have gone on in the soil for thousands of years, we would have neither food crops, oil nor coal. If we destroy it, our most advanced technology will become useless and any economic and political system that depends on it will founder. The environmental crisis is a signal of this approaching catastrophe.[5]

If this description is accurate, and if we are to respond to the crisis, we need to free ourselves from the alienated way of life and move around and see and experience ourselves and our social system as embedded and ourselves as engaged participants in (damaged) relationships of interdependency and cosustainability within the ecosystems in which we live and breathe.[6]

We need to think not of independent social systems and ecosystems but of interdependent social and ecosystems: that is, *ecosocial systems*. Rather than independence we need to think about sustainability in terms of mutually interactive and interdependent cosustainability of all life systems, human and nonhuman.

Three more important distinctions between the alienated and unsustainable way of life and the unalienated and sustainable way of life follow.

1. The distinction Anthony Laden makes between ways of conceptualizing reasoning in his workshop presentation and his book on Social Reasoning maps onto the alienated/unsustainable versus unalienated/sustainable distinction of Bilgrami. Laden's conception of "reasoning and acting *with* others" and "connecting" describes the ways humans interact reasonably in mutually sustainable ways.[7]

2. Spatially, the alienated way of life consists of independent entities in (efficient) causal or consensual relationships; temporally, they are in linear or developmental time.

3. Spatially, the unalienated way of life consists of interdependent beings as always already in relationships of mutual support, networks, or webs of life; temporally, they are in cyclical time.

That is, we need to think of ourselves, as Aldo Leopold put it in "The Land Ethic," as "plain members and participatory citizens" of the "commonwealth of all forms of life on earth" with responsibilities not only to sustain ourselves, but always to sustain ourselves in such a way that we also reciprocally cosustain all the other forms of life on which we are interdependent and which cosustain us.[8] This is an unalienated way of thinking about our responsible capabilities as sustainabilities.[9]

And to do this we need to learn from the living Earth how living systems sustain themselves over billions of years and use this knowledge to transform our unsustainable and destructive social systems into sustainable and symbiotic systems within systems.

On this view symbiotic ecosystems are the ground of life and socioecosystems are the ground of human life. That they sustain life on Earth is their value and the ground of all human value.

This is how Fritjof Capra puts the challenge:

The key to an operational definition of ecological sustainability is the realization that we do not need to invent sustainable human communities from scratch but can model them after nature's ecosystems, which are sustainable communities of plants, animals and micro-organisms. Since the outstanding characteristic of the Earth household is its inherent ability to sustain life, a sustainable human community is one designed in such a manner that its ways of life, businesses, economies, federations, physical structures, and technologies do not interfere with nature's inherent ability to sustain life. Sustainable communities and networks evolve their patterns of living over time in continual interaction with other living systems, both human and non-human.[10]

I would like to respond to this challenge in the following way. In section 1 I set out what I take to be four central features of sustainable living systems according to the life and Earth sciences. In section 2 I set out what I take to be the main features of our unsustainable social system that cause damage to the ecosphere on the one hand and give rise to the illusion of independence from it on the other hand. In section 3 I then turn to several ways of responding to the sustainability crisis that are informed by this way of thinking about our interdependent relationship in and with ecosocial systems. These are ways of disengaging from our unsustainable practices, beginning to engage in practices of reengaging and reconnecting socially and ecologically, and thus beginning to bring into being unalienated and sustainable ways of life on Earth. If the symbiotic interdependency thesis of section 1 is true, then participants in these connecting practices should be empowered in reciprocity by the interdependent relationships they connect with, and thus initiate expanding virtuous cycles. This reciprocal empowerment is discussed in section 4.

I. Symbiosis and Interdependency

Let's begin with Commoner's description of how life sustains life:

> There is an important lesson here. In the form in which it first appeared, the earth's system had a fatal fault: the energy it required was derived from the consumption of a non-renewable resource, the geochemical store of organic matter. Had this fault not been remedied, the rapid self-propagated growth of life would have consumed the earth's original organic "soup." Life would have destroyed the condition for its own survival. Survival—a property now so deeply associated with life—became possible because of a timely evolutionary development: the emergence of the first photosynthetic organisms. These new organisms used sunlight to convert carbon dioxide and inorganic materials to fresh organic matter. This crucial event reconverted the first life-form's waste, carbon dioxide, into its food, organic compounds. It closed the loop and transformed what was a fatally linear process into a circular, self-perpetuating one. Since then the perpetuation of life on the earth has been linked to an essentially perpetual source of energy—the sun.

Here in its primitive form we see the grand scheme which has since been the basis of the remarkable continuity of life: the reciprocal interdependence of one life process on another; the mutual, interconnected development of the earth's life system and the nonliving constituents of the environment; the repeated transformation of the materials of life in great cycles, driven by the energy of the sun.

The result of this evolutionary history can be summarised in a series of propositions about the nature of life and its relation to the environment.

Living things, as a whole, emerged from the nonliving skin of the earth. Life is a very powerful form of chemistry, which once on the earth rapidly changed its surface. Every living thing is intimately dependent on its physical and chemical surroundings, so that as these changed, new forms of life suited to new surroundings could emerge. Life begets life, so that once new forms appeared in a favorable environment, they could proliferate and spread until they occupied every suitable environmental niche within physical reach. Every living thing is dependent on many others, either indirectly, through the physical and chemical features of the environment or directly for food or a sheltering place. Within every living thing on earth, indeed within each of its individual cells, is contained another network—on its own scale, as complex as the environmental system—made up of numerous intricate molecules, elaborately interconnected by chemical reactions, on which the life-properties of the whole organism depend.[11]

Of course, there are important differences between human social systems and ecosystems. However, there are similarities as well. We are, after all, earthlings in the first instance. Following the earlier quotation from Capra, there are three kinds of education we can gain from the study of nonhuman life systems: (1) how they manage to sustain and complexify life over 3.5 billion years; (2) how humans are interdependent participants in ecosystems, and thus how we must design our social systems so that they support rather than undermine them; and (3) how to design or transform social systems so they are similarly self-sustaining. We need to acquire this education to engage in the practices of cultural and ecological reconciliation. Over the last forty years the ecological, life, and Earth sciences have advanced three hypotheses concerning the ways life sustains life that are essential for these reconciliation practices.

Interdependency and Symbiosis: Life Sustains Life

First, sustained life is a property of ecological systems rather than of single organisms or species. No individual organism can exist in isolation. Animals depend on the photosynthesis of plants for their energy needs; plants depend on the carbon dioxide produced by animals, as well as on the nitrogen fixed by the bacteria at their roots; and these living entities are interdependently coupled with abiotic rocks, atmosphere, and waters.

The major factor in the coevolution of forms of life is "symbiosis": the ways that organisms and ecosystems "live together." This refers to the way relations of interdependency within and among organisms and ecosystems are *mutually supportive* of the members, for example, the ways that animals depend on photosynthesis of plants, plants depend on carbon dioxide produced by the animals, and so on, in a circular manner.

Here is an example from Michael Simpson, a permaculture expert:

> Living systems do not only reproduce themselves. Their very life processes nourish their habitat and strengthen the conditions of life around them. They thereby create an organism that is larger than themselves or their individual species. When a forest is growing back from a disturbance, herbaceous (non-woody) plants are the first to move in. These plants exude sugars that attract bacteria around their roots. The bacteria in turn exude an alkaline "bioslime" that creates a favorable habitat for themselves as well as for the pioneer plant species. The alkaline condition of the bioslime also allows the bacteria to break down ammonia in the soil into nitrates that are taken up by plants allowing them to grow vegetatively. This cycle of life creating the conditions for more life continues as the forest gradually grows into a rich, biodiverse ecosystem (ecological succession).
>
> In short, living systems are not only self-regulating but they are relational in so far as they build the conditions of life around them. Hence, the organism is not something independent in its own right which then adapts itself to its environment; on the contrary the organism adapts a particular environment into it.[12]

The central point is this: in the activities of sustaining themselves, living systems also cosustain the conditions of life around them, and vice versa. This symbiotic relationship of reciprocal sustainability among interdependent living systems is how life sustains life.

It is called a "virtuous" cycle because it reproduces itself, becoming more complex. Moreover, it gives rise to symbiogenesis: the emergence of new systems of life out of the background symbiotic relationships and their complex relationships, creating the immense biodiversity of forms of life over billions of years.[13]

Thinking about living systems in this symbiotic way brings to awareness Arne Naess's "ecological self," in contrast to the independent, ego-self of the dominant way of life.[14] We realize that if we wish to live well we should live in such a way that our way of life supports the ways of life of those with whom we are related and that they should do the same in reciprocity.

That is, Gaia citizens gradually come to identify with the interdependent members of the commonwealth of all forms of life. They see themselves as citizens of this commonwealth with reciprocal responsibilities of mutual recognition and sustenance as the condition of sustaining life on Earth. They also realize that if they are suffering, it is probably because they are not living in ways that support such mutually supportive networks. This way of life is neither altruistic nor egoistic, for that debilitating distinction rests on the presupposition that organisms are independent and self-sufficient to begin with.

Despite the individualistic and competitive relationships of the dominant economy, we humans are participants in multiple social systems of this symbiotic or gift-reciprocity-gift kind. Relationships within families, neighborhoods, communities of practice with fellow workers, and an array of social networks are often symbiotic. Many psychologists argue that symbiotic relationships are the bedrock of communities and mutual well-being, unnoticed by the dominant competitive ethos, yet necessary to hold societies together.[15]

Symbiogenesis and Gaia Hypothesis

Second, the evolution of all the complex symbiotic feedback loops among life systems, atmosphere, rocks, and water has symbiogenetically given rise to the biosphere or ecosphere as a whole, that is, the evolving, self-regulating ensemble that has maintained habitable conditions on the surface of the planet over vast stretches of geological time. Earth-systems scientist James Lovelock named this the Gaia Hypothesis in the 1960s.[16] As Lynn Margulis commented, it is symbiosis and symbiogenesis from a planetary perspective.[17] The Gaia Hypothesis has since sustained several tests and is now accepted, in one form or another, by a wide range of scientists and members of the IGCC.

"Gaia" is the Greek term for the Earth as a living or animate Earth. The idea that the Earth is alive (animate) is a view shared by Indigenous cultures and many religious traditions, as well as by the Greeks and early modern Europeans. For the last four hundred years the dominant Western view was that the Earth is inanimate, a mechanism of some kind, and that the animate view was nothing but a premodern primitive superstition. With the Gaia Hypothesis, there is now a growing convergence and conversation in theory and practice between leading trends in Western sciences and the complementary fields of Indigenous sciences. For example, Stephen Harding writes: "The key insight of the Gaia theory is wonderfully holistic and non-hierarchical. It suggests that it is the Gaian system as a whole that does the regulating, that the sum of all the complex feedbacks between life, atmosphere, rocks and water gives rise to Gaia, the evolving, self-regulating planetary entity that has maintained habitable conditions on the surface of the planet over vast stretches of geological time."[18] And Lynn Margulis: "Gaia is not an 'organism' but an emergent property of interaction among organisms. Gaia is the series of interacting ecosystems that compose a single huge ecosystem at the earth's surface. Gaia is symbiosis on a planetary scale."[19] *Homo sapiens* does not control these systems, but is rather an active participant within and with them, as both subject and agent.[20]

Dynamic Resilience, Tipping Points, and Responses

Third, complex, overlapping, nonlinear systems of symbiotic and symbiogenetic relationships are neither stable nor harmonious. Their interactions are cooperative and competitive, and often far from equilibrium. They change and transform continuously as they interact and readjust to the interactions of their neighbors: that is, living systems are dynamic systems that are resilient—they respond to disturbances in complex ways that systems theorists call positive and negative feedback loops.

In negative feedback, the initial change is counteracted. In positive feedback the initial change is amplified. Feedback loops are the "circles of participation" by which living systems bring about constancy or change.[21]

Now, it is *always* possible that a positive feedback loop that causes the system to move away from dynamic equilibrium may begin a series of loops that move the system further away until it reaches a tipping point and transforms into an unsustainable system: that is, the system transforms from its conciliatory feedback loops of sustainable interaction (virtuous cycles) into unsustainable relationships by a series of positive feedback loops (vicious cycles) and eventual systemic collapse.

Systems theorists explain the last four centuries of developmental modernization as the rise of a viciously cyclical global system—the Anthropocene—that is reaching, or has reached, a catastrophic tipping point in our time. They describe the overall crisis in two steps and the appropriate two steps of dealienation and reconnection as follows:

1. The human enterprise is structurally and functionally inseparable from nature: that is, the human enterprise is a fully embedded, totally dependent subsystem of the ecosphere—people live within socioecosystems. Human activities can therefore significantly affect the integrity and behavior of supportive ecosystems and these changes immediately feed back to affect the state of the human subsystem. We can no longer understand the dynamics of either the natural system or the human subsystem in isolation without understanding the dynamics of the other component.

2.   Linked/integrated/interdependent and interactive socioecosystems are constantly changing in response to both internal and external forces—they are dynamic complex adaptive systems. The changes within these systems are not linear, smooth, or predictable, particularly outside the systems' "normal" regime. Indeed, under sufficient pressure, critical systems variables may "flip" (cross a threshold or tipping point) into a different regime or alternative stable state. In other words, like natural ecosystems, socioecosystems also have multiple possible equilibria, some of which may not be amenable to continued human use or existence (remember the collapse of the North Atlantic cod fishery).[22]

3.   The sustainability of the human enterprise on a crowded and resource-stressed planet depends on our ability to conserve the resilience of socioecological systems. In this context, resilience defines the capacity of the system to assimilate disturbances without crossing a threshold into an alternative and possibly "less-friendly" stable state. A desirable socioecological system characterized by high resilience is able to resist external disturbance and continue to provide biophysical goods and services essential for a satisfactory quality of life.

4.   For sustainability, resource-management efforts must shift from reshaping nature for the purpose of satisfying human demands to moderating human demands so that they fit within biophysical limits. They must do this in a way that is consistent with both the productive and the assimilative capacities of ecosystems, and in a way that enhances the long-term resilience of the integrated socioecosystem.[23]

II. The Hegemonic Unsustainable and Vicious Social System

It is my working hypothesis that the current destructive and unsustainable relationship to the living Earth that is causing climate change and the ecological crisis developed in Europe and spread around the globe by European imperial expansion and modernization. It is the driving force of the Anthropocene Age. It is a vicious and unsustainable system.

It is this social system that brings about and reproduces the alienated way of life Akeel Bilgrami describes and sustains the illusionary worldview of independence. This "illusion" of independence and linear temporality is grounded in the basic structure of the system, which constantly works to "disembed" humans from the ecosphere.

One way of *seeing* this unsustainable relationship is to reflect on a saying Indigenous people across Canada use to distance themselves from it. Since the 1960s Indigenous people have been saying the following: "the land does not belong to us; we belong to the land." If our mode of being in the world is one of being engaged participants in living interdependent and symbiotic ecological and sociological relationships that mutually sustain life, then the true expression of this mode of being is "belonging to the land": that is, as we have seen, we belong to this great commonwealth of earthlings and have shared responsibilities to sustain it and our fellow members.

From this orientation, it would be absurd to say that we stand in a *relationship of ownership or property* to the living Earth. Ojibway Elder Basil Johnston says this is like having property in your mother: mother earth. Yet, this relationship of the land belonging to humans has displaced the earlier relationship of belonging to the land and has become the basis of our modern way of life and the cause of the crises.[24]

For 95 percent of its time on Earth *Homo sapiens* lived in predominantly sustainable social systems and understood itself to belong to the ecosystems in which it dwelled. Although things began to change with the sedentary agricultural revolution eleven thousand years ago, the "great transformation" of the world as a whole to the orientation that the Earth belongs to humans occurred over the last four hundred years and accelerated exponentially in the twentieth century. There are many accounts of this, but among the best was advanced by Karl Polanyi in *The Great Transformation* (1944) and those who have learned from his analysis.[25]

The Great Disembedding and Reembedding in the Global Social System

Polanyi argued that during this great transformation humans have been disembedded from participation in the interdependent ecological and social relationships that sustain life and reembedded in abstract and competitive economic, political, and legal relationships that are dependent on, yet destructive of, the underlying interdependent ecological and social relationships. This "great disembedding" takes place in the following three steps.

First, the peoples who are embedded in symbiotic ecological and social relationships are dispossessed of this way of life and the territories in which it is carried on, first by the enclosure of the commons in England and then by the dispossession of Indigenous peoples throughout the non-European world (the "second enclosure").

The second step is to impose an ownership relation to the land by the spread of Western legal systems of private property and so to transform "earth into property," as Anthony Hall puts it.[26] Polanyi describes the privatization of land as a "fictitious commodity" because land is not a commodity produced for sale on the market. What we now call commodifiable and exchangeable "natural resources" are, as we have seen, interdependent coparticipants in the symbiotic webs and cycles of life that sustain life on Earth.

Relating to the living Earth as a storehouse of commodifiable resources disembeds them from these interdependent ecological relationships and reembeds them in the abstract and competitive relations of the global market system. The ecosystems in which resources are embedded are then treated as "external" to the global system of commodification and radically changed. The result of "development" under this system is the destruction of the webs of interdependent ecological relationships that sustain the natural and human world, giving rise to the environmental crisis and climate change.

Once the means of the reproduction of human life are placed under the ownership of independent corporations, the third step is to treat the productive capabilities of human beings as commodities for sale on the labor market by the spread of Western contract, labor, and corporate law. This kind of commodification disembeds human producing and consuming capabilities and activities from the surrounding social and ecological relationships in which they take place and reembeds them in abstract, competitive, and nondemocratic global market relationships. Polanyi describes the commodification of the productive capabilities of individual humans as the second "fictitious commodity" of modernization.

It is fictitious because abilities to work together and sustain ourselves are not commodities made for the market. These capabilities are, as we have seen in the previous section, the cooperative response-abilities and sustain-abilities through which we humans participate in the social and ecological systems that conciliate and sustain life on Earth. They are the capabilities through which we "belong to the land" and are grounded in it. Yet, they are now treated as abstract capabilities that we as separate individuals "own" (self-ownership); and, by selling the use of these abilities to a corporation, they become the means by which we insert ourselves in the global market system. The underlying social systems that producers and consumers live in and that sustain them—such as families, communities, First Nations, networks, and so on—are treated as "external" to the market system. The result of "development" under this system is the destruction of the webs of interdependent social relations of mutual aid that sustain human communities, giving rise to the well-known forms of social suffering of modern life: alienation and anomie, the horrendous inequalities in life chances, and the planet of slums and gated "communities" in which we find ourselves.

In 1944 Polanyi predicted that the result of this "great transformation" would be disastrous:

> To allow the market mechanism to be sole director of the fate of human beings and their natural environment . . . would result in the demolition of society. Robbed of the protective covering of cultural institutions, human beings would perish from the effects of social exposure; they would die as the victims

of acute social dislocation through vice, perversion, crime and starvation. Nature would be reduced to its elements, neighbourhoods and landscapes defiled, rivers polluted, military safety jeopardized, the power to produce food and raw materials destroyed.[27]

The Global Vicious Cycle

Despite Polanyi's warning and hundreds of others, this global system of double commodification in which the human species is reembedded continues to unfold as he predicted. It is now a deeply entrenched *vicious cycle*, what global systems theorists call an "automaton."[28]

Briefly, they argue that corporations are caught up in a competitive system in which they must continuously extract and exploit natural resources and human resources (capabilities) at the lowest price and at maximum speed in order to make a profit or go under. Any damage to the environment and communities is treated as external and off-loaded to governments. Governments and communities are constrained to compete for these corporations, because they fund their campaigns, bring jobs to the electorate, and provide the taxes that enable governments to provide basic services and repair the damage they do to social and ecological systems. If governments try to internalize the externalities and regulate and charge the corporations, corporations move to more compliant countries. Governments are also constrained to give military and economic support to companies that extract resources in foreign countries, giving rise to the huge global military network and the wars over scarce resources that follow. These degrade further formerly self-sustaining human communities and ecosystems. As nonrenewable resources become scarce, it becomes more expensive and destructive to extract and exploit them. And it becomes more difficult to regulate the "race for what's left" that we see today in the North, in fracking, in the Northern Gateway Pipeline, and other examples. Even "renewable" resources, such as fisheries, are exploited to extinction, because the temporality of market competition is faster than the natural cycles that *renew* fish, forests, and oceans.

Finally, in contrast to sustainable cycles, in which the "waste" of one organism is used up by another so there is zero waste or emissions overall, the global market system produces commodities in a liner, noncircular way, so they rapidly become waste, and require new commodities to replace them, as in automobiles or terminator seeds. And new commodities are required to repair the damage they cause during their short life span. It is a cradle-to-grave system of production rather than a cradle-to-cradle system.

If the real costs of this global system were taken into account it would collapse under its own economic irrationality. As Lester Brown points out:

As the world economy expanded some 20-fold over the last century it has revealed a flaw—flaw so serious that if it is not corrected it will spell the end of civilization as we know it. The market, which sets prices, is not telling the truth. It is omitting indirect costs that in some cases now dwarf direct costs. Consider gasoline. Pumping oil, refining it into gasoline, and delivering the gas to US service stations may cost, say, $3 per gallon. The indirect costs, including climate change, treatment of respiratory illnesses, oil spills, and the US military presence in the Middle East to ensure access to oil, total $12 per gallon. Modern economic thinking and policymaking have created an economy that is so out of sync with the ecosystem on which it depends that it is approaching collapse.[29]

As a result, this is how the dominant system looks from the resiliency perspective of section 1: Contemporary resource-management approaches typically attempt to maximize one or a few desirable systems components at the expense of other species and systems functions—think agricultural or forestry monoculture. Diversity plummets and functions are lost. The managed system becomes inflexibly brittle and vulnerable to unexpected external shocks.

Now, consider the form of contemporary global economic development. The emphasis here is on maximizing economic growth by exploiting the efficiency gains conferred by local specialization and global trade (comparative advantage). The approach tends to maximize resource exploitation and material dissipation (pollution), both of which simplify ecosystems, undermine life-support functions, and erode systems resilience. The global economy becomes dominated by a few global enterprises (and their numbers continue to shrink with each merger or acquisition). The sheer economic power of these monster corporations stifles meaningful competition and blocks new players from entering the market: both local diversity and global diversity plummet. Meanwhile, the economy and society have become dependent on a few declining energy sources (petroleum) and on energy-intensive systems (e.g., the global transportation systems and even the Internet).[30]

Moreover, when we participate in this vicious system in our business-as-usual activities we *overlook* the destruction it is causing to the symbiotic social and ecological relationships on which humanity and other forms of life depend. When we can no longer overlook, deny, and discount the problem and must confront the underlying relationship that causes it, the response is to try to regulate the system: to repair damage and compensate stakeholders. Most of these responses take place within and reproduce the vicious system because we continue to see the problem through the lens of the artificial system in which we are reembedded. And it creates the illusion of independence, as Commoner's quotation earlier mentions.

There are three ways in which the crisis is denied or discounted: (1) the boiled frog syndrome—the brain functions so that slow changes, long-term implications, and multiple connections cannot be easily seen; (2) mental apartheid—the psychological barrier between modern humans and the rest of reality, perceptual dualism since Descartes; (3) the idea of the tragedy of the ungoverned commons, from Hobbes's state of nature to Garret Hardin's analysis, makes it appear that the only alternative is privatization (and this becomes our present tragedy). So, there appears to be no other possibility.[31]

I think all three of these strategies can be seen as consequences of the alienated form of life.

However, the really basic one is that our alienated way of life causes us to overlook the living Earth and to represent it as resources for production and consumption, whether capitalist or Marxist, and this because of the sense of ourselves as independent and the sense of the only alternative to it as being a kind of servility, dependency, or heteronomy. The third possibility of being agents within and with the living Earth disappears from the picture.

III. Responses of Reengagement and Reconciliation: Realizing Interbeing

Aldo Leopold was one of the first people to point out the way in which our modern mode of life diverts our attention from its destructive effects and to suggest an appropriate response:

> One of the penalties of an ecological education is that one lives alone in a world of wounds. Much of the damage inflicted on land is quite invisible to laymen. An ecologist must either harden his shell and make believe that the consequences of science are none of his business, or he must be the doctor who sees the marks of death in a community that believes itself well and does not want to be told otherwise.[32]
>
> All ethics so far evolved rest upon a single premise: that the individual is a member of a community of interdependent parts. His instincts prompt him to compete for his place in the community, but his ethics prompts him also to cooperate (perhaps in order that there may be a place to compete for).
>
> The land ethic simply enlarges the boundaries of the community to include soils, waters, plants, and animals, or, collectively: the land.
>
> In short, a land ethic changes the role of Homo sapiens from conqueror of the land-community to plain member and citizen of it. It implies respect for his fellow-members, and also respect for the community as such.
>
> In human history, we have learned (I hope) that the conqueror role is eventually self-defeating. Why? Because it is implicit in such a role that the

conqueror knows, ex cathedra, just what makes the community clock tick, and just what and who is valuable, and what and who is worthless, in community life. It always turns out that he knows neither, and this is why his conquests eventually defeat themselves.[33]

An ethic . . . presupposes . . . some mental image of land as a biotic mechanism. We can be ethical only in relation to something we can see, feel, understand, love, or otherwise have faith in.[34]

Practices of Dealienation, Reconnection, and Reengagement

Now, let's turn to the practices of freeing ourselves from captivity to the alienated and unsustainable way of life, of beginning to reconnect and engage with the living Earth in regenerative rather than extractive-dominative ways, and thereby of beginning to see our place in the world differently and being moved by the animacy of Gaia, as Leopold recommends. There are countless ways in which this transformation can take place in everyday life. Here are several examples.[35]

The most obvious activities of reconnection are those engaged in by citizens and elected officials in the representative institutions and courts. These are indispensable yet insufficient. As we have seen, the social system as a whole places severe limits on what can be done for all the reasons discussed in the previous section. When Gaia citizens run up against these limits they turn to more direct forms of transforming unsustainable and destructive social systems.

In response to five hundred years of dispossession we are witnessing in our time a reclaiming of the local commons as a worldwide phenomenon. From the renaissance of over 350 million Indigenous people, to global movements such as the Food Sovereignty movement, and on to the return to local food production and distribution in the global North, people are not only repossessing privatized commons. They are also reasserting local knowledges of how to use and sustain the living Earth, knowledges that were swept aside as primitive and replaced by modern scientific agribusiness and synthetic fertilizers, yet are grounded in thousands of years of practical knowledge of the renewability conditions of local resources. In Boaventura de Sousa Santos's famous characterization of this renaissance of local practices and local knowledges, there is no global justice without local epistemic justice.[36]

In response to the commodification of labor power, democratic, cooperative citizens refuse to comply with this undemocratic mode of production and consumption. As much as possible, they reappropriate their producing and consuming capabilities from commodification and exercise their capabilities "in common," in democratically run cooperatives and community-based organizations that are reembedded in social and ecological relationships. Such grassroots democracies then produce and distribute the basic public goods that are privatized under the dominant form of democracy: food, shelter, clothing, health care, clean water, security, and so on. These social and economic democracies are linked together by global networks of fair trade relationships that are also under the democratic control of the producers and consumers subject to them.

This famous response to the injustices of the privatization of labor power has given rise to the tradition of cooperative democracy throughout the world. From Robert Owen, William Thompson, and Peter Kropotkin in Europe, to Gandhi, Richard Gregg, Fritz Schumacher, and the Swaraj and Swadeshi movements across Asia and Africa, to food sovereignty in Latin America, there is a turn to local food production, microcredit, democratic cooperatives, and Indigenous and non-Indigenous community-based organizations of diverse scales and types. These are then linked together by global networks of fair trade and self-reliance. These cooperative practices generate social capital and realize social and economic justice directly, by bringing the local and global organization of economic activities under the democratic cooperation and mutual aid of all subject to and affected by them.

This is the cooperative citizenship response to the global problem of economic inequality and exploitation. It is important to note that despite the global spread of the institutional module of modern representative government and civil citizenship, poverty and hunger persist on this "planet of slums": "963 million people go to bed hungry every night. One billion people live in slums. One woman dies every minute in childbirth. 2.5 billion people have no access to adequate sanitation services and 20,000 children a day die as a result."[37]

Cooperative citizens offer a response to this *glocal* injustice that is more immediate and perhaps more lasting than representative responses because the victims of hunger, starvation, and poverty become the agents of grassroots democracy and economic self-reliance.

Another response is ecological or Gaia citizenship. In response to the third process of the commodification of the living Earth, civic citizens withdraw their capacities from activities based on the commodification of the environment and develop a responsible way of relating in and to it. They reembed natural resources and the human uses of them into their place within ecological relationships and cycles of no waste and regeneration. They see the webs of ecological relationships as a living commonwealth of all forms of life. They derive the fundamental duties, responsibility, and rights of democracy in the first instance from their membership in the webs of ecological relationships in which democracy takes place and on which all forms of life depend. This natural gift economy, as Vandana Shiva puts it, is for them the true

mother of democracy.[38] The norms of ecological well-being govern economics, not the other way round.

This revolutionary response to the injustice of privatization of the natural
world has given rise to the great cooperative and community-based ecology
movements. From Aldo Leopold, Rachel Carson, and Vandana Shiva to the
Chipko Movement in India and Asia, and on to Japanese fishing cooperatives,
the water justice movement, Food Sovereignty, and everyday ecological footprint initiatives, millions of Gaia citizens are reclaiming the commons and
exercising their capabilities democratically in ethical relationships of stewardship in the commonwealth of all forms of life. These experiments in ecodemo-
cracy and cyclical economics are responses and alternatives to the idea of
unlimited linear development that gives rise to the ecological crisis.

Finally, two of the foundational premises that justify the imposition and
continuation of the unsustainable processes of modernization are that humans
are naturally antagonistic, and thus that they need an authoritarian master
who coercively imposes a structure of law over them as a socializing pre-
condition of peace and democracy. These are foundational presuppositions of
modernization in theory and practice since Hobbes. As we can see from the
examples I have just given, Gaia citizens reject these premises.

From Kropotkin, through Gandhi and Richard Gregg, to millions today,
these citizens argue in contrast that humans are self-organizing and self-
governing animals. Autopoiesis—self-organization, cooperation, and nonviolent contestation and dispute resolution—is a more basic condition of human
evolution than antagonism, violent conflict, and the hegemonic relationships
of command and obedience that violence aims to establish, yet it tends to generate counterviolence. If this were not the case, if Kantian antagonism and
Hobbesian war of all against all were primary, the human species would have
perished long ago. We overlook this pacific feature of our everyday activities
precisely because it is so commonplace and familiar.[39]

Humans are not unique in this respect. The hypothesis holds for all forms
of life and for the ecological relationships in which they all live. This was put
on scientific footing in the 1960s by James Lovelock in the Gaia Hypothesis. It
is widely endorsed by biological, ecological, and climate scientists today.[40] This
view that the ground of our being as earthlings is ecological and sociological
relationships of mutual interdependence and support is also widely endorsed
by many of the spiritual traditions of the world. This helps to explain the powerful attraction of cooperative citizenship to people from such different secular and spiritual traditions.

It follows from this scientific revolution in the way we think about our place
and roles in the ecosphere that there are two gift-reciprocity or mutually cosustaining socioecosystems. The first is the natural ecosystem that sustains all
forms of life. The second is that there are informal, and often unnoticed, social

networks of mutual sustenance that sustain the day-to-day lives of members in any community of practice. These informal networks of mutual aid underlie the dominant, formal, and unsustainable social systems. The dominant systems are parasitic on these relationships of "social capital," as Polanyi argued, in the sense that they need these symbiotic systems to sustain the members that they simultaneously separate, exploit, place in competition, and disemploy. This is a complex dynamic because the more concerned citizens repair the social and ecological networks that sustain life, the more attractive they become as resources for recommodification.[41]

Another important local and global initiative is the turn to zero waste, and cyclical, nonviolent technology, architecture, and urban planning. These movements find their inspiration in Gandhi's constructive programs and Fritz Schumacher's cyclical economics and appropriate technology. More recently, McDonough and Braungart explain their "cradle-to-cradle" version in the following way:

> Nature operates according to a system of nutrients and metabolisms in which there is no such thing as waste. A cherry tree makes many blossoms and fruit to (perhaps) germinate and grow. This is why the tree blooms. They fall to the ground, decompose, feed various organisms and microorganisms, and enrich the soil. Around the world, animals and humans exhale carbon dioxide, which plants take in and use for their own growth. Nitrogen from wastes is transformed into protein by microorganisms, animals, and plants. Horses eat grass and produce dung, which provides both nest and nourishment for the larvae of flies. The earth's major nutrients—carbon, hydrogen, oxygen, nitrogen—are cycled and recycled. Waste equals food.
>
> This cyclical, cradle to cradle biological system has nourished a planet of thriving, diverse abundance for millions of years. Until very recently in the earth's history, it was the only system, and every living thing on the earth belonged to it. Growth was good. It meant more trees, more species, greater diversity, and more complex, resilient ecosystems. Then came industry, which altered the natural equilibrium of materials on the planet. Humans took substances from the earth's crust and concentrated, altered, and synthesized them into vast quantities of materials that cannot be safely returned to soil. Now material flows can be divided into two categories: biological mass and technical—that is, industrial—mass.
>
> From our perspective, these two kinds of material flows on the planet are just biological and technical nutrients. Biological nutrients are useful to the biosphere, while technical nutrients are useful to what we call the technosphere, the systems of industrial processes. Yet somehow we have evolved an industrial infrastructure that ignores the existence of nutrients of either kind.

Humans are the only species that takes from the soil vast quantities of nutrients needed for biological processes but rarely puts them back in usable forms.[42]

Another important example is the way green legal theory and practice are presenting alternatives to the modern legal commodification of land and labor under national and international law. The role of green legal theory is to develop a way of thinking about law that cognizes the cyclical and symbiotic interdependency and cosustainability of "natural resources" in ecosystems, and "human resources," technology, and "communities of practice" (cooperatives and corporations) in socioecosystems.[43]

Finally, one of the most important movements over the last few decades is the attempt to bring together and integrate the various fields of knowledge, research, and teaching that I have drawn on in describing our predicament and responses to it.[44]

The challenge today is thus not to dream up responses to climate change or to seize power. Rather, it is to find ways to democratically coordinate the multiplicity of such practices of dealienation and reconnection around the world so they grow and gradually hollow out, displace, and transform the unsustainable systems. The tragedy of the present is that we who inhabit the dominant social systems and view the world through their theories of transformative change by means of reform or violent revolution do not see this alternative, nonviolent way of systemic transformative change from below, yet it is the way damaged life systems repair and regenerate themselves.[45] If it took five hundred years for the dominant system to reach "peak everything," perhaps we should think of this new "great transformation" as also taking centuries of sustained and sustaining practices of regeneration.

IV. Reanimation

Perhaps one of the most important practices of sustainability is the everyday individual and cooperative practices of engaging with the living Earth through the primacy of dialogical perception. This consists in disengaging our perception/cognition (senses and synthesia) from the built environment and reconnecting with the animate Earth sensuously, as our nervous system does naturally. This is how David Abram, following Maurice Merleau-Ponty, describes it:

The event of perception, experientially considered, is an inherently interactive, participatory event, a reciprocal interplay between the perceiver and the perceived.

Perceived things are encountered by the perceiving body as animate, living powers that actively draw us into relation. Our spontaneous, pre-conceptual experience yields no evidence for a dualistic division between animate and inanimate phenomena, only for relative distinctions between diverse forms of animateness.

The perceptual reciprocity between our sensing bodies and the animate, expressive landscape both engenders and supports our more conscious, linguistic reciprocity with others. The complex interchange we call "language" is rooted in the non-verbal exchange always already going on between our own flesh and the flesh of the world.[46]

Reengaging with the natural world in this way reconnects us with our interdependent "ecological self" and overcomes the separation of culture and nature in our dominant identity formation. Interacting in this way "brings forth the world" cognitively as interdependent and coevolving and, in so doing, brings us into connection with the suffering of the world. It enables us to see our responsibility for and interdependency on the living world. Yet, even more importantly, it connects us with the animacy of life itself (*anima mundi*) that empowers us.[47] As Joanna Macy and Chris Johnstone argue, the gratitude we experience for the invaluable gifts of the life-sustaining goods and services that the living Earth gives us every second motivates and moves us to reciprocate.[48] It is animacy of life itself that empowers us to continue to act in reciprocally sustaining ways.[49]

Notes

This chapter follows from the first chapter on this theme at the 2013 NOMIS Workshop: *Life Sustains Life 1*. I would like to thank the participants in both the 2013 and 2014 workshops for their helpful comments on my presentations, especially the late Jonathan Schell, Akeel Bilgrami, Charles Taylor, Anthony Laden, David Kahane, and Pablo Ouziel. I would also like to thank Michael Carpenter for assistance with editing both chapters. I have discussed the topics in this chapter further in James Tully, "Reconciliation Here on Earth," in *Resurgence and Reconciliation: Indigenous-Settler Relations and Earth Teachings*, ed. Michael Asch, John Borrows, and James Tully (Toronto: University of Toronto Press, 2018), 83–132.

1. Rachel Carson, *Silent Spring* (Boston: Houghton Mifflin, 1962); Barry Commoner, *The Closing Circle: Nature, Man, and Technology* (New York: Knopf, 1971); Donella H. Meadows, Dennis L. Meadows, Jørgen Randers, and William W. Behrens III, *The Limits to Growth: A Report for the Club of Rome's Project on the Predicament of Mankind* (New York: Universe, 1972).

2. Ugo Bardi, *The Limits to Growth Revisited*, SpringerBriefs in Energy (New York: Springer, 2012); *Climate Change 2014: Synthesis Report. Contribution of Working Groups I, II and III to the Fifth Assessment Report of the Intergovernmental Panel on*

Climate Change (Geneva: Intergovernmental Panel on Climate Change, 2015); Lester R. Brown, *World on the Edge: How to Prevent Environmental and Economic Collapse* (New York: Norton, 2011); Craig Dilworth, *Too Smart for Our Own Good: The Ecological Predicament of Humankind* (Cambridge: Cambridge University Press, 2009).

3. For example: Stephen Emmott, *Ten Billion* (New York: Vintage, 2013); Kenneth S. Deffeyes, *Beyond Oil: The View from Hubbert's Peak* (New York: Hill and Wang, 2005); Paul Crutzen and Hans Günter Brauch, eds., *Paul J. Crutzen: A Pioneer on Atmospheric Chemistry and Climate Change in the Anthropocene* (New York: Springer, 2016). For a history of sustainability, see Jeremy L. Caradonna, *Sustainability: A History* (New York: Oxford University Press, 2014).

4. Akeel Bilgrami, *Secularism, Identity, and Enchantment* (Cambridge, MA: Harvard University Press, 2014), 101–116; and see Tully, *Life Sustains Life 1*.

5. Commoner, *The Closing Circle*, 14–17.

6. David C. Korten, *The Great Turning: From Empire to Earth Community* (Oakland, CA: Berrett-Koehler, 2007).

7. Anthony Simon Laden, *Reasoning: A Social Picture* (Oxford: Oxford University Press, 2012).

8. Aldo Leopold, "Land Ethic," in *The Sand County Almanac and Other Sketches Here and There* (Oxford: Oxford University Press, 1949).

9. Ellen LaConte, *Life Rules: Nature's Blueprint for Surviving Economic and Environmental Collapse* (Gabriola Island: New Society, 2012).

10. Fritjof Capra, *The Hidden Connections: A Science for Sustainable Living* (New York: Anchor, 2002), 230.

11. Commoner, *The Closing Circle*, chap. 2.

12. Michael Simpson, PhD student, University of British Columbia, personal correspondence with the author.

13. Lynne Margulis, *Symbiotic Planet: A New Look at Evolution* (New York: Basic, 1998).

14. Arne Naess, *The Ecology of Wisdom: Writings by Arne Naess*, ed. Alan Drengson and Bill Devall (Berkeley: Counterpoint, 2008).

15. See further section 3.

16. James Lovelock, *Gaia: A New Look at Life on Earth*, 3rd ed. (1979; Oxford: Oxford University Press, 2000).

17. Margulis, *Symbiotic Planet*.

18. Stephen Harding, *Animate Earth: Science, Intuition and Gaia* (White River Junction: Chelsea Green, 2013).

19. Margulis, *Symbiotic Planet*, 119.

20. Harding, *Animate Earth*, 62–85.

21. Harding, 62–85.

22. William E. Rees, "Thinking 'Resilience,'" in *The Post Carbon Reader: Managing the 21st Century's Sustainability Crises*, ed. Richard Heinberg and Daniel Lerch (Healdsburg: Watershed Media, 2010), 32.

23. Rees, 32.

24. Robin Wall Kimmerer, *Braiding Sweetgrass: Indigenous Wisdom, Scientific Knowledge, and the Teaching of Plants* (Minneapolis: Milkweed, 2013).

25. Karl Polanyi, *The Great Transformation: The Political and Economic Origins of Our Time* (1944; Boston: Beacon, 2004).

26. Anthony J. Hall, *Earth Into Property: Colonization, Decolonization, and Capitalism* (Montreal: McGill-Queen's University Press, 2010).

27. Polanyi, *The Great Transformation*, 197.

28. Capra, *The Hidden Connections*, 129–157.

29. Brown, *World on the Edge*, 8.

30. Rees, "Thinking 'Resilience.'"

31. Mathis Wackernagel and William Rees, *Our Ecological Footprint: Reducing Human Impact on the Earth* (Gabriola: New Society, 1996).

32. Leopold, "Land Ethic," 197.

33. Leopold, 239–240.

34. Leopold, 251.

35. I have discussed these in more detail in James Tully, *On Global Citizenship: James Tully in Dialogue* (London: Bloomsbury, 2014); and Tully, *Imperialism and Civic Freedom* (Cambridge: Cambridge University Press, 2008).

36. Boaventura de Sousa Santos, *Epistemologies of the South: Justice Against Epistemicide* (London: Taylor and Francis, 2015).

37. Irene Kahn, *The Unheard Truth: Poverty and Human Rights* (New York: Norton, 2010).

38. Vandana Shiva, *Earth Democracy: Justice, Sustainability, and Peace* (Berkeley: North Atlantic, 2015).

39. Richard Gregg, *The Power of Nonviolence* (New York: Schocken, 1966); Mark Engler and Paul Engler, *This Is an Uprising: How Nonviolent Revolt Is Shaping the Twenty-First Century* (New York: Nation, 2016).

40. Fritz Capra and Pier Luigi Luisi, *The Systems View of Life: A Unifying Vision* (Cambridge: Cambridge University Press, 2015).

41. Michael Klare, *The Race for What's Left: The Global Scramble for the World's Last Resources* (New York: Picador, 2012); Dilworth, *Too Smart for Our Own Good*.

42. Fritz Schumacher, *Small Is Beautiful: Economics as If People Mattered* (New York: Harper, 2010); William McDonough and Michael Braungart, *Cradle to Cradle: Remaking the Way We Make Things* (New York: North Point, 2010), 92–93, 96.

43. Cormac Cullinan, *Wild Law: A Manifesto for Earth Justice*, 2nd ed. (Cape Town: Green, 2011).

44. See, for example, Capra and Luisi, *The Systems View of Life*; Sean Esbjorn-Hargens and Michael E. Zimmerman, eds., *Integral Ecology: Uniting Multiple Perspectives on the Natural World* (Boston: Integral, 2009).

45. See the quotation from Michael Simpson at note 12.

46. David Abram, *The Spell of the Sensuous: Perception and Language in a More-Than-Human World* (New York: Vintage, 1997), 89–90.

47. Harding, *Animate Earth*.

48. Joanna Macy and Chris Johnstone, *Active Hope: How to Face the Mess We're in Without Going Crazy* (Novato: New World Library, 2012).

49. Kimmerer, *Braiding Sweetgrass*.

CHAPTER 14

THE VALUE OF SUSTAINABILITY AND THE SUSTAINABILITY OF VALUE

ANTHONY SIMON LADEN

Human activity is causing the planet to rush headlong into a climate catastrophe. Our combined ways of living here are not sustainable. What reasons, then, do we have to change how we live? To anyone who believes the first two sentences, the question in the third one will seem ridiculous. It is the sort of question that gives philosophers a bad reputation. After all, there are some pretty obvious answers: it is in our interest to do what we need to do to keep the planet habitable. (As the slogan goes, there is no planet B.) Moreover, on our way toward the uninhabitability of the planet, many people and other living things will suffer. Among those who are likely to suffer the most are those who are already badly off. So we have reasons of prudence, morality, and justice to change course. Notice, however, that none of these reasons, as powerful as they are, makes direct reference to the unsustainability of our way of life. The sustainability of our activities, as itself a value, as a reason to act one way rather than another, is hard to see clearly. It turns out that there are some deep reasons for this. Working through them opens up a different way of answering the question I began with. Or so, anyway, I argue here.

The argument unfolds in three broad steps. First, I do some methodological ground-clearing in order to clarify one role I think philosophy can play in addressing real-world problems. This work makes room to suggest that some of our failures to act in environmentally sustainable ways might admit of helpful philosophical treatment. The second step provides the bulk of that treatment by suggesting that we can bring considerations of sustainability more clearly into view by shifting our conceptual frameworks for thinking about action. This shift provides an improved account of the value of sustainability. The final step involves drawing out some further consequences of that shift for thinking about the relation of action to value more generally, and in particular, the role our action, suitably conceived, plays in sustaining, rather than producing or expressing or being guided by, value.

While the importance of thinking through the relation of action, value, and sustainability is given its urgency by the onrushing climate catastrophe we face and the urgent need to find environmentally sustainable practices, and I will throughout take environmental sustainability as one of my main cases, the interest of the project does not stand or fall with its use in helping us solve our environmental predicament. As I hope to show along the way, considerations of sustainability arise in all sorts of practical contexts, and yet many of our habitual ways of analyzing and thinking about those contexts make those considerations hard to see properly.

Types of Problems

We can categorize problems along two dimensions, in terms of the kinds of solutions they require and in terms of the contexts in which they arise. Along the first dimension, there will be properly philosophical problems if there are problems whose solution requires the tools of philosophy. Along the second dimension, what are often called "real-world" problems (in contrast to "theoretical" or "philosophical" problems) are those that we encounter in the course of living our lives. Philosophy can usefully contribute to solving real-world problems only if there are some real-world problems (as understood on the second dimension) that are also philosophical (as understood on the first dimension). Here is one way that can come about: sometimes we conceptually frame a problem in a way that obscures possible solutions or diverts our sense of what the problem is. We can, for instance, conceive of a problem as of one type in a way that obscures the need for other types of solution. We make this kind of mistake when we treat a problem as merely technical, and thus pay insufficient attention to the shifts in collective will that will also be necessary to solve it. For example, we might take the problem of climate change to be entirely a matter of how much investment in clean energy sources is necessary to lower emissions levels, without thinking about whether this change will also require a change in political will or some people's conceptions of what a decent life looks like.

In what follows, however, I focus on a different kind of conceptual distortion. In this second kind of case, the very terms we use to frame a problem within some particular category block or blur our vision of what the problem is.[1]

Let me try to make that thought a bit clearer: our concepts are the tools we use to make sense of the world and our lives; they function as something like lenses through which we look out at the world. And like any set of lenses, concepts serve to make certain features of the landscape clear and perspicuous while occluding or otherwise obscuring other parts. So a problem can be philosophical along the first dimension when our habitual concepts hamper our seeing clearly some feature of the world that we need to see clearly in order to grasp and respond to a given situation.

Here is an example of how that can work. We can, and often do, conceive of equality in terms of nondiscrimination. When we do so, our attention is directed to features of our social and institutional lives where people are treated differently and to the purported justifications for that differential treatment. We expect inequality to be manifest in things like hiring and admissions decisions and in the distribution of resources and opportunities. Determining when these differences count as injustice will lead us to articulate criteria for drawing distinctions among people that have a rational basis, and thus to try to explain what features of the world and of people justify what sorts of treatment. But we can also conceive of equality in terms of the structure of our relationships to one another, such that equality obtains where there is no oppression, where no one can dominate another. Conceiving of equality this way will direct our attention to various background features of our society and culture, and how these create and assign us to roles within various hierarchies. It may reveal to us that by the time we get around to asking if a certain decision to treat two people differently had a rational basis, most of the work of creating inequality has already happened. If it turns out that we live in a society where people are situated in these oppressive relationships by various background norms and roles, then conceiving of equality as nondiscrimination may obscure our ability to see clearly certain sources of deep inequality. In that case, addressing inequality becomes in part a philosophical problem, although clearly not merely a philosophical one.[2]

Determining whether making this kind of conceptual shift in response to a philosophical problem has any value will always be importantly tentative, pragmatic, and incomplete. To see this, think through the analogy with lens-crafting more closely. The quality of the lens that gets shaped to correct your vision will depend on the accuracy of the diagnosis of your initial vision problem. If the optometrist mismeasures your vision, then the specifications she sends to the lens-maker will result in an unhelpful pair of glasses. Even if the diagnosis is spot on, however, the value of the lenses can't be determined merely by attending to the quality of the manufacturing process: we will need to try them out. This is a general point, but is made more acute by the further point that any pair of lenses will be better suited to some activities than others. A pair of glasses that lets you pick out fine details in the far distance in low light may make reading in full daylight unpleasant and difficult. So, the lenses are only going to be good if they serve the purpose of the eventual user, not those of the lens-crafter. Similarly, philosophical solutions to philosophical problems can go wrong in a number of ways: they can be responses to a misdiagnosis of the problem, they can be badly constructed, they can fail in the field, and they can fail to serve the purposes of those who need them. Moreover, there is no way, ahead of time, and by philosophical work alone, to avoid all failures of at least the last two kinds. In other words, philosophers can listen and diagnose and think hard and well, but then can do no more than propose conceptual shifts and invite others to try them on and see if they help. This relates to another key feature of treating a problem as philosophical. Technical solutions work around agents. Political and institutional solutions aim to manage and otherwise guide agents. The political or institutional actors who need new institutions or a different collective will need not recognize that this is the problem they face for its solution to be effective. In contrast, philosophical solutions of the sort I am proposing here need to address the agents to which they hope to apply. If a problem is philosophical, then those who face it need to be treated as agents who are guided by their conceptualization of what they are doing. Thus, they will also have to recognize the solution as a solution to their problems. (This is, I think, a fundamental difference between the role of philosophy sketched out here and its role if we think of it as theory construction.)

Note, also, that this suggests a way of understanding the philosopher's use of "we" when she makes a diagnosis: it is an invitation to join her in her vision of the problem. So understood, it has the form of "I do, don't you?"[3] It asks us to think about whether our own ways of conceiving of a certain domain have the characteristics being pointed out.

Finally, sometimes seeing what is being pointed out can be made more difficult because of the ways that our concepts fit together. Oftentimes, particular concepts work together to reinforce one another. We might say that concepts that fit together in this way form a picture. When they form such a picture, it can be hard to change only one feature of our conceptual outlook at will, as it is held in place by a further network of other conceptual commitments. When this happens, merely flipping one concept will be unstable or strike us as unmotivated or ruled out by other features of our conceptual outlook. Thus, we may need to investigate the hold of a whole set of interrelated concepts if we are to shift our thinking about an issue.[4] In such cases, we cannot just flip a single feature of our conceptual outlook at will. This interweaving of our conflicts is then one of the things that make some philosophical problems not only hard to solve, but hard to even see as problems.

Even with all these difficulties, however, philosophy can help us address real-world problems by helping us to see where and how our habitual concepts obscure or distort our vision. It is this kind of work that I undertake in the rest of this paper.[5] My claim is that our habitual ways of thinking about action obscure the importance of acting sustainably and thus at least one type of reason to change our actions when they are not sustainable.

Thinking Differently About Action

I invite you, then, to consider how we commonly think about action. In particular, note that the most ready-to-hand conceptual frameworks for thinking about action picture actions as episodic and goal-directed. Although these two aspects of our conception of action work together to form a picture of action, it helps to treat each in turn as if it were an independent element. My aim in each case will be not to argue for one conceptual framework over another, but to make room for the possibility and attractiveness of alternatives to these dominant and habitual frameworks for thinking about action. The point of doing so is to make more clearly visible the claims that acting sustainably has on us, and so one way to test the value of the work being done as we go along is to ask whether it is bringing something into focus that you have a tendency to otherwise lose sight of.

Action as End-Directed: Ends as Products vs. Ends as Wholes

It is a commonplace of action theory that action, or perhaps intentional action, is behavior that is directed at, and thus in some deep sense shaped by its relation to, an end. There are, however, two broad ways to understand that idea. The first seems to find its natural home in the case of productive activities: action that aims at producing something (an object, a state of affairs) that is separate from the action itself. Here we picture action as a kind of machine that generates a product: baking a cake, building a house, walking to the store, or playing a sonata all seem to be actions that we can characterize in terms of what they aim to produce—a cake, a house, my being at the store, a performance of the sonata. As the last two cases suggest, we can expand the notion of a product beyond its perhaps natural boundaries to make this conception of action rather widely applicable, though doing so does put a certain amount of pressure on the original picture. We can see that pressure by asking what precisely is produced when I play the sonata, such that what I produce in baking a cake and what I produce in playing a sonata could be members of the same category: Is it a set of sounds, a performance, a state of affairs that includes the sonata having been played here and now by me?

What I want to highlight about this picture of action, however, is not its potential weaknesses but its underlying structure: we picture some behavior or activity that is organized and rationalized by something outside or beyond it toward which it aims—the end here is like a target at which we aim our arrow or the object that comes off the assembly line, and the action is the drawing and loosing of the bow or the series of steps that happen along the line. We picture the action as organized and rationalized by the product or goal at which it aims in the sense that if we ask why the action is performed a certain way, or what justifies various steps within the activity, our answers are grounded in the aim of producing the product or the goal, or producing it well: you need to beat the eggs like this to create enough air bubbles to generate a light crumb.

On the second picture of what an end is and how it structures and rationalizes action, we conceive of the end as a kind of whole whose nature structures and generates norms for the parts. Though, as I will argue later, we can describe productive activities this way, this structure serves most naturally to describe and analyze activity that is done for its own sake, whether Aristotle's examples of virtuous activity and contemplation, or more mundane activities like casual conversation. Part of what makes a casual conversation casual is that it has no particular goal or product. Nevertheless, for it to be a conversation and not something else (a series of monologues, a browbeating, an argument, or a verbal food fight), it must exhibit certain characteristics and these set up norms for the particular actions that make it up. If I ask you why, given your greater knowledge of the topic at hand and the clearly right position on the question under discussion, you didn't press that advantage and keep at it until you had persuaded your interlocutor of your position, you can say, "I didn't want to browbeat him, we were just chatting." Here, you refer to the nature of the activity as a whole to explain why you undertake some part of it in a certain way in a manner similar to the baker explaining why the eggs need to be beaten just so. The difference is that the structure of that explanation goes from whole to parts, not product to means. Your reason for not pressing your advantage is not that then you will fail to produce a conversation, and this is your aim, but rather that this is what having a conversation is or involves, and that is what you are doing. Here, as with goal-directed actions, the value of the action comes from the value of the end: explaining why you are not brow-beating your interlocutor by saying you are having a conversation only succeeds if having a conversation is a worthwhile thing to be doing here and now, with him. But the relation of the particular action to the broader activity and thus the way that value is transmitted from one to the other are structurally different: from whole to parts via the constitutive structure of the whole rather than from goal to means via the logic of production.[6]

One of the features of actions that light up when they are conceived as wholes in this way is the work that might need to be done in order for the activity to keep being the kind of activity that it is. What makes the conversation a conversation is that its participants are guided in their participation by the norms of conversation. But that is just to say that among the things they must pay attention to as they enact this conversation is whether the conversation will continue on as a conversation. In other words, one of the tasks that the end of conversing with each other sets them is to sustain the conversation.

The contrast I am drawing here is between two ways of conceiving the nature of action, rather than two classes of action or activity. I can apply the second picture to productive activities and the first picture to actions done for their own sake, and doing so will change how I describe them, what features are lit up and what obscured, and how I understand their structure, the source of their value, and the kinds of value they realize. So, for instance, I can picture the activity of baking a cake as a whole, the parts of which are arranged and oriented around the production of a cake, but that take their point from the fact that I am *baking a cake*, rather than that I am baking *a cake*. Doing so might draw my attention to features of the activity that make the activity as a whole worth doing that are not reducible to the quality of the product and the process that leads to it: the joy I get from making a mess in the kitchen or performing a complex series of steps to produce something elaborate out of basic ingredients or trying out some cool new gadget I got as a present.

And similarly, I can picture a casual conversation as an interlocking set of goal-directed activities: chatting someone up, making a point, informing someone of something, telling a joke. Doing so may turn my attention away from the whole activity but will lead me to notice features of the moves within it. For my purposes in what follows, we need not assess which of these pictures is correct or valuable or gets right something deep about the structure of action or of a particular action. It is enough to recognize them as distinct pictures and notice how they shape our attention, highlighting some features and obscuring others.

Action as Episodic vs. Ongoing

The second contrast to bring forward concerns the temporal dimension of action. Our habitual ways of talking and thinking about action treat actions as episodic rather than ongoing. Actions are conceived as episodic when they are conceived of as having two related features: they are temporally bounded and that boundedness is internally related to their nature as actions rather than being an accidental and contingent feature. That is, I conceive of an action episodically when I think of it not only coming to an end in the sense of a stopping point, but as directed at that stopping point. So, for instance, if I think of baking a cake as action aiming at a baked cake, then when the cake is finished the activity of baking that cake comes to a stop. It is complete. Of course, it could be brought to a halt at some earlier phase, but this would only be related to the activity itself accidentally: if I am missing a key ingredient or I find my oven is broken, or I am called away by a philosophical emergency, then I will stop baking the cake, but in this case, we will say that I have been interrupted, rather than that I have finished or completed the task. The claim about the episodic nature of action is not that action can be interrupted but that it aims at being finished. In contrast, then, what is significant about actions conceived as ongoing is not that they are eternal or uninterruptable (they are neither), but that they do not aim at a stopping point: when they stop, it is not because they have been completed, but rather because they have been interrupted or in other ways wound down.

Many of our habits for thinking about action lead us to conceive of actions as episodic. And so while there are many familiar actions we engage in that are not episodic, they tend not to be our go-to examples when trying to make sense of action. It thus helps to start our analysis of ongoing action with some examples. Consider these:

- Casual conversations
- The unorganized play of children
- Marriages, friendships, and many forms of informal human relationships
- Contemplation, in the Aristotelian sense of reflexive attention to the world around us
- Human life on this planet

Although all five are meant to be cases where it is easier to see that they are ongoing, much analysis of them relies on characterizing them as a set of interlocking episodic actions. We often treat casual conversations as complexly interwoven attempts to inform and persuade and amuse, each part of which aims at a goal. We analyze the play of children as the playing of games and ask about the rules and aims and object and winners and losers, or treat it as a kind of developmental activity aimed at producing social skills or identities. We understand contemplation on the model of research, as the search for understanding or knowledge. And we think about human life on the planet as a set of interlocking end-directed activities that don't add up to a whole different from the sum of its parts.

If, however, we are to develop a different conceptual frame for thinking about such cases, then it helps to notice that though at least the first four all come to a close, that is not because they have been completed. They do not admit of completion in the way that goal-directed activities do. The conversation may wind down because we have run out of things to say or one of us is pulled away or the hour is late. The play ends when the bell rings or someone is called to dinner or a fight breaks out. Contemplation is hard work, and so we tire or become easily distracted and so break off. Nevertheless, in each case, while engaged in the activity, nothing internal necessarily leads it to one of these outcomes. We mark this fact in ordinary language by saying that they can be resumed after interruption without this being a matter of their being done again. Baking a cake ends when the cake is done, and I can bake another cake later, but I cannot resume baking this one (as I can if I get interrupted at certain points along the way). In contrast, though we can start a new conversation, we can also pick up the one we broke off last night or, indeed, many years before, and it is one of the great joys of seeing old friends when you find that you can pick up where you left off, even if you have not seen each other in years.

Some of these activities involve the pursuit of goals, but these goals are not definitive of the activity as such: they are rather connected to intermediate aims or best thought of as by-products. So in the course of our conversation, I may strive to make myself understood, which is to pursue a goal, but if I am conversing with you and not merely explaining myself, once you have grasped my meaning, we won't be done talking. And while conversation has its pleasures and fosters intimacy and may be a reliable way of spreading information, these are not what gives it direction. To see this, note that success at producing these results would not give us a reason to stop our conversation. If anything, it would be a reason to keep going.

The point of working out some of this material is to show what thinking about ongoing activities might look like if it does not resort to the structure of episodic action that makes it harder to appreciate the ongoing qualities of such activities.[7] I come back to these points in the final part of the paper when I turn to asking how we might think about our relation to values within a conceptual framework that makes proper room for ongoing activities.

First, however, it is worth noticing that once we have a way of thinking about ongoing processes on their own terms, we can more easily raise questions about whether we are sustaining our ongoing activities or not. Processes that are ongoing but not necessarily eternal require some sort of work to sustain them. Since they have no goal-determined termination point, they can be sustained, and since they are not necessarily eternal, they need to be sustained. What that work involves and whether it requires conscious effort and attention or comes effortlessly along with full-on engagement in the activity will depend on the activity and the circumstances in which it is undertaken. Some activities are easier to sustain than others, and when an activity begins to disintegrate or otherwise spiral out of control or wind down, it might require special effort and attention to restore to a sustainable footing.

Even if we grant those differences, however, it is still the case that insofar as these processes and activities realize some value, there will be value in sustaining them. Thus, one question we can easily ask about how we participate in such activities, systems, and processes is whether we do so in a way that contributes to sustaining them or inhibits or otherwise detracts from their continuing on. And it is very likely to turn out that among the things it means to engage in these activities well is to engage them in ways that sustain them. Think, for instance, of what makes a good friend or a good conversationalist. It is in large part the way they sustain the relationship or the conversation.

As should be clear from what I have said, conceiving of action as goal-directed and thinking of it as episodic fit well together. If I perform some action in order to produce some result, then the action has a natural, internally directed stopping point: the production of the result. Having aimed at that result, there is no reason to continue the action once the result is produced. Once the cake is done, if I was baking in order to produce the cake, it would be odd to keep on baking that cake (I might, of course, decide to bake another one or try my hand at a pie). If I am walking in order to get to the store, then when I get to the store I should stop walking there (I can, of course, continue walking around the store or change my mind and head off to the beach). If, however, I conceive of an action on the model of a whole done for its own sake, there is a less clear natural stopping point and, in fact, a reason to think that part of what doing that action involves is sustaining it as that very action going forward. It looks, then, as if shifting our conceptual framework for thinking about action brings more clearly into view the claim that considerations of sustainability have on our actions and deliberations about what to do.

Valuing Sustainability

To get the hang of this shift and see if it helps with matters relevant to climate change, it helps to start with a small example and move up. So begin with various things we aim to do when we are talking to one another. I may aim to persuade you of something or inform you of something or inspire a certain reaction in you. If these are my goals, and I conceive what I say as aiming to achieve that goal, that will structure how I think about what to say and how to say it. This is not a trivial fact or a straightforward one. The structuring of what I say and how I say it by my goal in saying it can be subtle and complex and it may make a big difference to what I can be understood to mean by what I say. But if we focus on these kinds of goals in interpreting or evaluating my actions, then we may miss other features of my actions: how what I say responds to you and what you have said, and perhaps how what I say leaves room or even invites you to respond.

If we shift our conceptual frame and think of the various things we might be doing while talking to one another as parts of a larger whole—the ongoing activity of conversation—then this shifts our attention in interesting ways. I may still be trying to inform or persuade you, but if this is a move in an ongoing conversation, then my success at informing or persuading will not be the only thing and perhaps not even the most important thing we need to pay attention to in evaluating what I say and how I say it. I can succeed in informing in a way that shuts down the conversation, silences or annoys you, or just leaves you feeling as if I am not worth talking to, and then I will have botched having a conversation with you.[8] So the shift in framework changes what we notice and in particular calls our attention to the ways that our actions sustain or fail to sustain the activity of which they are parts.

Moving to a larger stage, we can think about all sorts of activities that fall under the umbrella of politics as goal-directed actions: so campaigning aims at winning elections, legislative debate aims at cobbling together majority votes, and various forms of protest actions (marches, boycotts, direct action) aim at changing a vote or a decision or some particular governmental action or policy. Once again, getting precise about what one's goals are and how, given various other features of one's context, a particular action will advance those goals is difficult and much can be gained from seeing this terrain clearly. It is easy enough to lose sight of the particular goal you have and to confuse compromise that is a means to that goal and compromise of that goal if you are unclear in this way. But what can get lost in this way of understanding one's political actions is the larger, ongoing activity of politics of which all these are parts; that is, if we think of political actions as parts of the larger whole of the ongoing process of collectively working out the terms by which we live together, then part of what engaging in politics well will require is sustaining the possibility of that collective work. An effective campaign tactic or rhetorical move in a legislative debate may achieve one's goal, but if it simultaneously corrodes a culture of mutual respect and civic and civil disagreement, then it may also be an example of bad political action within this expanded frame. So, for instance, political speech can foster trust and trustworthiness among vulnerable citizens or it can destroy that trust, and which of these it does may not show up in its effectiveness in securing the immediate goals of the speaker.[9] In such cases, we will only clearly see the damage done by the trust-destroying speech or action if we widen our frame to a view of politics as an ongoing activity of living together of which this speech or action is a part.

We can similarly recast various goal-directed activities by seeing them as part of larger ongoing wholes: of certain traditions or institutions or movements, and in each case, I want to suggest, doing so will help us see why engaging in the various activities that make them up in ways that help to sustain those wider activities has value and what value it has. Note that in all of these cases, situating a given action or set of actions in a larger framework may also mean expanding what might be called the boundaries of our actions: turning them from very local and discrete sequences into parts of much wider structures involving potentially many more people. But while the boundaries of our actions expand when we reconceive of them in this way, there is also an important sense in which our actions themselves don't. Reframing my attempt to persuade you as part of our ongoing conversation or reframing my campaign for office as part of our ongoing political interaction is not to say that I should drop some selfish action for an altruistic one; it is merely to see the same action (persuading, campaigning) through a different lens that then highlights a different set of questions it is natural to ask about those actions, whether they are being done well, and the values they are realizing or thwarting.

If we can bring the activities that contribute to or detract from sustaining conversations and political cultures and institutions into view by this kind of reframing, can we also bring the environmental sustainability of our actions into view in this way? The shift I think we need to make to see the environmental impact of our activity as central to the value or disvalue of what we do is to think of our various activities as parts of the larger whole of living together with other life on this planet. This once again widens the boundaries of our action in important ways and it may be that it pushes those boundaries to a point where this kind of reconceptualization makes less sense. After all, the sense in which I am involved in an ongoing activity with you when we are engaged in conversation or in which the citizens of a country are engaged together in an ongoing activity through their political actions involves a certain kind of mutuality and reciprocity. It is precisely the fact that we are doing these things together that generates many of the norms that then structure what it is to do the parts of those larger ongoing activities well. But the planet does not act together with us the way our conversation partners and fellow citizens do, and even the other living things on the planet do not relate to us in quite that way. For one, they can't respond to or raise criticisms from this perspective, though they can be the grounds of our raising these criticisms to one another. I can't here resolve the question of what sort of difference this makes, but I want to at least raise it. It may be that in order to effect something like this boundary shift in our conception of our actions we need to also shift our conception of nature; that is, we may need to rethink not only our sense of the temporal boundaries of our actions and their relation to larger structures of action, but also that which we draw between an action and its context or environment. One way to do that might be to recover the idea of ourselves as embedded in nature and nature as a source of value that at least Europeans and their cultural descendants arguably lost in the early modern period as we shifted to a view of nature as natural resources.[10] This would suggest a different and perhaps complementary philosophical problem we need to solve in order to address climate change and see the value of environmental sustainability in our actions. These difficulties noted, however, it does seem as if something along the lines of the conceptual shift in our thinking about action discussed in this section can help us properly place the claims of sustainability, including environmental sustainability, with regard to our decisions about what to do and how to do it, and our evaluations of those decisions. It thus provides at least the sketch of an answer to the question with which I began: a new kind of reason to change our actions to make them more sustainable.

Sustaining Value

With the value of sustainability having been brought into view, it turns out that the resources developed to think about ongoing systems on their own terms also suggest a new picture of the relation of action to value. To see this, we need to return to a feature of the conceptual frame that pictures actions as episodic and goal-directed.

Imagining an action, process, or system as temporally bounded because goal-directed also generally involves thinking of it as having a starting as well as a stopping point. When we thus picture actions as both starting and stopping, we can be struck by human agency as the power of beginning or initiating various sequences of events.[11] So pictured, a central question to ask about human action is, what gets it going? And then there are roughly two types of answers: we are pushed from within or we are pulled from without. To explain action in terms of pushes is to point to psychological features like desires or intentions. To explain action in terms of pulls is to point to features of the world that call for or otherwise demand or entice action. In either case, we can raise human agency above the level of merely causal pushes and pulls by describing it as the capacity to set ends, and tying it to something like reason or the will. Note that setting ends is here something that we do in an important sense ab initio: even if we recognize that we set ends embedded in and influenced by all sorts of contexts, there is still the idea of a new beginning in the setting of an end or the performing of an action.

This can make some of the claims that get made on behalf of moral actions seem mysterious, even magical. Could the morality of an action set something in motion in the way that the setting of a productive end does? In a discussion of Gandhi's moral philosophy as a morality of exemplarity rather than a morality of criticism, Akeel Bilgrami relates the following story. As a boy, Bilgrami and his father came across a wallet lying in the road on a walk together. Bilgrami's father asked why they should not pick up the wallet and, when he does not get the answer he is looking for, declares, "If we do not pick it up, then no one will." Bilgrami says of this morality: "The romance in this morality is radiant. Somehow, goodness, good acts, enter the world and affect everyone else. To ask exactly how they do that is to be vulgar, to spoil the romance. Goodness is a sort of mysterious contagion."[12]

Notice, however, that Bilgrami's father's pronouncement and Bilgrami's analysis take the decision to pick up the wallet or not as a starting point, as if the world begins with this tableau: a wallet lying on the sidewalk as a father and son walk up to it. The romance in the moral vision is the thought that this is not an isolated, bounded episode, but that it will set a world in motion. But standing behind that romance is the assumption that the world begins here anew.

Though the details are different, a similar framework stands behind Kant's idea of treating the adoption of a maxim as legislative: in acting, I treat myself as setting something in motion, even if not in a causal or contagious sense. In both cases, what we are in danger of losing sight of is the fact that I am embedded in a temporal order and a set of activities that began before I acted and that I inherit.

Why does this matter? It turns out that how we think about values and our relation to them changes in interesting ways when we think of ourselves not as actors in a series of episodes but as participants in ongoing phenomena. In particular, our understanding of our placement in the world changes if we keep in mind what has already happened, how we got to this point. To see yourself as a participant in an ongoing process is to put yourself in a position to recognize that that process has already been sustained and thus that insofar as the process has or realizes value, you have benefited from it, and you may thus be called on to do what is necessary to sustain it going forward. The picture we need, then, is not of an agent standing still at the edge of a deliberative field and charting out a course to start moving. Rather, we should think of an agent already moving within a current, who needs to learn to steer among its ebbs and flows. Even this image is insufficient, as it leaves out the effects that our course going forward has on the current's strength and direction.

Let me begin to tease out the implications of this thought by returning to Bilgrami, his father, and the wallet. One might transpose his father's thought into a different temporal context: from the beginning of an episode to within an ongoing process. Doing so, the thought becomes not "If we don't pick it up, no one will" but "Since no one else has picked it up, neither should we." So put, it loses its radiance and also perhaps its romance. Goodness is no longer a mysterious contagion I unleash upon the world, but a tradition I sustain by participating in it.[13]

This points to a rather different way of thinking about value and our response to it. At least some values are realized by certain ongoing processes and activities. My use of "realized" here is meant to contrast with "being produced." The idea is that a value is made real or manifest, brought into being, in the course of the unfolding of the given activity or process, rather than by being produced or being aimed at. Not all such ongoing activities realize values or the same values, and so we need to ask, when thinking about such processes, what, if any, values they realize.[14] To see a practice or process or system as realizing a value is just to see it as having a claim on continuing, being worth sustaining. So when we conclude that some practice or system or process realizes a given value, this conclusion places a demand on its participants to act in ways that sustain it. As far as I can see, thinking about value this way does not require settling the metaphysical question of "where" values are to be found. Some values are realized through human practices, others are realized through practices that involve humans but are not artificial human creations, and some values will be realized through processes that don't involve humans at all. Since they need not set us in motion but only give us reason to keep moving, there is less of a need to determine whether they are pushes or pulls, in the same way there is less reason to worry about whether the eddy in the current is pushing or pulling me once I understand where it is moving me and what I need to do in response.

Perhaps more importantly, from the point of view of a participant in an ongoing process that she recognizes and thinks about as ongoing and worth continuing, the demands placed on her are not sui generis. They are more like the demands of gratitude: that is, once I see the practice that I find myself in as ongoing and value-realizing, I am led to appreciate the contributions to its sustenance that have come before, and how I have benefited from them: they stand toward me as a gift. Like any other gift, they call on me to respond in a certain way: to show gratitude in part by responding in kind. Recognizing the value of a world in which people do not take others' lost wallets, and recognizing the gifts others offer me in making this world such a place by not taking the wallet, I show my gratitude for that gift by reciprocating and also not taking it. In doing so, I, as it were, share in bearing the costs necessary to sustain such a world.[15]

Before turning back to the question of climate change, let me flag several points of interest that arise within this way of thinking about value in the course of ongoing activities. First, recognizing myself as the beneficiary of gifts is to recognize my own dependence and thus vulnerability. This changes the way I am likely to respond to the vulnerability of others and to the need on my part and others' for the kinds of conditions that make vulnerability bearable. I have in mind here things like trustworthiness and forms of security that are not understood as the absence of vulnerability and dependence.

Second, the normativity that is operative within this picture is not the normativity of law and command, or even of the peremptory demand that appears to be at work in Bilgrami's realism. It is also not the subjectivist normativity of desires or even that which follows the instrumental logic of goal-setting. It is, rather, the normativity that is involved in calling forth a response.[16] Moreover, because the response called for is one of reciprocity, the response itself must similarly call forth a response. Part of developing a better set of conceptual tools for thinking about ongoing processes would involve developing a better account of this kind of normativity, one that would allow us to understand how it plays out not only in human relationships but also perhaps in our relations with the nonhuman world. What goes on when some natural phenomenon or process calls for my response, and how can I respond in a way that calls forth a response that is not merely causal? Do we have examples of such interaction that are not merely mysteriously romantic? Perhaps they provide us with a conceptual model for thinking about climate change.

Third, once we picture our actions as parts of larger ongoing wholes that give them at least part of their points, it turns out that our actions are always contributing to keeping the activity going or detracting from that. Sustaining our activities is not some additional goal that we can choose to adopt or not, not some additional evaluative filter we can choose to adopt in deciding what to do. If we are not keeping these activities going, then we are contributing to their running down, deteriorating, or becoming corrupt forms of themselves. Of course, in some times and places and with certain activities, the required attention and effort to shifting from one mode of action to the other may be greater and more acute. And this is certainly the case when it comes to correcting our effects on the stability of the global climate.

Climate Change as a Problem of Sustainability

I have suggested that among the kinds of problems climate change poses for us are philosophical problems, and among those problems is one having to do with our difficulty in grasping the value of sustainability because of the unfitness of many of our conceptual tools for making sense of ongoing activities and systems. In response to that diagnosis, I then tried to sketch out some new resources that might be better suited to making sense of the role and demands of sustainability within ongoing processes. Can we now return to climate change to see what difference this might make?

A great deal of the popular and political discourse around climate change casts it as a problem we face as if we are setting out on a new episode of history: we face a set of established facts and now the question is what to do about them. Sustainable practices thus become one tactic in our arsenal for reducing emissions and limiting the damage already done and thus the catastrophic effects on our future. And given that they involve costs and sacrifices, and that we are unsure how much they will contribute and what other technological fixes might be on the way, it is perhaps no wonder that many of us find ourselves insufficiently motivated to bear the necessary costs.

But, armed with the skeletal conceptual resources developed in this paper, we might try to change that picture. Human (and nonhuman) life on the planet is an ongoing process that continues only because it has been sustained in the past. We find ourselves here due to the efforts of others in contributing to a system that has at least continued to this point (as well as actions that have undermined it). The changing global climate is not a new problem but a feature of an ongoing system that is no longer being sustained in the form in which it has been ongoing for billions of years. It is a system on which we are dependent and which we ought to regard as a gift. In this light, a failure to continue to sustain it or to take the corrective steps to respond to its spinning out of control is to show ingratitude to those who have sustained it until now. It is to squander their gift. In contrast, the adoption of practices that would return the global system to a sustainable footing is not a matter of altruism or long-term prudence or heroic sacrifice: these practices are merely the acts of reciprocal gratitude to which we are called.[17]

Notes

This paper was originally written for and presented at the NOMIS workshop on Nature and Value, in March 2016. I am grateful to Akeel Bilgrami and the NOMIS Foundation for the invitation to participate in this workshop, and a previous one in June of 2014. I also presented versions of the paper to the UIC philosophy department works-in-progress series in October 2016 and the Northwestern Practical Philosophy workshop in November 2016. I greatly benefited from feedback from participants in all three places. I am particularly grateful for detailed feedback from and conversations with Carole Rovane, Patchen Markell, and Will Small.

1. An issue arises here which I attend to later, about who the "we" is whose vision is thus blocked. Note here merely that this is not unique to philosophical problems: what *we* can do or are willing to do collectively is going to depend on who *we* are.

2. I take it that this general outline describes at least some forms of Marxist critiques of capitalist class structures, radical feminist accounts of sex inequality, and critical race theories of racial inequality. I lay out the feminist case in these terms in more detail in Anthony Simon Laden, "Radical Feminists, Reasonable Liberals: Reason, Power and Objectivity in the Work of MacKinnon and Rawls," *Journal of Political Philosophy* 11, no. 2 (2003): 133–152, and the general issue of different approaches to equality in Laden, "The Practice of Equality," in *Justice, Democracy and the Right to Justification: Rainer Forst in Dialogue* (London: Bloomsbury Academic, 2014), 103–126.

3. Stanley Cavell describes the authority of ordinary language philosophers' use of "we" when they "say what we say" in roughly these terms. See Cavell, *The Claim of Reason* (Oxford: Oxford University Press, 1979), 19.

4. I think this is what Wittgenstein has in mind when he talks of a "picture" holding us captive. Ludwig Wittgenstein, *Philosophical Investigations*, trans. G. E. M. Anscombe (Oxford: Wiley-Blackwell, 1991), §115.

5. I take it that this is the kind of work that Akeel Bilgrami's attempts to think through the shift in our concepts of nature, from the natural world to natural resources, is engaged in. See, for instance, the essays in Bilgrami, *Secularism, Identity, and Enchantment* (Cambridge, MA: Harvard University Press, 2014), as well as his contribution to this volume. In that sense, my project here and his are complementary, not competitive, even if they offer different diagnoses. I will have more to say about their possible relationship later.

6. The phrase "logic of production" is a bit obscure: I mean to capture something broader than the causal relations between means and ends, but to include such cases. Taking this route to the store is not the cause of my arriving at the store in the way that beating the eggs is the cause of the air bubbles in the batter, but the structure of rationalization follows a similar logic.

7. I take up the question of the logic and structure of ongoing activities in more depth, with a particular focus on thinking of reasoning as an ongoing activity that is a species of conversation, in Anthony Simon Laden, *Reasoning: A Social Picture* (Oxford: Oxford University Press, 2012).

8. In Laden, *Reasoning*, I begin with a discussion of the scene in *Pride and Prejudice* where Mr. Collins proposes to Elizabeth Bennett. One way to understand his failure to engage her is that he sees proposing as a goal-directed activity and she insists on a suitor who will see it as part of the "meet and happy conversation" that their marriage might be, and that requires proposing in a way that leaves space for and is responsive to a response. It is precisely this responsive sort of conversation at which he fails so spectacularly.

9. On some of these sorts of differences among rhetorical strategies and their importance for democratic politics in this broad sense, see Danielle Allen, *Talking to Strangers* (Chicago: University of Chicago Press, 2004).

10. This is where the project of this paper might benefit from being combined with Bilgrami's project.

11. Though this thought pervades Western philosophical thinking about human agency, it is brought to great clarity and force by Arendt, who takes it to be a kind of moral of the human condition: Hannah Arendt, *The Human Condition* (Chicago: University of Chicago Press, 1958). Interestingly, two of Arendt's categories of activity, labor and action, are conceived as ongoing and not directly goal-directed, and yet in both cases, she connects them to the idea of birth and the possibility of new beginnings. I am grateful to conversations with Patchen Markell for insight into Arendt.

12. Bilgrami, *Secularism, Identity, and Enchantment*, 114.

13. Though I have used Bilgrami's example as a foil here, I suspect that there is a fruitful way to bring the two perspectives together. In particular, both Bilgrami's realism and my analysis of ongoing processes involve the idea that value is a feature of our being embedded within something larger than ourselves and our episodes of action, and, as I suggested in the previous section, conceiving of an action as part of our living together with all life on the planet requires the shift in not only how we conceive of an action, but also how we think of our relation to the natural, nonhuman world.

14. It is the availability of this question, even from within our ongoing practices, that keeps this approach to thinking about value from being purely conservative.

15. This way of framing the issue is a lesson I have learned from James Tully's remarks on what he calls the gift-gratitude-reciprocity cycle in "A View of Transformative Reconciliation: Strange Multiplicity and the Spirit of Haida Gwaii at 20," the keynote address given at the "Indigenous Studies and Anti-Imperial Critique" symposium, Yale University, October 1–2, 2015. See also Robin Wall Kimmerer, *Braiding Sweetgrass: Indigenous Wisdom, Scientific Knowledge, and the Teaching of Plants* (Minneapolis: Milkweed, 2013). Such a willingness to bear the costs of sustaining a collective endeavor is what Danielle Allen calls "sacrifice" in her discussion of citizenship in *Talking to Strangers*.

16. In Laden, *Reasoning*, I describe this as the "authority of connection." See also Stanley Cavell, "Passionate and Performative Utterance: Morals of an Encounter," in *Contending with Stanley Cavell*, ed. Stanley Cavell and Russell B. Goodman (Oxford University Press, 2005), 177–198.

17. Note that if you take most of our current and recent past practices as part of the problem, as the source of unsustainability, you could run the argument just given the other way: whatever the productive power of these practices, when seen as part of the ongoing activity of life on the planet, they are unsustainable. This fact makes them a kind of poisoned gift, like the passing on of a debt incurred. Changing our practices into ones that recover the sustainability of life on the planet would amount to discharging that debt. We might still regard doing so as an act of gratitude, perhaps not toward the generations that incurred the debt, but reaching further back to the generations that sustained the values more recent generations have mortgaged.

CHAPTER 15

VARIETIES OF AGENCY

Comment on Anthony Laden

CAROL ROVANE

*S*ustainability has become something of a buzzword. It is a word that rightly makes us anxious, owing to all of the things we are currently doing that we recognize to be *unsustainable*. It is no surprise, then, that we find a lot of references to *sustainable practices*—sustainable energy practices, sustainable mining practices, sustainable manufacturing practices, sustainable building practices, sustainable farming practices, sustainable fishing practices, sustainable forestry practices, sustainable water practices, sustainable waste practices, sustainable development practices, and on and on. It is fair to say that a general consensus has emerged that no such practices will prove to be sustainable unless we find a way to halt, and perhaps even to reverse, climate change.

In the main, we have tended to view the problem that climate change presents as a *practical* problem whose solution calls for appropriate *action*. In that spirit, we have tended to approach it in the following sorts of ways, all of which come up for mention at one point or another in Laden's paper: as an engineering problem that would be solved by new technologies, as an institutional problem that would be solved by fostering suitable institutions, as a political problem that calls for relevant political accords (what Laden refers to as forging a "collective will"). To date, our efforts along these lines have been woefully inadequate. It is tempting to explain this disturbing fact disjunctively, by reasoning as follows: either the practical solutions we have attempted so far would in principle be adequate or they would not be; if they would be, then our failures to date have simply been failures to follow through on implementing the available solutions, whereas, if those solutions are in principle bound to fail, then we need to find better solutions, presumably in the form of better versions of the sorts of solutions we've already tried—better technologies, better institutions, better politics. But this is not the explanation Laden offers. He suggests instead that our failure may be due to our failure to see in the problem of climate change its specifically *philosophical* dimensions.

Laden has a great deal to say about the nature of philosophical problems, but here is what I take to be most central and important: a problem counts as philosophical when it is beset with conceptual obscurities or distortions that blur or block our vision of what the problem is; and such a problem calls for specifically philosophical work, which he portrays as *clarifying* the concepts involved, and then exploring whether certain conceptual *shifts* or *revisions* might aid us in seeing the problem better. He allows that philosophical problems can have practical dimensions as well, which call for practical solutions. He insists, though, that their practical solutions must wait until after we have brought the tools of philosophy to bear on them.

The particular concepts that Laden puts at the center of his philosophical investigation into the problem of sustainability all have to do, in one way or another, with the nature of *agency*. He draws many distinctions and connections between various *aspects* of our agency—or as he prefers to put it, between *various ways of conceiving* our agency. This is a rich field of investigation, in which he unearths many truths. But inevitably, my task as commentator will lead me to focus on claims that are less than fully convincing.

Let me begin by simply *listing* a number of the conceptual "shifts" that Laden recommends, with a view to solving the problems we now face with respect to sustainability. With respect to agency, he recommends that we shift our attention away from certain conceptions of our actions as productive, episodic, and structured by relations to external goals. With respect to the metaphysics of value, he recommends that we shift our attention away from two conceptions of values that philosophers generally take to be in competition with each other: as *pushes* from within, or as *pulls* from without. With respect to agents, he recommends that we shift our attention away from a conception of them as individual ends-setters who unleash their causal effects upon the world. What are we to shift our attention *to*? A conception of agency on which actions are nonproductive rather than productive, ongoing rather than episodic, structured by internal part-whole relations rather than by external relations to further ends; a conception of values that emphasizes reciprocal relations through which a whole system is sustained; a conception of agents as belonging to larger social systems that they together sustain.

The conceptual shifts that Laden is proposing rest on various *contrasts* that he draws, and much of conceptual/philosophical work he does in his paper is to explore various relations that he sees between the terms of these contrasts.

In connection with the three specific contrasts he draws regarding *actions*, he sees an affinity between their first terms, in which actions are conceived, respectively, as productive, episodic, and externally related to goals. Laden defines productive actions as actions that are undertaken for the sake of ends beyond themselves. Examples of such productive actions abound, for they have issued in every *thing* that any human being has ever intentionally *made*— every meal, every article of clothing, every building, every piece of furniture, every tool, every machine, and so on. He claims that such actions are naturally episodic, because they are expressly initiated for the sake of their intended outcomes, and then they are terminated as soon as those outcomes are achieved. He also claims that such actions are structured by their external relations to goals—though to my mind, this last claim barely adds anything to the very definition of productive actions, as aiming to produce outcomes distinct from them.

Laden also seems to see an affinity between the second terms of the same three contrasts, in which we conceive our actions respectively as nonproductive, ongoing, and standing in internal part-whole relations. He defines nonproductive actions as actions that are not done for the sake of any further end, but just for the sake of doing them. His examples of such actions include casual conversation, play, personal relationships, the life of virtue and contemplation as Aristotle understood them, and *human life on this planet*. It so happens that, when he first offers these examples, he offers them as examples of actions that are *nonepisodic*. To say that they are nonepisodic is not to say that they have no beginnings or ends, for they necessarily do—they are not eternal, after all. It is rather to say that this is not how we primarily conceive them, because we often find ourselves in the midst of them with no particular interest in how or why they first began, or when they will eventually end. It seems to me that this is what makes it natural to describe so many of our nonproductive actions as *activities* rather than as actions, to indicate that they are ongoing rather than episodic. Laden is especially interested in cases where these ongoing, nonproductive activities exhibit a complex internal part-whole structure, in which many component subactions together constitute a larger, unfolding whole.

Laden does not go so far as to claim that the three contrasts I have just laid out actually coincide. Such a claim would actually be mistaken, and it would have been helpful to this reader if Laden had clarified this. In any case, let me now clarify.

Laden does acknowledge that a great many productive actions exhibit the sort of internal part-whole structure that he finds in many nonproductive activities. But the point deserves much more emphasis. Consider the illustrative examples I myself offered earlier—the production of food, clothing, buildings, furniture, tools, machines. In *all* of these productions, we perform many component subactions that together constitute a larger process of production in which they are all internally related as parts to whole, even though they also bear external relations to their intended products as well. Furthermore—and this is something that Laden's discussion omits *entirely*—a very great deal of our production takes place within a larger, ongoing economic system, and as a result, production itself has come to be ongoing—which is to say, *not* episodic in the way he emphasizes. This is especially salient in capitalist economic systems, but I think it is also true of more traditional systems of petty production. I will return to these important points at the very end of my comment.

Similarly, I think it is misleading for Laden to emphasize the nonepisodic character of nonproductive actions to the extent that he does, given that they are often episodic, as his own example of playing a sonata attests.

Finally, although it is a matter of definition that nonproductive activities do not bear external relations to further ends beyond themselves, this does not guarantee that they possess the sort of internal part-whole relations that Laden emphasizes by contrast. The briefest reflection shows that a great many of them do not—think of lolling in bed, or basking in the sunshine. Of course, it is true that these activities are temporally extended, and on some philosophical conceptions it follows that they must have temporal parts. But this is not the sort of internal part-whole structure that Laden has in mind. The structure he has in mind arises when agents explicitly conceive their activities as constituted by component actions whose value derives from the fact that they are contributions to the larger whole that they together constitute—or, to put the point negatively, these component actions would not and could not have the value they have all on their own, were it not for the value of the larger whole for the sake of which they are undertaken. This is *not* true of the temporal parts of ongoing nonproductive activities like lolling in bed and basking in the sunshine—each moment of lolling and basking can be pretty good all on its own.

This last complaint does not really speak against Laden's suggestion that we can learn something by shifting our conceptual gaze to cases where nonproductive activities do exhibit a significant part-whole structure. But I must confess that I myself am more struck by the fact that we find this structure all over the place in human agency, in productive actions as well as nonproductive, and regardless of whether they are episodic or ongoing. In fact, I would argue that this structure is one of the most striking and distinctive features of human agency.

Here is why the three contrasts I have just reviewed are so central for Laden. What he seeks is a way to conceive—or, maybe, *reconceive*—human actions that affords a role for the *value of sustainability* that can be appreciated quite independently of the problems of unsustainability that we now face with respect to climate and the environment. And that is exactly what he takes himself to find in the wake of drawing the three contrasts. In his view, when we engage in certain ongoing nonproductive activities that exhibit a complex internal part-whole structure, we conceive the things we do as *sustaining* the larger whole, so that whatever value we attach to the specific things we do in the context of these activities is bound up with the value we attach to sustaining the whole.

Laden buttresses this finding with some further related arguments that have to do with the other conceptual shifts he is recommending. But before getting on to these matters, I would like to pause to observe that he isn't always explicit enough about what he has in mind. If he has in mind what I think he has in mind, then it is open to some criticisms that I'm about to share. If these criticisms are misplaced, then let the criticisms indicate that Laden leaves the reader with too much work to do in order to avoid, or, perhaps I should say, to clear up, confusion.

One wonders, why so much emphasis on *non*productive activities? After all, it is abundantly clear that we have come to a point in the history of our species when we literally cannot survive without producing a very great deal, including pretty much all of the things on the list of illustrative examples I gave earlier—food, clothing, shelter, buildings, furnishings, tools, and even machines. Our problem of the moment is not that we produce, but that we *over*produce, and in ways that have *bad effects* on the climate, and on the environment more generally—and also, on the quality of our work-life and social relations, as the young Marx observed so long ago. Prima facie, the only solutions worth pursuing are ones that show us how to produce *less*, and how to produce with *fewer bad effects*. I simply do not see how being reminded that we are capable of engaging in certain *non*productive activities that we value for their own sake is going to help on this front.

I gather that help is supposed to come when we see how the value of sustainability comes to the fore in connection with nonproductive activities— where the implicit suggestion seems to be that this value does not, and indeed cannot, come to the fore in quite the same way in connection with productive actions. But one cannot refrain from wondering, *why* does Laden think this, if he does? Does he think that if we do something for its own sake, then whatever value it has will be *extended* when the action itself is extended? In other words, does he think that if I value basking in the sunshine for its own sake, then I should value *sustaining* my basking? Since this thought is clearly mistaken, let us suppose not. Perhaps his thought is that the sort of extended value that can be achieved by sustaining something is at least *possible* in connection with nonproductive actions, whereas it is not so much as possible in connection with productive actions. The argument for this thought would be that once productive actions have produced their intended outcomes, they have no further value or point. Without quite knowing if this is what he actually thinks, let me register that I am not sure that we should agree. As I've already observed, much of our *current* production takes place in an ongoing way, within an ongoing economic system, and it is unclear to me that we currently see its value as confined to its products in the way that Laden seems to be suggesting—but I'll leave further comments on this matter for the end.

At this juncture, I'll simply note that Laden wants to explore what can happen in *various* domains if we shift our attention in such a way that we see the many things we do as contributing to, and thereby *sustaining*, an unfolding whole that has no definite end, rather than as discrete and episodic doings. It is striking that the *wholes* that interest him are all *social*. This shows up right at the start in his initial examples of ongoing activities—conversation, personal relationships, play, the life of virtue and contemplation that Aristotle urged upon us, and *human life on this planet*. It might seem that the life of virtue and contemplation stands apart as an exception. But it is not, for Aristotle explicitly portrayed it as a *social achievement* that is born out of a right upbringing in the right social conditions. In any case, en route to discussing the last example on his list, which is his real quarry, Laden discusses other cases of ongoing social activities that become increasingly large in scale.

As anyone can see, it is never up to any one *individual* to sustain a large-scale social activity. Individuals can only do their small parts with an eye to the larger relationships and practices that they thereby help to sustain. Laden notes that this can lead to feelings of vulnerability, insofar as the value and indeed the very meaning of our individual contributions are contingent on their relation to the contributions that others make as well. But he doesn't see this vulnerability as entirely negative. For it can give rise to feelings of gratitude when others are doing their parts, or have already done their parts, for this is what makes it possible for our own contributions *to be what they are*. Laden concludes that when we find ourselves embedded in such a large-scale social activity, we should see the value of sustaining it as paramount, because of the way in which the practice itself drives and organizes all within it. This is the internal part-whole structure that I've already discussed, now writ large across a temporally extended social whole.

Laden invites us to view whole social *traditions* along the lines of this argument—so that the value of a tradition becomes inseparable from the value of sustaining it. But I must confess that I am strongly inclined to reject this invitation. I do not believe that it has been typical in human history that the actions through which individuals have sustained their traditions have been explicitly conceived by them in the way that Laden suggests, as deriving their value from the fact that they are the means by which a larger whole is sustained. In the main, I think that the participants in traditions have primarily focused on *preserving knowledge*, in the form of collective memories of a common past, along with shared truths about how to live. They have conceived this inherited knowledge as a *good* that is worth passing on, and what they directly and explicitly aimed to do was to pass their knowledge on to the next generation or two, leaving it to them to do with it what they will. This is profoundly different from conceiving a tradition as a whole that will extend far into the future as well as the past, and conceiving one's present actions as the means by which this will happen. In fact, it seems to me that the explicit aim to sustain a tradition further than the next generation or two generally arises only after it has been made vulnerable by the effects of large-scale migration. Here I have in mind both the migrations of powerful colonizers who impose their own practices and values wherever they go, destabilizing local traditions in the process, and the migration of less powerful people who try to carry their inherited traditions to distant places that are not hospitable to them. In both sorts of cases, these threats to the survival of a tradition generate a new kind of concern for it, and it is at this point that thoughts and efforts directed at sustaining it *as a whole*, far into the future, arise. Alas, I think it is quite typical that once a tradition requires such an explicit aim to sustain it in the face of challenges, it is already starting to feel a bit artificial to its participants. And even if their aim should still be realized, there will be other problems that follow upon the attempt to sustain a tradition in such circumstances, for it will require insulating itself from, and disengaging from, the surrounds that threaten to intrude upon it.

When traditions are not threatened in these ways, they carry on in the much more myopic way that I just described earlier, by passing inherited knowledge on to the next generation or two, without any explicit aim to sustain a temporally extended whole that will reach far into the future. I think this *myopia* is more or less imposed by a genuine difficulty that stands in the way of knowing *enough* about the course of future events to allow us to effectively direct our efforts at *very* long-term effects. This difficulty is one thing that makes it hard to take utilitarianism as a serious guide in moral life: How can we coherently *aim* to satisfy the principle of utility if we cannot know very much about how our present actions will affect aggregate utility in the long term?

But let me return to the larger points that Laden wants to drive home, through his discussion of traditions. He presents traditions as examples of ongoing, large-scale social activities that are governed by an *aim* to sustain them, because sustaining them is recognized to be *valuable*. Even if I am right that this is not a good way to view traditions, it may be a good way to view *some* large-scale social activities. So what are the larger points that Laden aims to drive home in this connection? They concern the other conceptual shifts he is recommending, which I've not yet addressed, concerning our metaphysical conceptions of value and of agents.

Laden expresses dissatisfaction with two metaphysical conceptions of value that now dominate in philosophy, and that are generally viewed as being in competition with each other. On one conception, values are *pushes* in the following sense: they are contributed to the world by agents, either in the form of desires or in the form of some more spontaneous activities.[1] On the other conception, values are *pulls* in the following sense: they are objective features of reality that exert a distinctive kind of normative force on agents. Laden leaves us a bit in the dark about what, exactly, is so unsatisfactory about these two competing conceptions of value. Speaking for myself, I can say that what is unsatisfactory about the *push* conception is that it deprives us of the genuine insight of the *pull* conception, which is that there are objective matters of value that *call* upon us, by exerting a *normative force* that is rather unlike the causal forces that science studies.[2] It bears mention that such objective matters of value can exert their peculiarly normative force only on *agents* who can *apprehend* that force, and thereby be *called* to *actively* respond. But this ineliminable role for the active responses of agents in the metaphysics of value does not give any support to the idea that agents somehow *make* value, or *contribute* value to the world, in the way that the *push* conception would have us believe.[3]

Laden himself leaves aside the *push* conception without much discussion, and goes on to offer two different lines of resistance to the *pull* conception, which he seems to see as going together. The first line is introduced when he claims that ongoing social activities are governed by *constitutive norms* that guide agents' social interactions as they sustain those activities—he seems to take it to be an obvious point that such norms do not fit realism. The second line is introduced when he claims that realism follows upon, or at any rate goes together with, the metaphysical conception of agents as *ends-setters*—here I find it less clear what is on his mind.

Laden introduces this second line of resistance to realism via a story that Akeel Bilgrami tells.[4] When he was still a child, Bilgrami went for a walk with his father and the pair encountered a wallet on the ground. His father asked him, "Should we pick the wallet up?" When Bilgrami answered *yes* his father then asked, *why?* Bilgrami's answer was, "Well if we don't pick it up then someone else will." To which his father retorted, "No, we should not pick it up, for if we don't pick it up then no one else will." When done in this spirit, the action of not picking up the wallet is *exemplary*—it displays a moral possibility that all of us can realize, when given the chance. But what Laden detects is the metaphysical conception of agents as ends-setters who thereby unleash their causal effects on the world—that is what he thinks was in play when Bilgrami senior urged his son to *initiate* something through his own agency, in the form of an exemplary act. In contrast, Laden would prefer that the father had invoked *tradition*, by saying, "Since no one else has picked up the wallet, neither should we." Thus, in place of the *spontaneous moral creativity* that he takes to be highlighted in Bilgrami's own telling of his story, Laden proposes a vision of morality as given with and through an *inherited and ongoing moral practice.* On the heels of that proposal comes a proposal to reject realism with respect to value in general. I gather that he sees this move as necessary for retaining his vision of the metaphysics of *moral* values in particular, which he proposes to model on the *constitutive norms* that govern some social activities.

I said that I do not believe that we have generally sustained traditions by explicitly conceiving the actions through which we sustain them as undertaken for the sake of sustaining them, or as deriving their value from the fact that they sustain them. I believe that, more often than not, we have sustained traditions through other aims—most centrally, the aim of preserving inherited knowledge by passing it on to the next generation or two. This belief reflects an implicit commitment to realism, which is to say, a commitment to the idea that there is *real* value that can come of what we do, and the further idea that matters of value are things that we can be right or wrong about. If we substitute a metaphysical conception on which we do things just because others *have already done them*, and so if we do them too, then *our practice will carry on*, then we shall have substituted commitments to sustaining social rituals in the place of genuinely *moral* commitments (along with other value commitments to nonmoral goods). So I do see it as crucial that Bilgrami's account of *exemplary* moral actions be informed by realism—so that they can be undertaken for the sake of a good that does not reduce to a fact about what others have done before, *or*, for that matter, to a fact about what will happen in the future. An exemplary action *recommends itself to future others*, not just on the basis of the historical fact that it happened, but on the basis of the *value* of what it recommends.

Why does Laden prefer the antirealist or, at any rate, *non*realist metaphysical conception of value instead? It's hard to say. He leaves the reader to connect the dots, and even to fill in some blanks. It seems clear that he finds the idea of *constitutive* norms helpful. But this idea seems clearest in connection with a kind of activity that is not a plausible or attractive model for a *moral* practice, namely, a *conventional* practice that is governed by *constitutive rules* that define what counts as a move within it—where neither the practice itself nor the moves within it aim to *answer to* any objective matters of value. Social rituals are a case in point, and so are games like chess and baseball.

A certain irony comes to light at this juncture in the dialectic. I have claimed that participants in a tradition generally do not explicitly aim to sustain their tradition, but have aims that presuppose a realist metaphysics of value, on which there are objective matters of value that call upon us. Laden sees Bilgrami's conception of exemplary moral action as involving this realist conception. By contrast, he recommends that we conceive moral actions as aiming to sustain a local moral practice that is governed by norms that are internal to it, and do not answer to objective matters of value beyond it. The irony is that, as I shall now argue, the *explicit aim* to sustain such a practice, *were it in place*, would require just the sort of *exemplariness* that Laden sees as going together with realism. Here is why: One cannot coherently *aim* to sustain a larger tradition by doing one's part without thinking that others will do their parts. But this cannot be seriously predictive or causal thought, to the effect that doing one's own part will actually *bring it about* that others do their parts, for as I've already pointed out, *one isn't generally in a position to know this*. There is more than one reason why this is so. One difficulty is that we don't have enough definite thoughts about enough definite others who might be affected by our actions. Another difficulty arises when we realize that we must think of the others who might be affected by what we do as *agents*, and to think of them in this way involves recognizing that we must *leave it up to them* to do their parts, or not, as they see fit. It follows that the *best* one *could* do, if one really did explicitly *aim* to sustain a moral practice by doing one's own part, would be to offer up the doing of one's own part as *an exemplary act* for others to follow. I'm not sure if the very idea of an exemplary act actually *requires* realism. But it seems that Laden thinks so. Insofar as he does, he probably ought to take realism on board—for, as I just explained, his own vision of how the value of sustainability functions within ongoing social practices requires him to take the idea of exemplary action on board.

Perhaps Laden is not trying to suggest that the way in which traditions were historically maintained was through explicit aims to sustain them. Perhaps he is suggesting that this is how we ought to employ our agency from here on in. If so, I see no need to ruin the suggestion by insisting that we should abandon a realist metaphysics of value. I have already raised a worry about how the explicit aim to sustain a tradition tends to arise when the tradition is already under threat, and has already begun to feel a bit artificial to its participants. I'll now add another worry, which is that this feeling of artificiality is bound to be exacerbated, when we ask the participants in traditions to conceive the acts through which they aim to sustain them as done *just for the sake* of keeping the tradition going, and not for the sake of passing on any genuine *knowledge* about *how to live*.

There is a confusion that is very widespread among philosophers who think about moral relativism, which I suspect Laden may be laboring under. A good way to think about what the moral relativist has in mind is that there is something *local* about moral values, in the sense that they do not give *universal* guidance to all agents in all times and places, because moral values must guide *particular* agents in the *particular* circumstances that they find themselves in—their particular opportunities, to act in particular ways, in response to particular others, in light of particular relationships to them. I suspect that most philosophers, like Laden, think that if moral values have only local normative force, then this force cannot be *objective*, because objectivity implies universality. But note that the idea of universality has a *social* reference—what it implies is that if something is true for one, then it is true for all, whereas the idea of objectivity has a metaphysical reference—what it implies is that if a thought or a claim is to be true, then it must answer to something that is independent of it, by getting it right rather than wrong. The possibility that is little noticed among philosophers who are interested in relativism is that there can be *local* truths that are at the same time *perfectly objective* even though they are *not universal*.[5] It is when we overlook this possibility that we will be tempted to think, as Laden seems to think, that if moral values apply only locally, within specific moral contexts, then they cannot be objective, and then it may well seem that we are forced to abandon realism in favor of some form of conventionalism—though I imagine that Laden would not welcome this way of characterizing his idea that social practices are governed by constitutive norms.

There is one matter about which I think my agreement with Laden runs deep, and it is a matter about which I've said little so far, at least in any direct way. This is his claim that our current conception of agency is dominated by a wrong kind of *individualism*. I agree with him that it is paramount to get away from a distorted conception of *agents* as metaphysically separate from one another, and as called upon by the world to exercise *their own* agency, seemingly in isolation from all others. This metaphysical conception becomes especially dangerous when we focus our moral attention on lines of individual responsibility that are supposed to line up with individual lines of causal control. For, then, even if we are prepared to focus our attention on the large-scale effects of what we are together doing, we will see no scope for effective ameliorative action—we will see nothing but how small the effects will be, of any given action on the part of any given individual. Correlatively, when we register the bad large-scale effects of what we are collectively doing on the environment, we will find it morally disconcerting that we cannot find an individual whom we might call to account—someone to blame for what has already happened, and someone to call upon to take whatever large-scale actions are now needed to ameliorate the situation, which lie beyond the power of mere human individuals to do on their own. It is no surprise, then, that a new subfield has emerged in contemporary philosophy, which goes by the name *social ontology*, that has tried to shift our conceptual gaze to such matters as collective agency, along with collective intentionality, collective rationality, and collective responsibility.

My own views on agency are revisionary in a different direction from the directions Laden is heading in with his various recommended conceptual shifts. But they do not follow the lines set out by this new literature on social ontology either. This already overly long comment is not the place to try to elaborate or defend them. But I can say that they proceed from the point of agreement with Laden that I've registered, that our current conception of agency is distorted by its commitment to *individualism*. Rather than try to bracket this aspect of our current conception, I have explored it from within, in order to press it to its ultimate logical and metaphysical limits. The upshot is a conception of the individual agent as something that is not metaphysically given with the individual human being, but as something that emerges through effort and will. Thus, individual agents come to exist when human agential capacities are exercised so as to forge the sort of rational and practical unity that is characteristic of individuality. It is already a revision of our current conception of agency to say that agents are self-constituting in this way. Further revisions follow, because it turns out that the requisite unity can be forged within different boundaries that need not coincide with the biological boundaries of a given human life—they can be smaller or larger than that, or in other words, there can be multiple agents within a single human life, and group agents comprising many human lives, and agents of roughly human size. In all cases, no matter what their size, individuals come to exist for the sake of ends that require the achievement of unity within certain boundaries rather than others.[6] Group agents in this sense, in which groups qualify as individual agents in their own right, are commonly confused with the sort of collective agency that arises when many distinct agents exercise their individual agency for common ends. But these two kinds of agency are profoundly different, and this can be appreciated when we consider the very different sorts of ends that can be accomplished by the one kind or the other. The achievement of some ends, like the development of the atomic bomb, require that a group achieve the sort of rational unity that characterizes individual agency— which is exactly how the Manhattan Project proceeded when it developed the bomb. Other ends, like winning an election or pushing a piece of legislation through, would not require achieving such rational unity at the level of a whole group—many distinct agents can collectively act together for such common ends for quite different reasons, and they can succeed in achieving those ends even if they remain in disagreement about a very great deal. It seems to me that we would do well to avail ourselves of *both* kinds of agency in response to the problem of climate change and related environmental problems— individual agency and collective agency. We might need to form group agents who function as (and therefore *are*) rational individuals in order to develop new technologies, in a way reminiscent of the Manhattan Project. We might need to form collectives who aim to lower aggregate energy consumption through mass campaigns and mass movements.[7] Individual agents of *human size* can do their part too, simply by lowering their own personal energy consumption. Such human individuals might also offer up their actions as *exemplary actions* in the sense explored by Bilgrami—if they should do so, then without knowing whether others will do the same, they can still reason,

If I don't pollute then no one else will. I myself think it would be quite wrong to underestimate the potential power of such exemplary action in our times, when it can be conveyed to vast numbers in the blink of an eye.

But let me return to the conceptual shifts that Laden is proposing. It seems to me that he could fairly respond to the new literature on social ontology by simply observing that, in spite of its effort to get away from an overly individualistic conception of agency, it simply does not take up the value of sustainability, whereas *his* way of getting away from individualism actually aims to provide us with some understanding of that value, by situating it in relation to certain large-scale social activities that are nonproductive and ongoing.

What I have so far left out of account in my comment are two related proposals in which Laden emphasizes these social contexts in which he sees the value of sustainability as playing a central role. One is that we should view our relation to nature itself on analogy with our relations to, and within, large-scale social systems that we actively sustain. Then we shall keep our eye on how reciprocal relations can keep the whole going, and on our reasons for gratitude for the part that others play in keeping the whole going. But as Laden himself says, this is not an entirely new proposal—it has already been made and pursued, as he acknowledges, by James Tully, for example.[8] The other proposal is that when we seek to model our relations to nature on social relations, we should bear in mind that the best models on which to draw for this purpose are not ones in which agents proceed from their own individual points of view, but ones in which agents proceed from the point of view of a larger social whole. This is not really a new proposal either, and in fact it is very much the sort of thing that Bilgrami has in mind in his account of the unalienated life.[9] Still, it is good to have Laden's essay integrating with Tully and Bilgrami in the same volume.

I would like to close by returning to the matter I flagged earlier, concerning how, in the context of capitalism, productive action has come to have many of the features that Laden associates with large-scale nonproductive social activities. Each participant in the capitalist system does things that would have little value or point unless others did their parts. This is true of every productive action on an assembly line, and it is true of every transaction—whether it be labor in exchange for payment, or payment in exchange for commodities, or stocks, or currency, or whatever. Increasingly, the participants in this global system have become aware of this feature of it, and as a result, even when they play the role of producers within the system, their role is very much driven by the thought of making a sustaining contribution to an ongoing system, as opposed to carrying out a productive action that will be finished when the product is produced. So I really do think that Laden's efforts to connect the *ongoing* character of many of our activities with their being *nonproductive* rather than *productive* are quite misguided.

In my opening remarks, I noted the many ways in which we have come to be *anxious* about the sustainability of this or that human practice, because of its effects on the environment. I think we should add the practice of capitalism—by which I mean, the *global capitalist system*—to the list. I am moved by the claims of some Marxist economists that the capitalist system is not sustainable *in principle*, because it requires never-ending economic *growth*. In the past, such growth has been made possible not just by innovation, but by the existence of ever-new sources of cheap labor along with new sources of raw materials to extract. But although these new sources of labor and resources were available to capitalism in its imperialist phase, they are not endlessly available. And when they cease to be available, what we get is a different kind of so-called growth, which takes the form of speculative bubbles, which are bound to burst in the true manner of a bubble.[10] But let us suppose for the sake of argument, or perhaps out of deference to a prevailing consensus that I myself think is misguided, that these Marxist economists are wrong—that is, let us suppose that there is no such source of unsustainability that will inevitably come from *within* the capitalist system, due to its very nature. Even so, it is becoming abundantly clear that there is another source of unsustainability anyway, in the form of large-scale environmental disasters, which have come to be the familiar and frightening by-products of the system. I'm afraid that no one really knows how to put any real brakes on the capitalist system, and few would want to put an end to it. But it really is terribly unclear how it can be sustained, given its negative environmental impact, which can be predicted to increase because of its requirement of never-ending *growth*.

Does Laden's philosophical work help on this front? By his lights it does, insofar as it manages to clear away some conceptual obstacles and distortions that would otherwise impede any other efforts that we might make on this front. I have my doubts. I have argued that the recommended conceptual shift away from productive activities to nonproductive activities is simply beside the point, given that our very survival now *depends* on production, and given that we have created a productive *system* that we don't know how to control. I have also argued that the recommended conceptual shift away from episodic to ongoing activities is beside the point, given that production has itself already shifted from an episodic to an ongoing form. And I have argued that the recommended shift away from a realist metaphysical conception of value is not attractive, and indeed throws up a new conceptual distortion—for it obscures the way in which traditional social practices incorporate and pass on knowledge of real truths about how to live. And finally, I have argued that his recommended shift toward a vision of agents as embedded in large-scale social activities that they explicitly aim to sustain fits the case of capitalism surprisingly well. What I think we need to direct our attention to is the nature and effects of that system, as it relates to the large-scale environmental problems that we now face.

These are, no doubt, points of substantial disagreement, but they are registered by me here within a deeply sympathetic concern shared with Laden about the importance of considering questions of sustainability side by side with questions of agency; though he has not carried me with him in his arguments and conclusions, I remain very grateful to him for the stimulus provided by his admirably imaginative discussion of the relevance of the nature of human action to understanding our place in nature.

Notes

1. This *push* conception was famously pushed by David Hume, and has now become known as the *Humean* view. Central representatives of contemporary Humeanism include Michael Smith (see Smith, *The Moral Problem* [Hoboken, NJ; Wiley-Blackwell, 1992]) and Simon Blackburn (see Blackburn, *Ruling Passions* [Oxford: Oxford University Press, 1998]).

2. The *pull* conception is advocated by some so-called moral realists. The most prominent defense of moral realism along these lines was given by Derek Parfit in *On What Matters* (Oxford: Oxford University Press, 2011).

3. See Akeel Bilgrami, "The Wider Significance of Naturalism: A Genealogical Essay," in *Naturalism and Normativity*, ed. Mario De Caro and David Macarthur (New York: Columbia University Press, 2010), for further discussion of this point, as well as Carol Rovane, "Self-Constitution, Reductionism, and the Metaphysics of Value," in *Derek Parfit's Reasons and Persons: An Introduction and Critical Inquiry*, ed. Andrea Sauchelli (London: Routledge, forthcoming).

4. Akeel Bilgrami, "Gandhi, the Philosopher," in *Secularism, Identity, and Enchantment* (Cambridge, MA: Harvard University Press, 2014).

5. See Carol Rovane, *Metaphysics and Ethics of Relativism* (Cambridge, MA: Harvard University Press, 2013), for an account of relativism as denying *only* the universality of truth and *not* objectivity. The account eschews the prevailing consensus, according to which relativism arises in domains where antirealism holds because in those domains there can be so-called *irresoluble disagreements* in which all parties are right. Instead, the account seeks to elaborate and improve an admittedly underdeveloped idea that governed the main twentieth-century debates about relativism, according to which relativism arises with the possibility of so-called *alternative conceptual schemes* that cannot be embraced together, not because they are in disagreement but because they are *incommensurable* with one another.

6. See Carol Rovane, *Bounds of Agency: An Essay in Revisionary Metaphysics* (Princeton: Princeton University Press, 1998), for a book-length treatment of these issues.

7. For an account of the moral and metaphysical importance of the distinction between group agency and collective agency, see Carol Rovane, "Group Agency and Individualism," *Erkenntniss* (2014); Peter A. French and Howard K. Wettstein, eds., *Forward-Looking Collective Responsibility*, Midwest Studies in Philosophy (Boston: Wiley, 2014); Carol Rovane, "Is Group Agency a Social Phenomenon?," *Synthese* (2017).

8. See James Tully's essays in this volume.
9. See Akeel Bilgrami, "Gandhi and Marx," in *Secularism, Identity, and Enchantment,* as well as his essay in this volume.
10. Utsa Patnaik and Prabhat Patnaik, *A Theory of Imperialism* (New York: Columbia University Press, 2017).

CHAPTER 16

NONHUMAN AGENCY AND
HUMAN NORMATIVITY

NIKOLAS KOMPRIDIS

The Unsettling Time of the Humans

It is not without some considerable irony that as we enter the Anthropocene,[1] the era named after us, we humans are becomingly increasingly uncertain about what and who it is we are, about whether we can, and whether we should, distinguish ourselves from what and who we are not. Old distinctions and boundaries between human and animal, human and nonhuman, and human and machine have lost their firmness and certainty, and have come under renewed suspicion and critical pressure. Some of this is due to the enormous, ongoing revision of our knowledge and understanding of nonhuman life in general, making untenable distinctions long taken for granted. We are coming to see, finally, how complexly entangled and interdependent are human and nonhuman forms of life. The day has passed when questions of justice and the good life could be posed only in relation to human forms of life. Those questions have become all the more difficult and complex with the recognition of entanglement and interdependence. Making sense of those questions as we extend them outward to embrace the nonhuman alongside and amid which we dwell represents a conceptual and normative challenge we have never had to face before.

It is an unsettling time, because some fundamental questions seem to be open again, and it is unclear how we go about answering them, or where the answers will take us. One would think that rethinking formerly settled distinctions (and convictions)—be they based on cognition, consciousness, or language, inter alia—would bring about a renaissance in thinking about both the question of the human and the question of the nonhuman. After all, if who and what we take ourselves to be now depends on an unsustainable contrast with and opposition to the nonhuman, one would expect some revision of our self-understanding as humans. But it would appear that all the energetic rethinking is taking place in relation to the nonhuman, which, up to a point, is entirely understandable.[2] Much of this is due to a justly motivated endeavor to dislodge the dominant position of the human subject and the human species in our thinking—in short, to overthrow human exceptionalism. But there may be better and worse ways to go about this, and although it may appear counterintuitive, displacing the question of the human exclusively in favor of the nonhuman might unwittingly retrench rather than dislodge human exceptionalism (as I will show in the next section).

In the early years of this century, for a very brief time the question of the human was a very live question. The successful mapping of the human genome at the close of the previous century produced a series of anxious reflections by philosophers and social theorists in response to the potential of genetic engineering and the new technosciences to alter fundamentally and irreversibly the very basis of what counts as a human being, and what counts as a human form of life. In quick succession, Jürgen Habermas, Jacques Derrida, Francis Fukuyama, and Michael Sandel, among others, published widely read and intensely discussed books and articles on how and why the emergent technoscientific revolution posed a threat to what is peculiarly human. Suddenly, a question that until the very recent past had been treated as arcane, passé, a question that only thinkers with a conservative, essentialist bent would regard as philosophically obligatory, is "taking on, here, now, a terribly concrete and urgent form at an infinitely accelerated rate."[3] But what do we say, what can we say about what is peculiarly human, when the conceptual boundaries between human and nonhuman, person and thing, have become fluid?

A decade later, rather unnoticeably and without, so far as I am aware, any compelling explanation, this pressing question, a question absolutely pressed for time, seems to have lost its urgency.[4] It has been displaced and overtaken by a question to which an even-greater urgency is assigned: the question of the nonhuman, the animal, the vital materiality of things and nonhuman forces, the question of Gaia, and the emergence of the Anthropocene. You might say that ecology has trumped ontology, or, if you are especially persuaded by the work of Bruno Latour and Jane Bennett, that ecology has flattened our ontology, in more than one sense. Only a decade later, those panicked debates and concerns that coalesced around technoscientific threats to the very idea of the human seem very remote from our present concerns. What happened? Why?

I think there are a few reasons, one of which is the simple fact that we were, and still remain, largely unaware of how inseparably and fatefully linked the question of the human is to the question of the nonhuman. One won't find in the reflections of Habermas, Sandel, and Fukuyama any concern whatsoever with the fate of the nonhuman. Nor will one find any acknowledgment that human and nonhuman forms of life are entangled with one another. It was still possible a decade or so ago to write about the question of the human in such debates without any mention of how interdependent and entangled are human and nonhuman forms of life. By neglecting the question of nonhuman forms of life, those panicked inquiries foreclosed themselves, and came across as defensive and alarmist, concerned only with the "right understanding of cultural forms of life,"[5] as though the place of the nonhuman in human forms of life was a secondary or irrelevant question.

Another reason is that we are quickly realizing that we are no longer on Holocene time; the time of the Anthropocene is altogether different. It is not just human life that is pressed for time; all life is pressed for time. As humans carry on somnambulistically as though it's business as usual on the planet, a sixth mass extinction of nonhuman animal life proceeds apace. We are facing anthropogenic mutations of the living systems of the planet on a scale that no human civilization has faced before, mutations that are unpredictable, unprecedented, possibly ungraspable in their multiple aspects and implications—call it the Anthropocenic sublime—and just as uncontrollable. Hardly the time to be preoccupied, rather narcissistically, with the question of what it means to be human in order to reassert some version of human exceptionalism. We are told repeatedly that we need to act, and act quickly, to make a difference now to the scale of the catastrophes to come, to make them less unjust, less destructive, to save what we can. But at the same time, the "we" who needs to do all this cannot remain as it is or was. To respond to the challenges of living in and with the Anthropocene, this "we" needs to cultivate "a new mode of humanity."[6] But such cultivation at this scale and breadth takes time; it can only happen slowly, if it can happen at all. So there are competing and conflicting temporalities in play, and at stake, in the judgments we make about what "we" are called upon to do.

I want to adduce a further reason for the turn toward the nonhuman in the humanities and social sciences. The highly dramatic thematization of the question of the human in the first decade of the century was in fact a deviation from the norm, a blip, in a very long history of suppression or of indifference. Thereafter, the subsequent emergence of a wide and rapidly increasing array of "posthumanist" inquiries into the nonhuman is consistent with the antihumanist temper of the humanities and social sciences since the 1960s. It is no accident that the concerns of the present seem better and more urgently expressed by people like Bruno Latour, Donna Haraway, and Jane Bennett, whose aim is to diminish, not rescue, the difference between human and nonhuman. There are powerful moral motivations driving their efforts, of course, not least of which is the honorable and timely effort to overcome human exceptionalism, to which is so rightly attributed so much of the destructiveness that has brought into being the era of the Anthropocene.

But that's not the whole story. Decades of antihumanism have left their mark on us as thinkers, and on what we think is thinkable. It is not just about closing the door to any return to humanist comforts and pieties. It is about what we allow ourselves to imagine as possibilities beyond humanism and posthumanism. There is in the air, and has been for some time, a general sense of exhaustion with the human, a sense that the time has come to supersede the human and humanism, and you can find this expressed in various discourses throughout the humanities and social sciences. This thought was paradigmatically expressed five decades ago in the closing pages of Michel Foucault's *The Order of Things*. As Foucault took great pains to show, in what may be the most extraordinary intellectual performance of the twentieth century, the human being that became the epistemic object of the "human sciences" is "an effect of a change in the fundamental arrangement of knowledge,"[7] the very change through which "man" came into being as a living, laboring, and speaking subject.

> As the archaeology of our thought easily shows, man is an invention of recent date. And one, perhaps, nearing its end.
>
> If those arrangements were to disappear . . . if some event of which we can at the moment do no more than sense the possibility . . . were to cause them to disappear, as the ground of Classical thought did, then one can certainly wager that man would be erased, like a face drawn in sand at the edge of the sea.[8]

Of course, Foucault's closing statement has a different ring to it in the age of anthropogenic climate change. There are peoples and cultures living on islands in the Indian and Pacific Oceans who are about to disappear, whose history and ways of life will literally be erased as the ocean levels rise. Of course, Foucault was talking about a certain idea or figure of the human, not about the human species, not about cultures and peoples literally being erased. Certainly, the dangers and challenges of the Anthropocene were not on the horizon of Western intellectuals back then. The event of the Anthropocene is not the event that Foucault had in mind, or the kind of event he had in mind. But there is nonetheless a smug confidence in Foucault's statement that expresses our antihumanist temper at the point of its ascension to the status of taken-for-granted doxa about humanism—an unquestioned and unquestionable starting point for any sensible critical position.

I don't think Foucault would write that sentence today. It was easier to talk like that then; we were still on Holocene time, and could take for granted the stability of the climate systems and life systems of the planet. We're not on Holocene time anymore. We might have been rushing headlong into the future on modernist time, recklessly, not knowing where we were headed; but we're on Anthropocene time, now, and the future is rushing headlong at us, and we're by no means ready for it. The unprecedented challenges of the Anthropocene era are not just about the looming catastrophes threatening the Earth's living systems and all life on the planet, human and nonhuman. They are also challenges to the idea of the human, and to human responsiveness to the nonhuman. We very much need to cultivate different modes of our humanity if we are to be properly responsive to the nonhuman. The question I want to pose here is whether human responsiveness to the nonhuman is enabled or impeded by our antihumanist temper.

The final pages of *The Order of Things* certainly presaged the sense of the exhaustion of human possibility. My worry is that the exhaustion of the sense of human possibility might lead to the exhaustion of the possibility of the human. Today, in some sectors of the academy, it is easier to affirm the radically other, currently manifested in the figure of the "animal," than it is to affirm the human. This is not just because of a change in the arrangement of knowledge; it is also because of a mounting, completely understandable disappointment with the human. Given all that we have irresponsibly laid to waste, and what we continue to lay to waste, what kind of faith can we have in ourselves? As Stanley Cavell has shown, disappointment and skepticism are interlinked.[9] Our disappointment with the human feeds on our skepticism toward the human, and our skepticism feeds on our disappointment. But what if our disappointment with the human does more than prop up our antihumanist temperament; what if it also stands in the way of our responsiveness to the nonhuman?

The Question of Nonhuman Agency

In her widely read and admired book *Vibrant Matter*, Jane Bennett offers an account of nonhuman agency that is consistent with our antihumanist temper and, at the same time, in conformity with a very standard, taken-for-granted conception of human agency. Drawing on Spinoza and Latour, Bennett extends and distributes to things, and to nonhuman life, a form of agency previously identified, almost strictly, with human agency. The ontological and analytical priority Bennett further attributes to nonhuman agency flattens, deliberately, the difference between human and nonhuman agents. Indeed, Bennett wants entirely to "bracket the human" and the question of human agency in order to make room for the agency of the nonhuman. For her, any attempt to distinguish human from nonhuman agency will always over-shadow the vitality and agency of the nonhuman.

> To attempt, as I do, to present human and nonhuman actants on a less vertical plane than is common is to bracket the question of the human and to elide the rich and diverse literature on subjectivity and its genesis, its conditions of possibility and its boundaries. The philosophical project of naming where subjectivity begins and ends is too often bound up with fantasies of human uniqueness in the eyes of God, of escape from materiality, or of mastery of nature; and even where it is not, it remains an aporetic or quixotic endeavor.[10]

Bennett therefore has no interest in pursuing the question of human agency: "Sooner or later" the topics of subjectivity or of what distinguishes the human from the nonhuman "will lead down the anthropocentric garden path, will insinuate a hierarchy of subjects over objects, and obstruct freethinking about what agency really entails."[11] Apparently, our thinking about nonhu-man agency is much less encumbered by what are essentially fruitless episte-mological and ontological questions. "No one really knows what human agency is, or what humans are doing when they are said to perform as agents. In the face of every analysis, human agency remains something of a mystery. If we do not know just how human agency operates, how can we be so sure that the processes through which non-humans make their mark are qualita-tively different?"[12]

But if we do not "really know" enough about human agency—the agency, which is, after all, our own, that tells us something about what and who we are—on what basis can we be confident that our grasp of nonhuman agency is any more secure? In what way can it be less mysterious to us than we are to ourselves? We are seeking to understand the agency of an "other" we have pre-viously ignored or underestimated, an "other" whose otherness raises all kinds of epistemic and ethical issues for us. And yet Bennett doesn't pose, let alone answer, the question of what it is that makes nonhuman agency less mysteri-ous and more conceptually accessible than our own. The problem is not only that Bennett is overstating the mystery of human agency. The problem is that she disallows any meaningful basis of comparison between human and non-human agency. Although one term of the comparison is rendered conceptu-ally opaque and mysterious, it nonetheless provides the vocabulary on which the intelligibility of the other term depends. How can the conceptual language of human agency make the explanatory object of that language mysterious without also making nonhuman agency at least as mysterious? Why should we expect that an avowedly flawed language of human agency can illuminate nonhuman agency?

Let me further clarify what is at issue here. Bennett espouses unquestioningly a view of agency that is indistinguishable from the standard view of agency as a doing or acting, through which doing or acting an agent brings about an effect of one kind or another. In other words, what we have here is a causal picture of agency, a view of agency that is synonymous or identical with causality. The agent is understood, and understands herself, as the efficient cause of something, acting on that something (any something), producing an effect that affects, alters, or redirects that on which it acts. This picture of agency is shared by diverse philosophical positions, both naturalistic and antinaturalistic. To regard oneself as an agent is to recognize oneself as a locus of causality. One therefore attributes to oneself a causal power, the power to act on oneself, on others, and on things in accord with one's beliefs and desires.

One post-Kantian line of thinking about agency from Fichte through Kierkegaard to Tugendhat has drawn on this causal picture of agency to explain how it is that one comes to be the individual or person one is, and thereby to give an account of one's practical identity as self-determination.[13] One's practical identity is the outcome of one's own activity: the "I" that I am is both the cause and effect of itself.[14] One can find versions of this picture of agency in a wide range of philosophical views, even views that see the agent as constituted by forces and powers not her own—for example, Nietzsche, Sartre, and Foucault. Bennett is not interested in this particular tradition of thought on agency, not least because it is a distraction from her primary concern, which is to amplify nonhuman agency. To do so she takes over Bruno Latour's concept of "actants"—a concept that puts human and nonhuman actors on the same ontological level, and the same level of conceptual analysis, thereby neutralizing the difference between human and nonhuman agency. But the definition of what makes agency agentic, so to speak, remains one that nonetheless conforms to rather than departs from the standard causal picture of what agency is.

Contrary to what Latour and Bennett believe, they are not neutralizing the distinction between human and nonhuman so much as they are extending to the nonhuman domain an already flawed (and much contested) picture of human agency. They do not challenge this picture; they only reproduce its shortcomings. Notice the definition of agency that is supposed to render neutral the difference between human and nonhuman agency: an actant "is a source of action that can be either human or nonhuman. It is that which has efficacy, can do things, has sufficient coherence to make a difference, produce effects, alter the course of events."[15] Thus, what was originally a form of agency that distinguished human from nonhuman agents reappears as a form of agency that makes human and nonhuman agency indistinguishable from each other. But if this view of agency is correct, why then does Bennett claim that no one really knows what human agency is? Why does it remain a mystery, best left undisturbed? Clearly, it would seem more consistent for Bennett to argue that we do know what agency is, and the problem has been that we have been too stingy and anthropocentric to share it with nonhuman beings. To be an agent simply is to be a causal force, a locus of causality with the power to affect things, to produce effects, to alter the course of events. And so, contrary to what we have heretofore thought, it is not a power that is restricted to, or the special property of, specifically human agents.[16]

The argument stated this way would be more clearly aligned with Bennett's hope that by seeing the active, agentic materiality of things, we will become more responsive to the nonhuman. This is after all the whole point of her endeavor, one that, I should state clearly, I wholeheartedly endorse. "Because my hunch is that the image of dead or thoroughly instrumentalized matter feeds human hubris and our earth-destroying fantasies of conquest and consumption. It does so by preventing us from detecting (seeing, hearing, smelling, tasting, feeling) a fuller range of the nonhuman powers circulating around and within human bodies."[17] That hunch, then, supposes that "if matter itself is lively, then not only is the difference between subjects and objects minimized, but the status of the shared materiality of all things is elevated."[18]

But there is a false equation here between shared materiality and shared moral status. The latter does not follow from the other, logically or morally. We can recognize the vibrancy and vitality of the nonhuman, and quite simply carry on with fantasies of conquest and consumption and, of course, with actual conquest and consumption. We can continue to eat animal flesh made easily and attractively available to us through mass-scale industrial livestock farming and still (perversely) appreciate the vibrant vitality of the animal life that we crave and consume. Green consumerism could easily accommodate vibrant materialism: the more vibrant, the more desirable; the more desirably consumable, the more expensive; and the more expensive, the more desirably consumable. Perversely, that very vitality may feed further fantasies of conquest and consumption. Practices of torture whether performed on human or nonhuman animals need not be blind to the vitality of the beings that are tortured; indeed, there is evidence that their vitality (as well as their resistance) is recognized, and that torture is partly a response to that vitality. People who hunt for sport, for example, have no trouble recognizing the vitality and vibrancy of the animals they hunt; that's what makes it a sport, a contest, a source of excitement and satisfaction. Vital nonhuman life can incite a violent response in virtue of its vitality.

Bennett supposes that the recognition of the vital materiality of the nonhuman will induce in humans a change of normative attitude, on whose basis humans will be inclined to treat nonhuman creatures and bodies as subjects rather than as objects. Is such a supposition warranted? Is this all it would take to change our normative responses? Is our perception of their vitality, their status as actants, enough to bring about a change in our normative attitude? Perhaps Bennett underestimates how much we already grasp the vitality of nonhuman life, and yet remain unmoved to change our normative attitude toward it. I do not think that it is merely our failure to recognize the material vitality of the animals we eat and breed for this very purpose that stands in the way of rethinking our attachment to the consumption of animal flesh. From childhood on, these animals richly figure in our stories and cultural imagination in all their actual and anthropomorphized vitality, and yet we slaughter them in the billions for our consumption. Recognizing the vitality of human life has not stopped human beings from torturing and executing fellow human beings by the millions. What makes us think that recognizing the vitality of nonhuman animal life will make us act any differently? So what normative force does the notion of material vitality actually possess? It is not a little puzzling that Bennett wants to bring about a change in the moral status of nonhuman life through the deployment of nonmoral concepts. If a change in moral status is her goal, why does she wish explicitly to privilege and give priority to "physiological over moral descriptors"?[19]

The answer to these questions lies in Bennett's commitment to a naturalistic ontology and metaphysics that she absorbs through the influence of Spinoza and Deleuze. Indeed, much of what she says about agency could be regarded as a straightforward statement of ethical naturalism and moral noncognitivism as they are understood in contemporary analytic moral philosophy. Bennett states her allegiance to Spinoza's naturalism from the very outset: "I invoke [Spinoza's] idea of conative bodies that strive to enhance their power of activity by forming alliances with other bodies, and I share his faith that everything is made of the same substance. Spinoza rejected the idea that man 'disturbs rather than follows Nature's order,' and promises instead to 'consider human actions and appetites just as if it were an investigation into lines, planes, or bodies.'"[20]

However, Bennett does not fully clarify just what this allegiance to naturalism commits her to. In his *Ethics*, Spinoza argues deductively that since the "laws and rules of nature, according to which all things happen, and change from one form to another, are always and everywhere the same," the "way of understanding the nature of anything, of whatever kind, must also be the same, viz. through the universal laws of nature."[21] Such a view, consistent with naturalism, is one that must accord ultimate epistemic authority to the natural sciences, for these "universal laws of nature" are not only best stated in the vocabularies of the natural sciences, but what counts *as* nature is empirically exhausted by these vocabularies. Obviously, to be consistently naturalistic, the normative dimension of human life must be placed within a naturalistic framework, and thereby naturalized, making it explicable exclusively in terms of the same "universal laws of nature."

It is rather odd to find this convergence of continental thought with the predominantly naturalistic bent of analytic philosophy, particularly since post-Kantian continental thought has been decidedly antinaturalistic. However, the Deleuzian side of continental thought is one through which the naturalism of Hume and Spinoza flows unobstructed, and thereby brings it into a strange proximity with the dominant side of analytic philosophy, namely, its scientism. This has been much too little noticed, and too little commented upon.[22] In the present context, it impinges on the question of whether Bennett's espousal of naturalism undermines or weakens the normative project of transforming human attitudes to the nonhuman. Alongside this question is the question of political theory's relation to the natural sciences. If all things are to be understood from within a single monistic ontology, as things of the same kind, and grasped as the things they are in one and the same way, how does this metaphysical and epistemological commitment redefine the role of political theory? For naturalistically oriented philosophers, the role of philosophy is to clarify the key concepts of science, which is to say, its role is to provide philosophical translations of scientific knowledge, which is taken to be the most basic and primary form of knowledge. Bennett doesn't write as though she believes political theory is limited to this translational role. However, it seems to be a direct implication of the naturalism she espouses. As she makes clear, she forthrightly endorses the naturalistic view that human behavior and action should be analyzed and understood by the same methods that are used to analyze and understand "lines, planes, or bodies." It is hard to tell whether Bennett has thought through this implication of her naturalism. It is certainly not thematized in *Vibrant Matter*, surprisingly, given its significant import.

Bennett is clearly committed to analyzing politics through a naturalistic lens. However, the question of how much such a framework illuminates and how much it mystifies in the realm of ethics and politics is insufficiently addressed. The following statement of Bennett's purpose makes this issue all the more conspicuous: "The task at hand for humans is to find a more horizontal representation of the relation between human and nonhuman actants in order to be more faithful to the style of action pursued by each."[23] But what does it mean to be more "faithful" here without begging all the questions? Which kind of epistemic representation would be most faithful? Which human and nonhuman "style of action" would be most definitive? In which epistemic language would it be most authoritatively couched? Biology? Physics? Neuroscience? Certainly, if one were to be consistently naturalistic, the authoritative language of representation could not be moral or political theory, and any use of normative language would have to be of a kind reducible to the relevant scientific language. By definition, it could not be irreducibly normative. So at the very least, there is a question here of whether the more horizontal mode of representation called for by Bennett is stipulated by her naturalistic commitments or by a "faithful" representation of "the style of action pursued by each." But what if being faithful to at least one if not both sides of that relation required irreducibly normative language?[24] How could we possibly justify such a language from within a naturalistic framework that explicitly denies its validity?

Consider the way Bennett describes the underlying goal of what is undeniably a normative project: "The ethical aim becomes to distribute value more generously, to bodies as such."[25] This statement not only exposes the tension between Bennett's ethical motivation and the naturalistic language she deploys; it also exposes a further tension between Bennett's interest in reenchanting nature and her naturalistic framework. Since the reenchantment is undertaken in naturalistic rather than normative language, it is not surprising that the redistribution of value is one-directional: from humans to the nonhuman. Once one abandons normative language, privileging "physiological over moral descriptors," the natural or nonhuman world is deprived of any value properties of its own. If it is to have value at all, value must be bestowed upon it by human beings. As Akeel Bilgrami has argued, in this naturalistic (Humean) view, value is understood essentially as an extension of human desires and preferences. And it is a view that complements and reinforces a view of nature as wholly disenchanted and devoid of value properties of its own: "Since nothing intrinsic in nature is valuable, nothing in nature can constrain our desires, utilities, and moral sentiments."[26] There is, in other words, nothing "out there" for and to which we are accountable other than the desires and preferences we ourselves project.

Of course, the very point of Bennett's work is to get us to notice the nonhuman, and to take note of its *own* vibrancy—and, moreover, to become more accountable to it. But these necessary and urgent goals are undermined rather than advanced by the naturalism she espouses, which, contrary to her self-understanding, is also implicated in the human exceptionalism she seeks to overthrow. As she concedes, but only partially, her attempt to reenchant and revitalize the nonhuman world may render her open to the accusation that she is committing a performative contradiction, a concern she raises a couple of times in her text.[27] The performative contradiction that worries her arises from the circumstance that it is a "self-conscious, language wielding human who is articulating this philosophy of vibrant matter."[28] But the performative contradiction cuts deeper than Bennett realizes, for in engaging in value redistribution, all good intentions notwithstanding, she is presupposing the absence of intrinsic value in the nonhuman world. If, by acknowledging the liveliness of all matter, we dissolve the distinction between subjects and objects, we can, as a consequence, attribute value to that from which we withheld value. But that supposes and concedes that *we* are the value-distributors. It is we humans who have the power to more generously and more democratically distribute value. Values are ours to redistribute as we desire, to whatever we desire to attribute value. We might even say, in accordance with Spinoza's view of the agency of conative bodies, that our power to distribute value is life-enhancing when human bodies align their actions with the actions of nonhuman bodies. But we cannot say that those nonhuman bodies possess their own value, and are themselves value-distributing bodies. So we, we humans, could choose to withhold value from nonhuman bodies, arbitrary or capriciously, or for apparently good internal reasons of our own. It doesn't matter. We are the ones with the power to generate and distribute value. The beneficiaries of our value-redistribution scheme, having no intrinsic value of their own, depend on us for the acquisition of value. Unavoidably, then, if we rely on a naturalistic framework, we can't but subjectivize value; we entrench it in the value-conferring human subject, and paradoxically reinforce rather than overcome human exceptionalism.

Bennett thus seems to be paradoxically captive to a picture of human agency that is continuous with human exceptionalism. But I think it is also a corollary of contemporary skepticism about and disappointment with the human. Throughout the text she asks, somewhat peevishly, why it is that we think we need a human-specific account agency, discontinuous with the nonhuman: "Why are we so keen to distinguish the human self from the field? Is it because the assumption of a uniquely human agency is, to use Kantian language, a 'necessary presupposition' of assertion as such? Or is the quest motivated by a more provincial demand that humans, above all other things on earth, possess souls that make us eligible for eternal salvation?"[29]

But are these the only possibilities? Must we be so skeptical, not in the sense of critical, but simply suspicious ex ante of any inquiry that seeks a differentiated account of human and nonhuman agency? Why couldn't we say that we humans are unavoidably self-interpreting animals without diminishing the agency and value of nonhuman animals? Or as, Heidegger put it in the more ontological register of *Being and Time*, human Dasein is the kind of being whose being is an issue for it. This hermeneutic or ontological distinctiveness does not underwrite human exceptionalism or license human beings to rule over other beings, or to deny the existence of intrinsic value in the world, value that is therefore not at our subjective disposal either to distribute or to withhold. Neither our efforts to make sense of our kind of being nor skepticism about the kind of beings we are will come to an end so long as humans exist. That just may be the ontological condition of being human. However, our self-interpreting activities cannot go on as before. We cannot deny the entanglement of and interdependence between human and nonhuman life, and that too will affect our self-interpretations (in ways not yet foreseeable).

Although Bennett finds in Dewey and Adorno views of human/nonhuman entanglement and interdependence that approximate her own, they don't go quite as far as Bennett would like. And that is because neither Dewey nor Adorno would assent to the naturalistic premises of Bennett's view of agency. They thought it unwise make such a sharp cut between affect and reason, between responsiveness to affects and responsiveness to normative expectations. It would be wiser to find a way to gather all that together in a richer and more complex view that admits of both significant continuities and significant discontinuities. They would not favor a flattened ontology, simply because of their commitment to the irreducibility of the normative. As Adorno wrote in *Negative Dialectics*, a passage which Bennett herself cites, we should be disinclined "to place the object on the orphaned royal throne once occupied by the subject. On that throne the object would be nothing but an idol." To avoid the mistake of fetishizing the object in order to correct the mistake of fetishizing the subject, Adorno recommended thinking both in dialectical relation, but with a normative orientation to the other—whether it be human or nonhuman.[30]

We no longer want an outlook that posits a radical and total discontinuity between human and nonhuman. That is no longer an option. But neither do we want an outlook that erases all discontinuity, in principle. We want and need multiple lines of continuity and discontinuity in all their human and nonhuman complexity. But these lines can never be drawn once and for all; they will be contestable and revisable in practice. In *The Public and Its Problems*, a text from which Bennett also draws inspiration, Dewey acknowledged the interconnection and interdependence between human and nonhuman life without erasing significant discontinuities: "In describing what is common in human and other animal junctions and consolidations we fail to touch what is distinctively human in human associations. . . . Physics and chemistry as well as zoology may inform us of some of the conditions without which human beings would not associate. But they do not furnish us with the sufficient conditions of community life and of the forms which it takes."[31]

Agency as Normative Responsiveness

As much as she expresses confidence about the view of agency she attributes to nonhuman agents, Bennett also registers that there is something not quite right about her approach, but she can't quite identify what that is: "It seems necessary and impossible to rewrite the default grammar of agency, a grammar that assigns activity to people and passivity to things."[32] Why accept the limitations of this framework? Because she does not put into question the picture of agency on which she depends, and by which she is obviously constrained, Bennett cannot think agency anew. The only move open to her is to distribute causal agency to the nonhuman.

That move does not only follow from her naturalistic premises. It arises from a normative impulse, which, however, Bennett cannot express in normative terms. She claims that things call for our attentiveness, or even "respect" (provided that the term is stretched beyond its Kantian sense).[33] But what does it mean to say things "*call* for our attentiveness, or even 'respect'"? Let's pause right here for a moment, and take a closer look at what is involved in understanding ourselves as called, as the addressees of a call. This call for attentiveness or respect is not just any kind of call; it is a normative call, a call demanding normative response. Why? Because it calls us to attend to how we're treating an "other," to how we're living rightly or wrongly in relation to it, potentially calling us to live differently. Moreover, it calls to a particular kind of being, not just any kind of being. It calls to a being capable of normative response, capable in the sense that it understands itself as called, and therefore as answerable, even if it might not, to begin with, understand what the call is about or what it actually calls for in response.

Curiously, in this respect, her book begins with just such a call for attentiveness and respect, provoked by the sight of miscellaneous objects contingently drawn together and caught in the grate over a storm drain—a dead rat, a stick of wood, a plastic glove, some oak pollen, a plastic bottle cap. "Struck" by the sight of these things, Bennett comes to realize that things have a power or agency of their own, manifested in the effect they were having on her: "I caught a glimpse of an energetic vitality in each of these things, things I generally conceived as inert."[34] These objects, she came to see, were not just objects, but they suddenly possessed the dignity of things, and therefore called for attentive respect.[35] Indeed, through this encounter she "glimpsed a culture of things irreducible to a culture of objects."[36] That she responded this way cannot be captured in causal terms alone. To be moved to see mere objects as things that have revelatory power (to see a "culture of things" and not just a "culture of objects") is not merely an effect produced by some unexpected cause. There is a normative transformation of perception taking place, whereby objects become things, not merely through their "vitality" but through a normative revaluation, itself a consequence of a normative response to the call of things to attend to their thingliness, to see them as more than merely disposable, fungible objects.

Unfortunately, nowhere in her book does Bennett attend to the normative implications of this call, which presupposes a normative relationship between caller and called, and the normative being of the being who is called. For such being there is a question about how to respond, and whose response is open to judgment about whether it was the right or wrong response, an adequate or inadequate response, a morally sensitive or morally insensitive response. Bennett's naturalistic commitments preclude normative considerations of this kind. Bennett does speak of responsiveness in connection with Spinoza's account of affects, but the responsiveness she speaks of is construed natural-istically not normatively. She takes from Spinoza the idea of an affect as "the capacity of any body for activity and responsiveness."[37] But there are all kinds of ways for one body to respond to another without involving or calling for normative response. The latter is not just another effect brought forth by the causal action of one body on another. Inattentiveness to the normativity of the relationship between "call and response" is part of the reason why Ben-nett fails to see that there is no clear normative pathway from the recogni-tion of vibrant matter as such to a change of normative attitude toward the nonhuman.

Clearly, we need a different approach from the one Bennett or Latour offers, an approach that does not rest content with distributing a flawed conception of human agency to nonhuman beings. I understand and sympathize with the motivation, and I am also committed to the idea that the nonhuman has an agency of its own, but I am not persuaded by what Bennett and Latour have proposed, and am quite convinced that we have to think more deeply and more critically about agency, both human and nonhuman agency. Moreover, I think it is completely unacceptable to say that thinking about human agency is pointless and futile, an unavoidably aporetic enterprise, especially when such claims must rely on the very same deeply flawed conception of human agency to account for the agency of the nonhuman.

To get free of the inadequacies of the default notion of agency we need an altogether different starting point, one that does not unquestioningly begin from the premise that agency is a kind of causality. We need to think about agency as something other and more than a form of causality; we need to think of it as a form of normative responsiveness. What Bennett requires but cannot provide is a view of agency as a normative response to a call from something or from someone. She cannot provide such a view because she is so deeply committed to a naturalistic view of agency as causality, according to which what is significant about agency simply is its causal power, and for which success or failure is necessarily adjudicated by a performance standard. Successful agency means successfully affecting that on which your agency is exercised. And here too, by remaining captive to this picture of agency, we once again reinforce human exceptionalism. The performance standard that is built into this concept of agency will indeed lead inevitably to the primacy and superiority of human agency. Humans will always win at this causal power game, all the more so when all agents, human and nonhuman, are placed on the same level. The "age of the humans" is the age in which the human species has become a geological force—a superhuman agency whose actions as a spe-cies have brought about the end of the Holocene era.[38] By subscribing to a causal conception of agency we allow humans to remain at the apex of agency, the very outcome that we wanted to avoid.

But what if we think of what it is to be an agent not in terms of causal power, but rather in terms of the capacity to be normatively responsive to something or someone, where being responsive means getting right—in some meaningful sense of getting right—what it is we are responding to, and how we ought to respond. Understanding that which calls us and understanding ourselves as called are continuous with each other, and, at the same time, are inseparable from a reflective awareness of and sensitivity to normative standards that are in play in how one responds. If we think of our agency as a form of normative responsiveness rather than as a special kind of causality, we are on the way to a very different picture of agency that captures something distinctive about human agency without reinstating human exceptionalism. And, I believe, it opens up the possibility of thinking of nonhuman agency in a way that is not simply the extension of the causal picture to nonhuman agents.

What I am suggesting here is part of an alternative picture of agency that has been proposed in different but overlapping ways in the tradition of philosophical romanticism, in Emerson and Heidegger, more recently in the work of Charles Taylor and Stanley Cavell, in a large body of feminist scholarship, and also, outside the academic world, in centuries-old Indigenous practices of caring for the human and nonhuman world.

As Charles Taylor puts it, "what is crucial about agents is that things matter to them."[39] Agency is thereby reconceived as kind of mattering, and we come to "understand an agent essentially as a subject of significance."[40] What is distinctive to human agency is "that there are matters of significance for human beings which are peculiarly human," which get expressed in peculiarly human kinds of mattering: "An agent can be a respondent, because things matter to it in an original way."[41] Contra Bennett, it is not the matter, vibrant or otherwise, but the *mattering* that is definitive of our responses to the human and the nonhuman. Human agency is distinguished not by some mark of superiority in the field of mattering, as though there could be a winner/loser contest over which species is better at caring about what matters to it. Unlike cases of cognitive and linguistic performances, there is no performance standard for mattering in light of which human agency could be tested. As a matter of normative responsiveness, human agency is not about cognitive and linguistic superiority vis-à-vis nonhuman agents, but about how agents let things matter to them, especially things that didn't matter before, but have come to matter now, and come to matter now because of the mattering things they have become.

Let me emphasize here that the responsiveness to what calls us is not exclusively to a human call; the call can come from anywhere, from anyone, from anything. What *may* be distinctively human is our responsiveness to what calls, or our failure to be responsive (experienced as shame, guilt, remorse, and so on), our failure to count what calls as one of our concerns, as that which matters or should matter to us, and thus to regard our failure to be normatively responsive in this sense as a failure to let something matter, as a failure to care about what should matter to us.[42]

Receptivity as Answerability

If we think of human agency exclusively in causal terms, we are unable to capture the kind of agency that is involved in a receptive answer to the call of the nonhuman. Conversely, to think of the nonhuman in causal terms also distorts the agency of the nonhuman, and thereby the responsiveness that is called for. My thinking about how the causal picture of agency distorts the nonhuman is still evolving, and I do not think I'm ready to articulate and defend my intuition about this here. However, I do want to say a little about my model of receptivity as one way to think about human agency as a form of normative responsiveness, which is far better attuned or attunable to both the call of the human and the call of the nonhuman than is any causal account.

By receptivity, I do not mean being open to someone or something; I mean becoming answerable to someone or something. I identify receptivity with answerability, not with openness. Being open and attentive to something or someone is a necessary but insufficient condition of receptivity in my sense. Thus, I conceive of receptivity as a form of normative responsiveness that is both spontaneous and reflective, which is to say, as a form of agency through which we are responsive to something or someone in an attitude of answerability. Conceiving of receptivity in this way allows us to think of our epistemic and normative agency, our mindedness, if you like, as involving and requiring exposure to human vulnerability—the vulnerability of a being that can be marked, struck, impressed by experienced reality, by what and whom it encounters in the world.

Thus, when we talk about empathy or compassion, about imaginatively entering into the being of another, whatever other that may be, we are presupposing receptivity to that other, for it is only when we allow entry of that other into our being—only when we allow ourselves to be marked, struck, impressed by it—that we are able to then enter into its being. That is why I prefer to think of receptivity as an unclosing of oneself. Unclosing involves cognitive and normative struggle—a struggle with ourselves to become receptive to that which we cannot know or understand without becoming answerable to it, whatever it may be. And this is why I do not think receptivity in my sense can be instrumentalized. We cannot instrumentalize what we cannot master. The struggle to become receptive to what we cannot know or understand without becoming answerable to it is not a struggle that comes to an end, is not a struggle that we can excel at, something over which we can gain mastery.

I have argued elsewhere that there is a class of recognition failures, cases of misrecognition that cannot be captured in the conceptual language of recognition.[43] That is because what we are dealing with in such cases are not failures of recognition, but rather failures of receptivity. Recognition presupposes receptivity, for the struggle to recognize and acknowledge the other—be it a human or nonhuman other—is first of all a struggle to become receptive to the other, particularly when the other's call for acknowledgment appears in an unfamiliar form, in a language we don't speak, or in a form for which the concept "language" does not apply. I have used as a test case the fictional character Elizabeth Costello in J. M. Coetzee's *The Lives of Animals*.[44] What makes her a salient example is that she is a figure of unintelligibility, both for the other characters in the novella in which she first appears and for the commentators invited to respond to it. As a figure whose intelligibility is constantly in question, Costello enjoins us to confront the question of how we are supposed to respond to something or to someone when it is not clear just what it is we are dealing with and what it is that is being asked of us. Can we speak meaningfully about responding rightly or wrongly in cases like this—as readers, as intimates, as citizens, as human beings? What can guide us, normatively speaking? How do we tell what form of response we owe an other—be it a human or nonhuman other—when that other appears inscrutable to us? Where do we begin? What might get in the way of a reply that is just, in every sense of just, not callous, insensitive, indifferent, destructive?

You could say that the picture I'm presenting here is one that conceives of agents as set within (or thrown into) a normative space of indeterminacy, striving to make intelligible what appears to be unintelligible. This differs markedly from the picture of agents as set within a normative space of reasons seeking to arrive at mutually acceptable reasons or of agents set within a normative space of unequal recognition seeking mutual recognition; rather, it is a picture of agents situated in a normative space of exposed intelligibility, a normative space in which their intelligibility and the intelligibility of what or whom they encounter are at issue. The point of politics would not be reducible either to practices of normative justification or normative recognition: the point of politics would be to enable practices of building receptivity to that which is voiceless or invisible, to that which is denied appearance, where building receptivity means building answerability.

Receptivity, then, is not merely a moral and political good, as in it is good to listen to others, to give them a hearing; it is both a condition of making sense of things and others and a normative demand we place both on ourselves and on others. We are always already receptive in a prereflective way. As Heidegger demonstrated in *Being and Time*, our prereflective stance of receptivity is an ontological condition of intelligibility in general. Receptivity is necessarily a selective stance, whereby—unjustly, unjustifiably—we are more receptive to some things, some persons, some concerns, more than to others, and so it is the normative task of our reflective stance of receptivity to attune ourselves to the selectivity of our reception, and to understand how this came to be so. This is not just a cognitive undertaking, for in becoming attuned to what and to whom we were unable to be receptive, we have also to acknowledge and take on, be answerable to, the normative demands that follow from such acknowledgment, even if, as it is quite often the case, it is not clear how to answer, or what to answer.

However, it is from just such normative responsiveness that we become aware of new normative demands, new claims laid upon us by something or someone, calling us to respond, not just with any kind of response, but rather with one that requires, manifests, a freer relation to ourselves. The freedom I am referring to comes into play when we make room for the call of an other, rendering intelligible what may have been previously unintelligible. Becoming receptive to such a call means facilitating its voicing, letting it become a voice that we did not allow ourselves to hear before, and responding to it in a way that demands something of us that we could not have recognized before. In responding freely to such a call, which means becoming answerable to it, we allow ourselves to be unsettled, decentered, thereby making it possible to occupy a potentially self-critical and illuminating stance. From such a stance it may become necessary to confront the possibility that we cannot go on as before, that some change is demanded of us, a change we may envision only in an indeterminate and inchoate way, but to which we feel nonetheless obligated to be receptive, answerable. Why? Because it is here, in this gap between who we have been and who we might be, that we see our own freedom is at stake, and here where we may "find" it manifested. Becoming receptive to the normative demands of change, to the work of going on differently in relation to an other or to others, is to manifest this freer relation to ourselves as agents for whom things matter.

I have been using the language of normativity a great deal but not in the way it is typically used. By normativity I do not mean the normativity of rules or law—I do not mean prescriptive or coercive norms. There is much more to normativity than what is captured in the idea of normativity as something rule-like or law-governed in nature.[45] Indeed, I'm of the view that this picture of normativity has distorted our understanding of the normative, and because it so dominates our thinking about what counts as normativity, it has had very destructive effects on our social practices and political institutions. We need to resist our tendency—whether we are critics or defenders of this picture of normativity—to identify it with normativity as such. But that is another story for another day.

Notes

1. In August 2016, the Working Group on the Anthropocene recommended to the International Geological Congress that the Holocene period should be regarded as over, and that we have already entered the Anthropocene period: Damian Carrington, "The Anthropocene Epoch: Scientists Declare Dawn of Human-Influenced Age," *Guardian*, August 29, 2016, www.theguardian.com/environment/2016/aug/29/declare-anthropocene-epoch-experts-urge-geological-congress-human-impact-earth.
2. I am thinking of the work of Latour, Haraway, and others in science and technology studies, work in neuroscience and cognitive science, such as the work of Andy Clark, and the now voluminous literature on the "animal turn."
3. Jacques Derrida, *Negotiations*, trans. Elizabeth Rottenberg (Stanford: Stanford University Press, 2002), 209.
4. Indeed, one might say that it got lost as a question—again. I say "again," because the question of the human is one that repeatedly gets lost, rather strangely, since so much is at stake in how we answer it—or, rather, to be more precise, in how we have answered and are always answering it.
5. Jürgen Habermas, *The Future of Human Nature*, trans. William Rehg, Max Pensky, and Hella Beister (Cambridge: Polity, 2003), 15.
6. Val Plumwood, "A Review of Deborah Bird Rose's Reports from a Wild Country: Ethics for Decolonisation," *Australian Humanities Review* 42 (August 2007).
7. Michel Foucault, *The Order of Things* (New York: Vintage, 1970), 387.
8. Foucault, 387.
9. See Stanley Cavell, *The Claim of Reason* (Oxford: Oxford University Press, 1979).
10. Jane Bennett, *Vibrant Matter* (Durham: Duke University Press, 2010), ix.
11. Bennett, 120.
12. Bennett, 34, my emphases.
13. See Ernst Tugendhat, *Self-Consciousness and Self-Determination*, trans. Paul Stern (Cambridge, MA: MIT Press, 1986).
14. I am not endorsing this view of agency. I am referring to it merely for illustrative purposes.
15. Bennett, *Vibrant Matter*, viii.
16. Although Bennett appeals to the idea of emergent causality, which she derives variously from Whitehead, Bergson, and Deleuze, the basic unit of agency remains unchanged. What she calls emergent causality is the confluence of diverse and heterogeneous agencies whose individual agency is nonetheless conceived in terms of efficient causality. In the end, what counts is "a doing and an effecting by a human-non-human assemblage." Bennett, 28.
17. Bennett, ix.
18. Bennett, 13.
19. Bennett, 12.
20. Bennett, x.
21. Baruch Spinoza, "Ethics," preface to part 3, in *The Collected Works of Spinoza*, vol. 1, ed. and trans. Edwin Curley (Princeton: Princeton University Press, 1985).
22. I do not have the space to do so here, and in any case it's not quite the right context, to properly discuss this convergence and its implications.
23. Bennett, *Vibrant Matter*, 98.

24. Of course, these are long-standing questions in the ongoing debates over the adequacy of naturalistic frameworks for explaining human action and human normativity. There is an enormous literature of arguments for the irreducibly normative character of human action, which cannot come into view and is therefore not accessible to third-person observation descriptions that naturalists privilege. What is frustrating and disappointing is that Bennett's argument for naturalism proceeds in the form of a testament and does not take account of antinaturalistic objections to the naturalization of agency and normativity, or of the available antinaturalistic accounts of agency and value.

25. Bennett, *Vibrant Matter*, 13.

26. Akeel Bilgrami, *Secularism, Identity, and Enchantment* (Cambridge, MA: Harvard University Press, 2013), 155. On pp. 161–162, he offers his own cogent objections to Bennett and Latour.

27. Bennett, *Vibrant Matter*, ix, 120.

28. Bennett, 120.

29. Bennett, 121.

30. Theodor W. Adorno, *Negative Dialectics*, trans. E. B. Ashton (London: Bloomsbury, 1981), 181.

31. John Dewey, *The Public and Its Problems*, in *Later Works*, vol. 2 (Carbondale: Southern Illinois University Press, 2008), 243.

32. Bennett, *Vibrant Matter*, 119.

33. Bennett, ix.

34. Bennett, 5.

35. Bennett seems to be invoking without mentioning Heidegger's distinction between mere objects and things, which, as distinct from mere objects, have the power to gather a world.

36. Bennett, *Vibrant Matter*, 5.

37. Bennett, xii.

38. Dipesh Chakrabarty, "The Climate of History: Four Theses," *Critical Inquiry* 35 (Winter 2009).

39. Charles Taylor, "The Concept of a Person," in *Human Agency and Language: Philosophical Papers*, vol. 1 (Cambridge: Cambridge University Press, 1985), 98.

40. Taylor, 102.

41. Taylor, 99.

42. I am not committed to a specifically and exclusively human notion of normativity. The wonderful thing about seeing normativity more capaciously is that it allows making distinctions between kinds of species-specific normativity which distinguish human and nonhuman animals from each other, but at the same time draws them together too. We need a differentiated understanding of agency and normativity, not one monistic conception distributed across altogether diverse kinds of agents.

43. Nikolas Kompridis, "Receptivity and Recognition: Forms of Normative Response in the Lives of the Animals We Are," *New Literary History* 144, no. 1 (April 2013). Reprinted in Nikolas Kompridis, ed., *The Aesthetic Turn in Political Thought* (London: Bloomsbury, 2014), 145–168.

44. John M. Coetzee, *The Lives of Animals* (Princeton: Princeton University Press, 1999).

45. See Nikolas Kompridis, "The Idea of a New Beginning: A Romantic Source of Normativity and Freedom," in *Philosophical Romanticism*, ed. Nikolas Kompridis (London: Routledge, 2006), 32–59.

CHAPTER 17

NATURAL PIETY AND HUMAN RESPONSIBILITY

DAVID BROMWICH

I will be pursuing some hints that follow from the idea of nature as a continuous process, and of human nature as an inseparable part of the same process. The idea has long seemed a necessary constituent of the complex planetary life we take for granted; but we live at a time when human nature is beginning to be discussed, instead, as a kind of baseline: a thing to be revised by the dictates of the market or the social will, and obedient to the new technologies of the self. The idea I am defending presumes that such revision cannot be a matter of will and choice without bringing in its train a radical disruption of our conception of humanity. Our self-definition as human creatures goes hand in hand with an experience of physical nature. To repeat: physical nature, and not a regulated simulacrum pliable to human will. For there is a moral distinction that turns on the experiential difference. One comes to know oneself, imaginatively, as part of nature; and nature is understood to have a dignity quite apart from its susceptibility to human control. It is the experience of human nature in that wider setting that sponsors the growth of conscience; and conscience, in turn, owes a responsibility to nature that is at once vivid and hard to specify. The sense of human life as a participant in nature also teaches humility in the face of willful experiments—whether the experiments are pursued for the sake of social improvement or self-expansion. My argument will draw heavily on William Wordsworth and Hans Jonas, and the resemblances here between a Romantic poet and a modern philosopher are fascinating in themselves. But I have been influenced by these writers, above all, because they reflect searchingly on the growth of the individual mind from the experience of something beyond its control: a source of wonder that always eludes complete comprehension.

Wordsworth thought of the poet as a *maker*. The poet makes things out of words that come from the imagination; and the imagination works on materials inseparable from the poet's experience of the world: sensations, impressions, ideas. As Wordsworth understands this process, there is no sharp distinction between feelings and thoughts: "our thoughts," he wrote in the preface to *Lyrical Ballads*, "are indeed the representatives of all our past feelings." He likewise enforced no separation between the blended elements of "all the mighty world / Of eye, and ear,—both what they half create, / And what perceive." The maker and the doer exist in the same person and are burdened by the same "daring sympathies with power": a power that the poet may belatedly recognize as having a correlative in himself. "Then your eyes shall be opened, and ye shall be as gods"—the text from Genesis is the almost quoted starting point of Wordsworth's reflections in *The Prelude* on the way a desire for knowledge is implicated in the drive toward an ever-expanding mastery of nature. The pursuits of art and science, the wish to make a thing that is beautiful and the will to discover an unknown element or property or possible combination in nature, proceed from the same restless instinct. The name, *sublimation*, which we have given to the refinement of this demand, is in one sense a misleading euphemism. All the productions of art and science originate in the same primitive and irreducible and by no means exalted motive.

This discovery may be deeply unsettling, but it carries with it an ameliorative possibility. In the autobiographical narrative of *The Prelude*, Wordsworth's gradual recognition of his self-sufficiency is shown to coincide with a realization that the world is a blessed place—a place that needs no God to assure us that it is good. Nothing, he says, makes the world a home for us, nothing for that matter gives unity to the world as we understand it, except ourselves and the environing conditions that make us possible. Yet in confronting this truth, we feel wonder and not despair. We see with gratitude that the world is a place adequate to our purposes:

> I was left alone,
> Seeking the visible world, nor knowing why.
> The props of my affections were removed,
> And yet the building stood, as if sustained
> By its own spirit![1]
> (*Prelude*, 1805 version, 2.292–296)

In an older form of theodicy, the building that is the visible world could only stand because it was sustained by the spirit of God. As Wordsworth comes to see, however, the objects of the world have a miraculous self-sufficiency; they do not require for sustenance the props of his affections or God's approval. The passage almost says outright that the visible world is superior to the invisible world of theological supposition.

While this discovery makes us lonelier than before, it also brings an immediacy to our contact with nature—as if nature with all its "finer influxes" were conjoined with our own life.

> All that I beheld
> Was dear to me, and from this cause it came,
> That now to Nature's finer influxes
> My mind lay open, to that more exact
> And intimate communion which our hearts
> Maintain with the minuter properties
> Of objects which already are beloved,
> And of those only.
> (2.296–303)

Wordsworth—still a boy, yet feeling the power of action belonging to later life—attends to all that can be seen and heard of nature. He hears it as an unfamiliar language, which nonetheless he finds that he can understand:

> I would walk alone,
> In storm and tempest, or in starlight nights
> Beneath the quiet heavens, and at that time,
> Have felt whate'er there is of power in sound
> To breathe an elevated mood, by form
> Or image unprofaned; and I would stand,
> Beneath some rock, listening to sounds that are
> The ghostly language of the ancient earth,
> Or make their dim abode in distant winds.
> Thence did I drink the visionary power.
> (2.321–230)

He attributes to "the ancient earth" characteristics we normally think of as uniquely human. It provides a home, an "abode" for the elements, and shows an integral continuity from its ghostly origins to the present moment.

Moments such as these, of identification with nature, become for Wordsworth a source of strength in his adult life, even when the constituents of the memory have faded. The moments are a pledge of continuity with himself, parallel to the influx he has felt from nature:

> I deem not profitless those fleeting moods
> Of shadowy exultation: not for this,
> That they are kindred to our purer mind

And intellectual life; but that the soul,
Remembering how she felt, but what she felt
Remembering not, retains an obscure sense
Of possible sublimity, to which
With growing faculties she doth aspire.

(2.331–338)

We are made to see that physical pleasure, derived alike from animate and inanimate nature, is vital to the growth of a poet's mind. (And Wordsworth knew this to be the true subject of his poem; he often referred to it simply as his "poem about the growth of a poet's mind.") He looks on his intensifying love of the physical world as a process inseparable from self-discovery; yet it is a process in which the resistant world of outward things must participate; for it involves all the textures and secreted stimuli of nature—"the hiding places of man's power"—which the poet heeds as an admonition and an obscure source of hope.

This way of thinking may border on animism. It certainly requires that we refuse to calculate the value of our relationship to nature from moment to moment. To reduce the ancient earth to an instrumental convenience—as we do when "we murder to dissect"—is to commit the error of denying the worth it carries as a whole. Further on in *The Prelude*, Wordsworth will describe the common work of a shepherd in a way that permits no such division of labor and quantification of value:

I remember, far from home
Once having strayed, while yet a very child,
I saw a sight, and with what joy and love!
It was a day of exhalations, spread
Upon the mountains, mists and steam-like fogs
Redounding everywhere, not vehement,
But calm and mild, gentle and beautiful,
With gleams of sunshine on the eyelet spots
And loop-holes of the hills, wherever seen,
Hidden by quiet process, and as soon
Unfolded, to be huddled up again:
Along a narrow valley and profound
I journeyed, when, aloft above my head,
Emerging from the silvery vapours, lo!
A shepherd and his dog! in open day:
Girt round with mists they stood and looked about
From that enclosure small, inhabitants
Of an aerial island floating on,

As seemed, with that abode in which they were,
A little pendant area of grey rocks
By the soft wind breathed forward. With delight
As bland almost, one evening I beheld,
And at as early age (the spectacle
Is common, but by me was then first seen)
A shepherd in the bottom of a vale,
Towards the centre standing, who with voice,
And hand waved to and fro as need required,
Gave signal to his dog, thus teaching him
To chase along the mazes of steep crags
The flock he could not see; and so the brute,
Dear creature! with a man's intelligence
Advancing, or retreating on his steps,
Through every pervious strait, to right or left,
Thridded a way unbaffled; while the flock
Fled upwards from the terror of his bark
Through rocks and seams of turf with liquid gold
Irradiate, that deep farewell light by which
The setting sun proclaims the love he bears
To mountain regions.

(8.81–119)

Sun, wind, and mist are blended here to create a texture of visibility that touches the edge of the visible world; the power of these glimpses of human work, in the mind of Wordsworth, depends on the thoroughness of the integration between landscape and figure. At his sight of the shepherd, the poet feels himself braced by a fellow intelligence, just as the shepherd is accompanied by the "Dear creature! with a man's intelligence." The poet's awareness of the ages of geological time required to form such "mountain regions" adds to the sense of a moment that is consecrated and timeless.

* * *

Episodes like this in *The Prelude* are to be understood in combination with those that recall Wordsworth's instinctive exertions of power: the prehistory of his natural piety. Such encounters arise from an adventurous will that must be chastened before it can deliver the poet as "a dedicated spirit." His passage on the labor of shepherds shows a happy discovery of beauty and the symbiosis of man and nature; and yet, Wordsworth says that his childhood imaginings were "Fostered alike by beauty and by fear." What does that mean? The fear was a rebuke to the egotism of his youthful exploits: taking eggs from a raven's nest, or stealing a boat and rowing far out on a lake. He feels stirrings then that seem to betoken a will in nature opposed to his own—the harsh wind, for example, which both exhilarates and threatens; or the mountain peak that reveals itself like a punitive shadow as he rows farther out, so that memories of the moment become "a trouble to my dreams." We are given to understand that Wordsworth in these interludes is morally unformed. He might grow up a monster of self-will or the reverse. But his sense of nature as a necessary limit—his suspicion that there is something he must give to it, as well as take— renders the stirrings of conscience finally stronger than the urgency of unchecked appetite.

The most vivid and directly argued of these encounters can be found in the fragment "Nutting"—a separate short narrative, most likely intended for *The Prelude* but never incorporated in the text. The boy Wordsworth, some way into a happy and aimless walk, lies down to rest in a bower by a stream over-hung with hazel trees. A nameless impulse takes him suddenly from passive contentment to action:

> I heard the murmur and the murmuring sound,
> In that sweet mood when pleasure loves to pay
> Tribute to ease; and, of its joy secure,
> The heart luxuriates with indifferent things,
> Wasting its kindliness on stocks and stones,
> And on the vacant air. Then up I rose,
> And dragged to earth both branch and bough, with crash
> And merciless ravage: and the shady nook
> Of hazels, and the green and mossy bower,
> Deformed and sullied, patiently gave up
> Their quiet being.[2]

A sense of self-rebuke and reflection follows almost immediately:

> Ere from the mutilated bower I turned
> Exulting, rich beyond the wealth of kings,
> I felt a sense of pain when I beheld
> The silent trees, and saw the intruding sky.

The instruction could not be clearer. We are not to blend our pleasure or our pride with the pain of anything that lives. Nature itself is a living entity.

A further axiom is implied by the importance Wordsworth confers on such "spots of time." The placement of man in nature involves a special trust, a responsibility for the powers (both known and potential) that human intelli-gence may engender. The keeping of that trust depends on the way *self*-trust is allowed to develop. As creative and destructive beings, Wordsworth believes, we act more potently than we can bear to acknowledge; and in the extraordinary passage that begins book 5 of *The Prelude*, he confesses the full extent of his intuition. It comes to him in a guilty dream; in the early version I will quote, he partly disowns the dream by attributing it to "a friend"; in the final version of *The Prelude*, he would acknowledge the dream as his own.

The poet begins by confiding to his friend that he is haunted by a curious wish or fancy. He would like to be sure that, if the earth itself were to perish, some means could be found to save from destruction the monuments of human intellect:

> One day, when in the hearing of a friend
> I had given utterance to thoughts like these,
> He answered with a smile, that in plain truth
> 'Twas going far to seek disquietude;
> But on the front of his reproof confessed
> That he, at sundry seasons, had himself
> Yielded to kindred hauntings. And forthwith
> Added, that once upon a summer's noon,
> While he was sitting in a rocky cave
> By the sea-side, perusing, as it chanced,
> The famous history of the errant knight
> Recorded by Cervantes, these same thoughts
> Came to him, and to height unusual rose,
> While listlessly he sate, and, having closed
> The book, had turned his eyes towards the sea.
> On poetry and geometric truth,
> The knowledge that endures, upon these two,
> And their high privilege of lasting life,
> Exempt from all internal injury,
> He mused, upon these chiefly: and at length,
> His senses yielding to the sultry air,
> Sleep seized him, and he passed into a dream.
> (5.49–70)

His friend—and it will matter in the following account that the friend is also a poet—cherishes a hope for the "privilege of lasting life," beyond the duration of the earth. Though the earth itself will be destroyed, this fantasy allows the great productions of the mind to stay "Exempt from all internal injury." Self-pity and ambition are equal partners in the formation of such a fantasy; and with that glimmer of self-knowledge, Wordsworth enters the dream recounted by his friend:

He saw before him an Arabian waste,
A desert, and he fancied that himself
Was sitting there in the wide wilderness,
Alone, upon the sands. Distress of mind
Was growing in him when, behold! at once
To his great joy a man was at his side,
Upon a dromedary, mounted high.
He seemed an Arab of the Bedouin tribes:
A lance he bore, and underneath one arm
A stone, and in the opposite hand, a shell
Of a surpassing brightness. Much rejoiced
The dreaming man that he should have a guide
To lead him through the desert; and he thought,
While questioning himself what this strange freight
Which the new-comer carried through the waste
Could mean, the Arab told him that the stone
(To give it in the language of the dream)
Was "Euclid's Elements"; "And this," said he,
"This other," pointing to the shell, "this book
Is something of more worth"; and at the word
The stranger, said my friend continuing,
Stretched forth the shell towards me, with command
That I should hold it to my ear. I did so
And heard that instant in an unknown tongue,
Which yet I understood, articulate sounds,
A loud prophetic blast of harmony;
And Ode, in passion uttered, which foretold
Destruction to the children of the earth
By deluge, now at hand. No sooner ceased
The song, but with calm look the Arab said
That all was true; that it was even so
As had been spoken; and that he himself
Was going then to bury those two books:
The one that held acquaintance with the stars,
And wedded man to man by purest bond
Of nature, undisturbed by space or time;
The other that was a god, yea many gods,
Had voices more than all the winds, and was
A joy, a consolation, and a hope.
My friend continued, "strange as it may seem,
I wondered not, although I plainly saw
The one to be a stone, the other a shell;

Nor doubted once but that they both were books,
Having a perfect faith in all that passed.
A wish was now engendered in my fear
To cleave unto this man, and I begged leave
To share his errand with him. On he passed
Not heeding me: I followed, and took note
That he looked often backward with wild look,
Grasping his twofold treasure to his side."
(5.71–120)

The Arab has entrusted to the poet the knowledge of the shell, identified with the prophetic power of imagination and with poetry. Yet that knowledge also encompasses—its existence may incite—a love of power that can lead to catastrophe. Though the poet thinks himself a natural historian of thoughts and feelings, and therefore a guardian of human nature against destruction, his imaginings pull him in the opposite direction. In the lines about following the stranger, one may notice that the poet suppresses any account of himself as an active force; rather, he portrays himself captive to the instructions of his guide; but the Arab looks "backward with wild look" and the look tells a different story, as if there were something about the poet that frightened his unsuspecting guide. Maybe it is the poet's claim of special and sublime knowledge (a knowledge as extrinsic to ordinary life as that of the scientist) that troubles the Arab. If one bears in mind the fear of himself that Wordsworth has recorded in "Nutting" and elsewhere, the dream at this moment suggests that the poet has failed the terms of his trust. He owed a debt to nature that he never fully grasped; and the consequence will appear in the dream, just before it breaks off.

As the poet resumes the narrative, we are told how the Arab, perceiving the effects of the "blast" of poetic prophecy, turned and rode quickly away:

"His countenance, meanwhile, grew more disturbed;
And, looking backwards when he looked, I saw
A glittering light, and asked him whence it came.
'It is,' said he, 'the waters of the deep
Gathering upon us'; quickening then his pace
He left me: I called after him aloud;
He heeded not; but with his twofold charge
Beneath his arm, before me, full in view,
I saw him riding o'er the desert sands
With the fleet waters of the drowning world
In chase of him; whereat I waked in terror,
And saw the sea before me, and the book,

In which I had been reading at my side."

(5.127–139)

The close of the dream signals the coming of a catastrophe. The forfeit of the world seems to be linked to the very success of imagination (or an acquiescence in the wrong kind of imagination). Looking back, we may see that a warning was implicit in the book the poet was reading before the dream, *Don Quixote*. It is a book, in fact, about fiction and poetry and the fantasies of power they may set in motion. Wordsworth understood that his own curiosity, like Quixote's, had its source in a generous-seeming interest in power. If we translate the insight into the allegory of the stone and shell, it suggests that the poet, as much as the scientist, may become an agent or instigator of destruction. So the dream is a cautionary fable. It warns against the excessive self-trust by which the mind may regard itself as a creator of value in the world. This warning applies not only to the technology of language, but to all possible technologies.

***

Natural piety had a complex meaning for Wordsworth. It referred to the connection I feel with my past self, and my gratitude toward the natural and human world in which I grew up. Each of these bonds seems sufficient only in the presence of the other; and the two together make a definition of integrity:

> My heart leaps up when I behold
> A rainbow in the sky:
> So was it when my life began;
> So is it now I am a man;
> So let it be when I grow old,
> Or let me die!
> The Child is father of the Man;
> And I could wish my days to be
> Bound each to each by natural piety.[3]

Notice that the sense of a bond—with one's own past and with nature—yields an idea of participation or integration. But to call it an idea may be too formal: Wordsworth declares his hope as a prayer—"Or let me die!" Why does the pledge of continuity require such vehemence of commitment? The reason lies in the power of a fantasy of independence, a fantasy of nonintegration and ascent above nature, which the prayer aims to counteract. Let me anticipate the argument that follows by calling it a fantasy of *secession* from nature. Through the exertions of imagination, assisted by technology, human beings have developed ever more sophisticated methods of creating a virtual climate, an environment of the mind, which provides its own envelope of stimuli, and in the presence of which the world of outward things becomes a secondary concern. Think of the temptations—possibilities that demand response with a pressing immediacy—created by our computer screens and handheld gizmos. By design, and with the aid of new-modeled habits, the effect is to cut us off from the experience of nature. Our best technology renders strange and possibly quaint the idea that it is right to feel an attachment to the actual world, to prefer the actual to the many virtual worlds in which we can work and play. Attachment to the physical world has become just one more option.

The use of natural scenes as a backdrop, a "screen saver" for computer adventures and word processing, illustrates the tendency clearly. While a hurricane outside destroys the natural environment, as part of the larger pattern of global climate disruption, you look into your screen and conduct a Skype interview or compose a second draft of a conference paper or make a personalized selection of clips to review the global news. This power to engulf ourselves in a chosen world, a substitute world, is something radically new. We have not yet begun to register its consequences for human consciousness. The allure of the computer screen, in particular, is new. And yet there have been warnings closer to our time than Wordsworth. In *The Human Condition*, Hannah Arendt pointed to a rupture in historical experience at the successful creation of the atom bomb. Man, in her phrase, for the first time was "acting into nature"; how far we would define ourselves by connection with nature was suddenly a matter of choice and volition. The coming of the personal screen inaugurated a second stage of the process; the effects are witnessed on a smaller scale and are apparently less astonishing; but, multiplied a billion times, in offices and cubicles and smaller spaces for clicking and sifting and watching, the inward secession takes on an appalling magnitude. In our lives online, we can create a thousand small modifications of our chosen psychological landscape or intellectual environment, and at the same time be reassured that none of this is necessary or permanent: we can always revise, rejigger, reconnect, and, if useful, exchange at pleasure all of the old relations for a new set. No condition seems intractable or stubbornly resistant.

This state of fulfillment is also a state of world-alienation. The chosen worlds we have created and are constantly modifying tend to become for us an indispensable resource and a super-induced primary need. To what end? We want, and we are being guided to realize as far as possible, the elimination from human life of accident and random experience. Earbuds are an apt emblem of the motive that drives this quest. For what purpose do they exist if not to seal off the world of outward things and to "program" a more desirable sound-world—without footsteps on the sidewalk, the rush of passing cars or trains, the barking of dogs, the chat and murmur of the couple at the next table. An intermittent and passive awareness of the crisis of global warming is the visible by-product of this self-isolation from nature. We are protected from regular reminders of the disintegration of nature by the ingenuity and ubiquity of our devices.

For anyone who, like Wordsworth, is convinced of the connection between personal identity, moral responsibility, and the placement of human beings in the physical world, the central question of ethics is how to think steadily about and somehow to counteract the temptation to secede from nature. And how to do so in a way that does not succumb to myth or superstition. This was the task Hans Jonas set himself in his great book *The Phenomenon of Life*. Jonas has come to be known to environmentalists for a later book, *The Imperative of Responsibility*, a work of polemical urgency first published in German in 1979 and widely credited as an inspiration for the founding of the Green Party in Germany. As Jonas interpreted it there, the imperative of responsibility to nature stands in contrast with the imperative of moral action toward persons above all, expounded in Kant's *Critique of Practical Reason*. Preserving and guarding nature become the duty that defines a fresh imperative, brought on by a situation that could hardly be foreseen in the eighteenth century. The resemblance to the later Heidegger's image of man as "the shepherd of Being" may offer a help to understanding what is at stake. Jonas's language, however, because it is less improvised and mysterious, seems to me less open to a reading that could evade its practical consequences.

The title *The Phenomenon of Life* was chosen deliberately to withhold the teleological consolation of Teilhard de Chardin's *Phenomenon of Man*. In that popular treatise, the reader was shown, says Jonas, "life's sure and majestic march toward a sublime consummation." He suggests, instead, that we look at the rising importance of human existence, in an environment that man both enlarges and disrupts, as "an experiment with mounting stakes and risks which in the fateful freedom of man may end in disaster as well as in success."[4] The imperative of responsibility contrasts with earlier interpretations of human history in another way. "Which is for the sake of which?" asks Jonas. "Contemplation for action, or action for contemplation. With this challenge to choice, biology turns into ethics."[5] I take it that contemplation for action denotes the use of nature instrumentally, for the purpose of extracting utilitarian goods or products to serve the short-term demands of society. The same motive may compel exertions that are world-altering in their effects: the splitting of the atom, for example, or genetic engineering. On the other hand, action for contemplation would treat the human stewardship of nature as an end that cannot itself become another means.

To excite the necessary commitment, Jonas asks that we appreciate the enormity of the transformation by which the elements of earth evolved from inanimate to animate matter. With time acting as one of the assistants, this was a miracle as great as any attributed to God: "It is in the dark stirrings of primeval organic substance that a principle of freedom shines forth for the first time within the vast necessity of the physical universe—a principle foreign to suns, planets, and atoms."[6] The gradual distinction of matter from living things was characterized, as Jonas puts it, by the evolution of three powers: perception, motility, and emotion. (Notice that these are the very traits remarked by Wordsworth, in the passage on the shepherd and his dog in book 8 of *The Prelude*.) Once we recognize that the acquisition of perception, motility, and emotion could follow only from the whole process of nature over time, we must wish to maintain the fabric and not just "samples" of nature as we have known it. We cannot pretend not to be informed by this prehistory of human thought and feeling; nor can we be expected to renounce the commitment we attach to it. Duties, therefore, arise from the fact that we live in the world: a point emphasized more recently by Akeel Bilgrami.[7] Human interdependence with nature is a compulsory and not an optional and occasional exercise. For we are not merely passers-by or spectators of nature.

Jonas's argument leads him to reconsider the human-centered account of morality that has prevailed in the West since the seventeenth century. "The contention—almost axiomatic in the modern climate of thought—that something like an 'ought' can issue only from man and is alien to everything outside him" must be understood as "more than a descriptive statement: it is part of a metaphysical position, which has never given full account of itself." Obligation, on the contrary, may be a discovery on a par with the discovery of gravity; and though it required the human mind to know this, the commandments that follow may be compulsory and affect man's relation to the nonhuman world. "Nor does it follow that the rest of existence is indifferent to that discovery."[8] Now, to speak of obligation as a discovery akin to those of physical science, or again, to speak of duties that arise immediately as a deliverance from our experience of value in nature itself: this is an unfamiliar way of talking. But there is a precedent for it in Wordsworth. Thus we find him saying in book 4 of *The Prelude*, about his sense of being companioned by society and nature: "I made no vows, but vows / Were then made for me; bond unknown to me / Was given."

* * *

I have borrowed and elaborated on Jonas's idea of a "secession" from nature with which the rapid advance of technology tempts the imagination. The suppressed matter always to be considered is that life "is exposed to the world from which it has seceded, and by means of which it must yet maintain itself."[9] Natural piety, however, remains a difficult and initially a counterintuitive achievement for a generation brought up with a constant reliance on technology and a minimum of exposure to nature: "*We know the thing at stake only when we know that it is at stake.*"[10] What, then, should be the change in our thinking once we learn that the world we care for might perish by our activity? That question is the starting point of *The Imperative of Responsibility*; and Jonas there propounds a guiding maxim: knowledge of the scope of the hazard, once absorbed, "can yield a not uncertain rule for decision making. It is the rule, stated primitively, that the prophecy of doom is to be given greater heed than the prophecy of bliss."[11]

The threat of destruction known to the poet—the prophecy "which foretold / Destruction to the children of the earth / By deluge, now at hand"—no longer appears an event in a dream. It is no longer a matter of imagination. "The privilege of freedom carries the burden of need and means precarious being," writes Jonas. "For the ultimate condition for the privilege lies in the paradoxical fact that living substance, by some original act of segregation, has taken itself out of the general integration of things in the physical context, set itself over against the world, and introduced the tension of 'to be or not to be' into the neutral assuredness of existence."[12] Those sentences accurately interpret the cause of the emotion of wonder that Wordsworth described in the first passage quoted earlier: "The props of my affections were removed, / And yet the building stood, as if sustained / By its own spirit!" In consequence of such an experience of wonder and gratitude, is there something that we owe to the world?

Jonas puts an answer in the form of an elemental precept. Man is "the executor of a trust which only he can see, but did not create."[13] The idea of human service as the executor of the whole estate of nature may help to clarify the otherwise abstract metaphor of stewardship. This is a form of service that stands against the endless process of human assimilation and rapacity. It can seem nothing but common sense for the prophecy of doom to override the prophecy of bliss; but when a way of life many centuries old is under challenge, what we *see* and *know* is hard to bring into line with what we are *willing to do*. Here again, Wordsworth seems in his way as farseeing as Jonas. One may take the dream of the stone and the shell to affirm that there are responsibilities that come with bad dreams. This is especially so when we know the dreams to be a direct issue of the ambitions and practices that govern our daily life.

The Phenomenon of Life—with its unusual scope and rigor as a philosophical, scientific, and imaginative work—seems to have been written to answer a broad question about human conduct. How can one *justify* the sensation of gratitude one feels for living in the world? And what follows from that justification? For wherever speculation may lead, we none of us live in "*a world*," a possible or constructed or future world. Our dwelling is "in the very world, which is the world / Of all of us, the place where, in the end, / We find our happiness, or not at all." Technology is the usable offspring of convenience that comes from the genius of science and art. But if the poet and scientist, the originators of the advances we know as technology, are to become at last the benefactors of humankind, they can have no higher purpose than to maintain the copresence of human nature with nature.

Expertise and innovation are beside the point. "The scientist," remarks Jonas, "is by his science no more qualified than others to discern, nor even is he more disposed to care for, the good of mankind. Benevolence must be called in from the outside to supplement the knowledge acquired through theory: it does not flow from theory itself."[14] He continues, with an insight that brings together the pagan myth of Prometheus and the Christian myth of the fall from Eden:

> The advent of man means the advent of knowledge and freedom, and with this supremely double-edged gift the innocence of the mere subject of self-fulfilling life has given way to the charge of responsibility under the disjunction of good and evil. To the promise and risk of this agency the divine cause, revealed at last, henceforth finds itself committed; and its issue trembles in the balance. The image of God, haltingly begun by the universe, for so long worked upon—and left undecided—in the wide and then narrowing spirals of pre-human life, passes with this last twist, and with a dramatic quickening of the movement, into man's precarious trust, to be completed, saved, or spoiled by what he will do to himself and the world. And in this awesome impact of his deeds on God's destiny, on the very complexion of eternal being, lies the immortality of man.[15]

The speculation is in some degree theological—inescapably so. For it is intended as a counterstatement against the Gnostic postulate of a spark of divinity in man that thrives only by hostility to the fallen world of created life. But supposing that God indeed has withdrawn from nature, his success or failure now resides with human beings who have it in us to complete or undo creation.

The injunction of natural piety once carried the force of a command from a higher power. It now draws credence from the recognition that human beings alone must make and keep the bond between humanity and the world; and with this goes our understanding of the punishment that negligence will incur: a penalty that human beings alone can inflict on themselves. In earlier ages, only the publication of Darwin's theory of natural selection afforded a comparable example of an epochal shift of human self-knowledge; but, as with that earlier discovery, so the present fears of human diminishment are fundamentally misjudged. "In the hue and cry over the indignity done to man's metaphysical status in the doctrine of his animal descent," Jonas comments, "it was overlooked that by the same token some dignity had been restored to the realm of life as a whole."[16] It is not a loss of dignity we face, but the emergence of dignity of an unfamiliar sort. Wordsworth, in the final lines of *The Prelude*, addressed his friend Coleridge to announce the mission they shared. Their poems would show how, in seeking the visible world, "the mind of man becomes / A thousand times more beautiful than the earth / On which he dwells." This can seem an anthropocentric conceit, and no doubt it bears a trace of the pride disclosed in Wordsworth's dream. But he concludes that the survival of the creative mind itself as a force for good is determined by the continuity of man's placement in "the earth / On which he dwells." This awareness, for those who feel it, can never be a source of resentment, as at an unwelcome limitation. We know it rather as the bond of experience that joins a distant past to a future beyond our imagining.

Notes

1. William Wordsworth, *The Prelude: A Parallel Text*, ed. J. C. Maxwell (Harmondsworth, UK: Penguin, 1971), 88. All subsequent references to the poem, given by book and line numbers, are to this text of the 1805 version.
2. William Wordsworth, *The Poems*, ed. John O. Hayden, 2 vols. (Harmondsworth, UK: Penguin, 1977), vol. 1, pp. 368–369.
3. Wordsworth, vol. 1, p. 522.
4. Hans Jonas, *The Phenomenon of Life: Toward a Philosophical Biology* (1966; Evanston, IL: Northwestern University Press, 2001), xxiv.
5. Jonas, 2.
6. Jonas, 3.
7. Akeel Bilgrami, *Secularism, Identity, and Enchantment* (Cambridge, MA: Harvard University Press, 2014), chap. 6.
8. Jonas, *Phenomenon*, 283.
9. Jonas, 5.
10. Hans Jonas, *The Imperative of Technology: In Search of an Ethics for the Technological Age* (Chicago: University of Chicago Press, 1984), 27, italics in original.
11. Jonas, 31. "The reproach of 'pessimism,'" he adds, "leveled at such partisanship for 'the prophecy of doom' can be countered with the remark that the greater pessimism is on the side of those who consider the given to be so bad or worthless that every gamble for its possible improvement is defensible" (34).
12. Jonas, *Phenomenon*, 4.
13. Jonas, 283.
14. Jonas, 195.
15. Jonas, 277.
16. Jonas, 57.

CONTRIBUTORS

Akeel Bilgrami is the Sidney Morgenbesser Professor of Philosophy at Columbia University. His latest book is *Secularism, Identity, and Enchantment* (Harvard University Press, 2014).

Jedediah Britton-Purdy is Professor of Law at Columbia University. His most recent book is *This Land Is Our Land: The Struggle for a New Commonwealth* (Princeton University Press, 2019).

David Bromwich is Sterling Professor of English at Yale University. His most recent book is *How Words Make Things Happen* (Oxford University Press, 2019).

Bina Gogineni is Assistant Professor of English at Skidmore College. She most recently coauthored "The Anthropocene's Dating Problem: Insights from the Geosciences and the Humanities" in the *Anthropocene Review* (2018).

Nikolas Kompridis is Research Professor in Philosophy and Political Thought and Director of the Institute for Social Justice at Australian Catholic University. His latest book is *Aesthetics and Political Theory* (Polity, 2020).

Anthony Simon Laden is Professor of Philosophy and Chair of the Department of Philosophy at the University of Illinois at Chicago and Associate Director of the Center for Ethics and Education. His most recent book is *Reasoning: A Social Picture* (Oxford University Press, 2012).

Kyle Nichols is Associate Professor of Geomorphology, Quaternary Geology, and Hydrology in the Department of Geosciences at Skidmore College. He most recently coauthored "The Anthropocene's Dating Problem: Insights from the Geosciences and the Humanities" in the *Anthropocene Review* (2018).

Joanna Picciotto is Associate Professor of English at the University of California, Berkeley. She most recently edited "Devotion and Intellectual Labor," a special issue of the *Journal of Medieval and Early Modern Studies* (2014).

Robert Pollin is Distinguished Professor of Economics and Codirector of the Political Economy Research Institute (PERI) at the University of Massachusetts, Amherst. His most recent book is *Greening the Global Economy* (MIT Press, 2015).

Sanjay G. Reddy is Associate Professor of Economics at the New School for Social Research.

Carol Rovane is Professor of Philosophy at Columbia University. Her most recent book is *The Metaphysics and Ethics of Relativism* (Harvard University Press, 2013).

Jonathan Schell was an author, journalist, and visiting fellow at Yale University and a writer for the *New Yorker, Newsday*, and the *Nation*. The author of numerous books, his *The Fate of the Earth* (Knopf, 1982) received the *Los Angeles Times* Book Prize, among other awards, and was nominated for the Pulitzer Prize, the National Book Award, and the National Critics Award. He died in March 2014.

James Tully is the Distinguished Professor Emeritus of Political Science, Law, Indigenous Governance, and Philosophy at the University of Victoria. His most recent book is *On Global Citizenship: James Tully in Dialogue* (Bloomsbury Academic, 2014).

Jan Zalasiewicz is Professor of Palaeobiology at the University of Leicester and Chair of the Anthropocene Working Group. He most recently coedited and coauthored *The Anthropocene as a Geological Time Unit: A Guide to the Scientific Evidence and Current Debate* (Cambridge University Press, 2019).

INDEX

Abram, David, 201–202

Action: basic, 86, 92; on climate change, 224; collective, 90–92, 95n6, 148; contrasts regarding, 226; conversation as, 215; as end-directed, 209–211, 226; as episodic *versus* ongoing, 211–214; exemplary, 232–235; intentional, 86; Kant on, 218; morality and, 217–218, 222n13, 231–235; nonepisodic, 226–227; nonproductive, 226; normativity, 219–220; political, 215; practical, 92; productive, 226, 228; social, 86; value and, 216–219

Activities: end-directed, 212–213; goal-directed, 215–216; nonproductive, 226, 228, 236; ongoing, 213–214, 216, 218–220; productive, 210–211, 236; sustainability of, 205, 213–214; value and, 213–214

Adamic man, 102–104

Addison, Joseph, 107

Adorno, Theodor, 97, 251–252

Aesthetic, the, 97–99

Agency, xiv, 110; Bennett on, 244–255; causality and, 254–256, 259n16; climate change and, 234–235; human, 244–245, 251, 254–255; nonhuman, 244–255; as normative responsiveness, 252–256;

practical, 69–70, 72–73; sustainability and, 225, 238; Taylor on, 255; value and, 69–73

Agents, 234–235, 244, 246–247

Agriculture, 149, 171, 192

Ahimsa, 174

Alienated form of life, 165–166, 182–183, 185, 196

Alienation, xiv, 76, 82, 85, 170–174; Bilgrami on, 92–93, 165, 167–168, 176–178, 182–183; dealienation and, 197–201; nonalienation and, 91–94

Alternative notion of consent, 78–79

Ancient Greece, 3, 14–15

Animacy of life, 202

Animals, 102–110, 128, 187–188, 244, 248

Answerability, 256–258

Anthropocene, 5, 17, 128, 153n11; action-guiding representation of, 164–165; biosphere in, 37; carbon dioxide levels in, 27; changes of, 41–42; climate and sea level in, 37; climate change and, 25–28, 152; colonialism and, 56–57; commencement of, 52–54, 56–57, 65–66; Crutzen on, 3–4, 15, 26, 32, 38, 47, 51; defining, 51–54, 58, 64–65; discounting, 168; Earth System in, 49–50, 54, 56–59,

Anthropocene (*continued*)
  65; economic activity in, 144; first-person approach to geosciences, for studying, 51; fossil record of, 41; Foucault and, 243; Gaia and, 174, 241; geology of, 32–35, 37–38, 40–43, 47–48, 50–53, 55, 57–58, 66–67; geoscientists on, 52, 55–56; global warming and, 25–28; the Great Acceleration in, 33–34, 36, 41, 54, 57; GSSA of, 53; GSSP of, 52–53; Holocene and, xi–xii, 3, 6–7, 15, 18, 21, 27, 32, 36, 54, 242–244, 254; human and natural history in, 63; human and nonhuman in, 240; human timescale of, 52; implications of, 56–59; life and, 166; mineraloids in, 34–35; nonhuman in, 240–241; nuclear weapons in, 57; Orbis spike in, 54; paradigm shifts, 54–56, 66–67; responses to, 168–170, 174, 242; Schell on, 163–164; science on, 54–55; Stoermer on, 38, 47, 51; strata of, 33, 35; stratigraphy of, xii, 16, 27, 32–33, 35, 38, 40–41, 52–53, 55, 57–59; time-rock unit of, 33; traditional geology and, 50; as unique age, 172; as vicious system, 191; world-disclosing representation of, 163–164; Zalasiewicz on, 52–53, 55–56
Anthropocene condition, the, 143
"Anthropocene" Working Group (AWG), 51–53, 55, 57, 59, 60*n*20, 66
Anthropocene Working Group of the Subcommission on Quaternary Research, 32
Anthropocentrism, 10, 55–56, 100–101, 103, 247
Anthropomorphism, 102, 248
Antihumanism, 242–243
Antipollution laws, 139
Antitrust law, 146–147
Appalachian Trail, 134
Aquinas, Thomas, 10
Archaeological time, 38
Arendt, Hannah, 10–11, 176, 222*n*11
Aristotle, 3, 14, 70–71, 87*n*1, 111, 226, 229
Art, science and, 274–275
Artificial Ground, strata of, 33
Ashworth, William, 103
Asteroids, 42–43

Astro-Theology (Derham), 109
Atmosphere, 26–28, 37, 47, 128
Attentiveness, respect and, 253
Auden, W. H., 4, 12
Automaton, 194
AWG. *See* "Anthropocene" Working Group

Bacon, Francis, 97–98, 101
Bangladesh, 148
Basic action, 86, 92
Beauty, 8–9
Being, Goodness and, 10
Being, knowing and, 176–178
Being and Time (Heidegger), 251, 257–258
Beliefs, 71–73
Belonging, social, 82–83
Bennett, Jane, 74, 244–255, 259*n*16, 260*n*24
Big Oil, 124–125
Bilgrami, Akeel, xii–xiv, 50, 95; on action, morality and, 217–218, 222*n*13, 231–235; on alienation, 92–93, 165, 167–168, 176–178, 182–183; on Bacon, 97; on exemplary moral actions, 232–235; on liberty, 91–92; on metaethical doctrines, 94; method of, 89; on possessive individualism, 90; project of, 164–165; Schell and, 165–167, 174–175; on unalienated life, 76, 82–87, 165–166, 175–178, 182–183; on unalienated worldview, 94; on value, 93, 250
Biocoup, 6–7, 18
Biology, 15
Biopolitics, 3, 14
Biosphere, 6, 18, 21, 25, 27–28, 37, 47
Biotechnology, open-source, 150
Boas, Franz, 169–170
Braungart, Michael, 200–201
Bromine, 21
Bromwich, David, xiv–xv
Brower, David, 136
Brown, Lester, 194–195
Buffet, Warren, 125
Bureau of Indian Affairs, 134

Califano, Joseph, 139
Canada, 191
Capital (Marx), 78
Capitalism, 13, 163–164, 168–169, 177, 221*n*2, 236–237

Capra, Fritjof, 173–174, 185–187
Carbon dioxide, 56–57; animals producing, 187–188; in atmosphere, 26–28, 47, 128; emissions, 117–118, 121–123; levels, in Anthropocene and Holocene, 27
Carboniferous Period and System, 31
Carbon Tracker, 124
Carson, Rachel, 131, 136–137, 175
Causality, 254–256, 259n16
Cavell, Stanley, 244, 255
CFCs. *See* Chlorofluorocarbons
Cheyne, George, 106–107
Chlorine, 20–21
Chlorofluorocarbons (CFCs), 20–21, 32
Christian apologetics, 99
Christian community, 109
Christo-Theology (Derham), 109
City, 3, 14–15
Clean Air Act, 131–132, 138–139
Clean renewable energy, 119–123, 125
Clean Water Act, 131–132
Climate: fluctuations, 26; greenhouse gases in, 27–28; progress, barriers to, 118–119; in Quaternary Period, 25–26, 38; science, xi; scientists, 48–49; sea level and, 25–26, 36–37; stabilization, 119–120, 124; warming, 27–28
Climate catastrophe, 205
Climate change: action on, 224; agency and, 234–235; agriculture and, 149; Anthropocene and, 25–28, 152; climate scientists on, 48–49; corporations on, xi; decadal, 49; degrowth agenda and, 122; discounting, 168–170; energy economy and, 150; Hansen on, 54; mass extinction and, 38; nuclear danger and, x–xi; as philosophical problem, 225; political economy and, xiii; as practical problem, 224; problem of, 206, 220–221, 224–225; responses to, 201; sustainable practices for, 220–221; timescale of, 48–49
Climate Change Conference, UN, 117
Club of Rome, 182
Coal, 31
Coetzee, J. M., 257
Collective action, 90–92, 95n6, 148
Collective cultivation of the commons, 80–81

Colonialism, 50, 56–57, 181
Command and control organizations, 172–173
Commodification, 167–168, 192–193, 198
Commoner, Barry, 184, 186–187, 195
Common law, 147
Commons, 80–82, 85–87, 192, 197–198
Conflict of temporalities, 64
Conscience, xiv–xv, 149–150
Consent, 78–79, 82, 85–86
Conservationism, 145
Constitution, U.S., 139–140
Constitutive infrastructure, 144–146
Constitutive norms, 231–232
Contract law, 145
Contractualism, 77–80
Conversation, 215
Cooperation, 142–143
Cooperative citizens, 197–199
Cooperative management of the means of production, 90–91
Copenhagen Climate Summit, 117, 122–123
Corporations: on climate change, xi; oil companies, 118–119, 124–125; in vicious cycle, 194
Courtemanche, Eleanor, 110
Creator, God as, 12, 99, 104–105, 109
Cretaceous Period, 31, 39
Crutzen, Paul: on Anthropocene, 3–4, 15, 26, 32, 38, 47, 51; on CFCs, 20–21; Stoermer and, 38, 47–48, 51
Cultures, 63–64

D'Alisa, Giacomo, 122
Darwin, Charles, 170–171, 275
Dasein, 251
Data, noise in, 48–49
Data collection, 49, 55
Dawkins, Richard, 6, 17
Dealienation, 197–201
Death, 11
Debt, owed to nature, ix
Deep time, of geology, 4–5, 15–16, 47–48, 52, 64
Defoe, Daniel, 107–108
Degrowth (D'Alisa, Demaria, Kallis), 122
Degrowth agenda, 119, 122–123
Deleuze, Gilles, 248–249
Demaria, Frederico, 122

Democracy, 136, 140, 148, 176, 181–182, 197–199

Derham, William, 100, 103–104, 107–109

Desacralization of nature, xii

Descartes, 176

Design, of nature, 100–101, 105–107

Desire, 70–73, 93

Destruction, 2, 5–7, 16–17, 168–170, 172–173

Detecting, policing and, 81–82

Devonian Period, 39

Dewey, John, 251–252

Disenchantment, xii, 50, 74, 97

Disparate impact, 139–140

Distributive dimension, of legal infrastructure, 146–148

Distributive justice, 131, 137–138, 147

Diversity, 195

Divestment, from Big Oil, 124–125

DNA, 5–6, 17–18

Dynamic resilience, 190–191

Earth: atmosphere of, 26–28, 37, 47, 128; crimes against, 19; domesticated land area of, 128; future of, 29; geological history of, 28; geologists on, 29–30; history of, 29–31, 37, 46, 51; human infrastructure on, 127; humanity and, 4; indigenous civilizations on, 176–177; living, 169–170, 184, 192–193, 196–198, 201–202; living systems of, 167–168, 170, 173, 186, 188, 244; minerals of, 34; as mother Earth, 176–177, 192; as property, 192; as singular, 19; sixth great extinction of life on, 1–2; surface of, 34, 143

Earth, life on, 15–16, 18, 20, 23, 172; conditions for, 167–169, 181; sixth great extinction of, 1–2; sustaining, 163, 166–167, 170–171, 174–177, 186–189, 191–192

Earth science, 46, 166, 184, 186

Earth System, 26–28, 34–35, 37–38; in Anthropocene, 49–50, 54, 56–59, 65; changes in, 48, 58; human beings in, 47, 49–50, 54, 57–58, 64–65; living networks comprising, 173

Ecclesiastes, Book of, 14, 21

Ecological crisis, xv; alienated response to, 170–174; biopolitics and, 3; discounting,

196; environmentalists on, 137; global, perplexity of, 151–152; history and, 14; human beings' place in, 2–3; human extinction, 7–8; IPCC on, 117–118; Muskie on, 138; nihilism in, 1–2; nuclear holocaust and, 22–23; responses to, 2, 10–12, 21–22, 151, 184; scales of, 151; Schell on, 163; traditional stories for, 14

Ecological destruction, 168–170

Ecological issues, social issues and, xiv

Ecological reconciliation, 187

Ecological self, 188, 202

Ecology, 4, 144

Economic inequality, 138–139, 141, 147

Economics, 4, 144. *See also* Green growth

Ecosocial systems, 185–186

Ecosystems, 166, 171–173, 181, 183–185, 187, 191

Egalitarianism, 90, 138–139, 141

Einstein, Albert, 17, 22–23, 76

Eisenstein, Charles, 177

Emergent causality, 259n16

Emotions, 72–73

Employment law, 145

Enchantment, xii–xiii, 12–13, 74

Endangered Species Act, 128–129, 132

Ends, as products or wholes, 209–213, 226, 231

Energy: clean, renewable, 119–123, 125; companies, 118–119; costs, 118–119; economy, 150; efficiency, 119–121, 123, 125; sector, 121–122; solar, 120, 122, 125

English Fenland, 36

Enlightened self-interest, 91–92

Enlightenment, the, 55, 74–76

Entrachrons, 39

Environment, 132–133, 136, 143–145

Environmental Defense Fund, 140

Environmental infrastructure, 144–145, 147

Environmentalism, 131, 132–137, 140–141

Environmental justice, 131–132, 138–140

Environmental law, 154n12; antipollution, 139; disparate impact and, 139–140; distributive justice and, 131, 137–138; environmental justice and, 131–132; environment in, 132; genealogy of, 131–141; the great forgetting and, 132–133, 137–141; green legal theory, 201; as infrastructure, 128–133, 141;

institutionalization of, 140–141;
    movement, 140–141
Environmental Law Institute, 132
Episodic action, 211–214
Epistemic authority, xii
Epistemologies, 49–50, 55, 65–66
Equality: freedom and, 91–93; liberty
    and, xii–xiii, 75–76, 83, 85, 94;
    nondiscrimination and, 207
Equal Protection Clause, of U.S.
    Constitution, 139–140
Ethical deliberation, 74
Ethics, 11–12; human-centered, 174–175;
    Leopold on, 175; metaethics and, 94;
    politics and, 249; of relations to nature,
    ix–x; scientific method and, 51; values
    and, 250
Ethics (Spinoza), 248
Europe, 31–32
Evaluation, perception and, 93–94
Evans, Gareth, 71–72
Evelyn, John, 97
Evil, 10–11
Evolution, 5; geology and, 15–17; history
    and, 7, 21; natural selection in, 4, 6, 15,
    18, 275
Exemplary moral action, 232–235
Existentialism, 3
Experimentalism, 97–99, 108–109, 111–112
Experimentation, 19–20
Extinction: human, 7–8, 11, 22–23; mass,
    37–38, 42, 164, 242; in Quaternary
    Period, 25–26, 38; sixth great, 1–2

Factual mentality, 97
Fate of the Earth, The (Schell), x
Feedback loops, 190
Fenland, the, 36
Fertilizers, 127
Festa, Lynn, 104
Fictional characters, 107–108
Financial Times, 122
First-person approach, to science, 50–51
First-person perspective, 69–71
Fisch, Harold, 98
Fishkin, Joseph, 156n31
Food production, 149, 151, 197–198, 226–227
Forbath, William, 156n31
Ford Foundation, 140–141, 159n69

Fortey, Richard, 29
Fossil fuels, 41, 57, 120, 122; companies,
    118–119, 124–125; workers, in industry,
    118, 123–124
Fossilized smoke, 41
Fossils: of Anthropocene, 41; graptolites,
    39–40; rocks and, 30–31, 39
Foucault, Michel, 243
Framework, of the Enlightenment, 75–76
Freedom, 90–93. See also Liberty
Free labor, 161n100
Free-rider problem, 90, 92–93, 95n3
Freon, 20
Freud, Sigmund, xv
Frischmann, Brett, 142, 159n76, 160n86

Gaia theory, 167, 170–176, 189–190, 197–199,
    241
Galbraith, John Kenneth, 139, 157n57,
    158n61
Gandhi, Mahatma K., 172, 175, 199–200, 217
Gaukroger, Stephen, 98–99
GDP, 118–123, 127
Generation Anthropocene podcast,
    52–53, 55
Genesis, Book of, 98, 108–109
Genocide, 11
Geography, 47
Geological epochs, 15–18; Cretaceous
    Period, 31, 39; Devonian Period, 39;
    Ordovician Period, 39–40; Paleogene
    Period, 39; Pliocene Epoch, 28;
    Quaternary Period, 25–26, 36, 38;
    Silurian Period, 39–40; Tertiary Period,
    35–36. See also Anthropocene;
    Holocene
Geological Time Scale, 27, 32, 36, 38–41, 51
Geologists, 3–4, 29–31, 33, 35, 46
Geology: of Anthropocene, 32–35, 37–38,
    40–43, 47–48, 50–53, 55, 57–58, 66–67;
    deep time of, 4–5, 15–16, 47–48, 52, 64;
    evolution and, 15–17; geosciences, Earth
    science and, 46; history and, 29–32; at
    human timescale, 52; nonhuman, 35;
    rocks in, 30–31, 38–39; silos in, 47–49;
    strata in, 33; traditional, 46–47, 50,
    58–59, 64–65
Geopolitics, 53
Geosciences, 46, 49–51, 54–56, 59, 64–67

Geoscientists, 46–47, 50, 52, 55–56, 64

Germany, 42

Gift: economy, 176–177, 198–199; life as, 177–178; of nature, xiv

Gift, The (Mauss), 177

Glasstone, Samuel, 20

Global average temperature, 25, 27–28

Global Boundary Stratigraphic Section and Point (GSSP), 52–53

Global capitalist system, 237

Global economy, 194–195, 198

Global emissions, of carbon dioxide, 117–118, 121–123

Global green growth program, 118

Global Standard Stratigraphic Age (GSSA), 53

Global vicious cycle, 194–196

Global warming. *See* Climate change

God, 10, 12, 99, 104–110, 153, 262

Gogineni, Bina, xii, 63–67

Golden spike, 39–40, 52–53, 58, 64

Goodness, 8–9, 218

Goodness, Being and, 10

Governing the Commons (Ostrom), 81

Governments, 194

Granger, James, 107

Grantham Institute on Climate Change and the Environment, 125

Graptolites, 39–40

Grassroots, 125, 198

Great Acceleration, the, 33–34, 36, 41, 54, 57, 127

Great disembedding, the, 192–194

Great forgetting, the, 132–133, 137–141, 156n31

Great Oxygenation Event, 34

Great Recession, the, 122–123

Great Transformation, The (Polanyi), 167–168, 177, 192–194

Greece, ancient, 3, 14–15

Green growth, xiii; climate stabilization by, 119–120; global program of, 118; jobs and, 121; "small is beautiful" and, 121–122

Greenhouse gases, 26–28, 37

Green legal theory, 201

Gregg, Richard, 199

Grew, Nehemiah, 106, 108

Grewal, David, 142–143, 149–150, 155n21

GSSA. *See* Global Standard Stratigraphic Age

GSSP. *See* Global Boundary Stratigraphic Section and Point

Hadot, Pierre, 111–112

Haff, Peter, 27, 33

Hale, Robert, 147

Hall, Anthony, 192

Halpern, Richard, 111

Hamilton, Alice, 135, 136, 157n47

Hansen, James, 54–55

Hardin, Garrett, 81

Harding, Stephen, 167, 189

Harrison, Peter, 99

Hathaway, Berkshire, 125

Hayes, Chris, 119

Heidegger, Martin, 251, 257–258

Hertz, Heinrich, 89, 95n2

Hidden Connections (Capra), 173–174

Hind Swaraj (Gandhi), 172

Hiroshima, x–xi, 2, 7, 14

History: of Earth, 29–31, 37, 46, 51; ecological crisis and, 14; evolution and, 7, 21; geology and, 4–5, 29–32; human and natural, 63; of science, 47

Hobbes, Thomas, 153, 160n87, 199

Holocene: Anthropocene and, xi–xii, 3, 6–7, 15, 18, 21, 27, 32, 36, 54, 242–244, 254; carbon dioxide levels in, 27; duration of, 36–37; the Fenland and, 36; GSSA of, 53; in Quaternary Period, 26, 36; start date of, 54; strata, 36

Homo sapiens, 164, 170–172, 174–175, 190, 192. *See also* Human beings; People

Horkheimer, Max, 97

Hueper, Wilhelm, 135–136

Human: agency, 244–245, 251, 254–255; extinction, 7–8, 11, 22–23; history, natural history and, 63; life, 167, 173, 183, 193, 221, 226, 229, 240, 248, 261; nonhuman and, 240–247, 252, 254; value, 250–251

Human agents, 246–247

Human beings: animals and, 102–110, 128, 244; biocoup of, 6–7, 18; biomass of, 128; in cities, 14–15; in climate catastrophe, 205; destruction by, 2, 5–7, 16–17; Earth and, 15; Earth's surface and, 34; in Earth

System, 47, 49–50, 54, 57–58, 64–65; ethics and, 174–175; feelings for, 104; in geosciences, 64; infrastructure of, 127; knowledge of, 18–19; nature and, 50–51, 55, 100; place in ecological crisis, 2–3; power of, 19; responses to ecological crisis, 2; responsiveness of, 255–256; as sentient beings, 9; stratigraphic record and, 46–47; survival of, 184; unsettling time of, 240–244; valuing by, 8–10; will of, 261. *See also* Anthropocene
Human-centered ethics, 174–175
Human civilization: in natural world, 13–14; science in, 17–18
Human Dasein, 251
Human exceptionalism, 55, 109, 241–243, 251, 254–255
Human history: natural history and, 63
Humanism, 243
Humanities, 54, 58–59, 63–64, 242
Humanity: of Adamic man, 103; constructive powers of, 7; destruction of, 172–173; Earth and, 4; infrastructure and, 151–153; natural selection and, 18; nature and, 19, 23; nuclear weapons and, 5; scientific knowledge of, 5; of scientists, 51; vulnerability to nature, 56
Human nature, 261
Human timescale, 46–47, 50, 52, 58–59, 64–65, 67
Hume, David, 249
Hydrosphere, 47

"I," 69–70, 246
Ice Age, 25, 28
Imperialism, 181
Incentivization of talent, 75
India, 42, 121–122
Indigenous civilizations, 176–177, 191–192
Individual intentionality, 86
Individualism, xiv, 90, 234–236
Individual liberty, 75, 83, 92
Individuals, 75, 81, 84, 92, 229
Individual self-governance, 75, 84
Industrial chemicals, 135–136
Industrial environment, 135–136
Inequality, xiv, 138–139, 141, 147
Infrastructure, xiii; as coordination regime, 142–143; distributive, 146–148; ecology

as incipient, 144; environmental, 144–145, 147; environmental law as, 128–133, 141; environment and, 143–145; human, 127; of humanity, 151–153; invisibility of, 152; legal, distributive dimension of, 146–148; legal, qualitative dimension of, 148–149; legally constitutive, 144–149; material *versus* legally constitutive, 144–146; natural systems as, 144; normativity in design of, 149; political, 142–143; politics of, 149–152; as resources, 142; scales of crisis and, 151; state and, 129–130; technosphere as, 143–145
Instrumental meaning, 101
Instrumental reason, 97, 101
Intelligibility, 257
Intention, 70–71, 73
Intentional action, 86
Intentionality, 86
Interbeing, 196–201
Interdependency, symbiosis and, 186–189
Interdisciplinarity, 56–58
Intergovernmental Panel on Climate Change (IPCC), 19, 117–118, 122
International Commission on Stratigraphy, 32
International Mineralogical Association, 34
International Renewable Energy Agency (IRENA), 120
Iowa caucus, 4, 16
IPCC. *See* Intergovernmental Panel on Climate Change
IRENA. *See* International Renewable Energy Agency

James, William, 176
Jesus Christ, 109–110
Job, Book of, 12, 107
Jobs, green growth and, 121
Johnson, Lyndon Baines, 139
Johnston, Basil, 192
Johnstone, Chris, 202
Jonas, Hans, xiv, 261, 273–276

Kallis, Giorgos, 122
Kalman, Laura, 159*n*71
Kant, Immanuel, 218

Kapczynski, Amy, 150
Keener, Frederick, 107–108
Klein, Naomi, xiii
Knowing, being and, 176–178
Knowledge: experimentation and, 19–20; human, 18–19; scientific, 5, 17–19, 249; tradition and, 229–230, 232–233
Koch, Charles, 118
Koch, David, 118
Kompridis, Nikolas, xii
Kropotkin, Peter, 171, 198–199
Kuhn, Thomas, 75–76

Laden, Anthony, xiv, 165, 224–227, 230–237
Land, 191–192
"Land Ethic, The" (Leopold), 185
Land ownership, 77–78
Larkin, Philip, 11
Latour, Bruno, xii, 74, 246, 254
Law. See Environmental law
Left, the, 119
Legal infrastructure, 144–150
Legal liberalism, 140–141
Legal theory, green, 201
Leopold, Aldo, 168–169, 175, 177, 185, 196–197
Liberalism, 76–78, 80, 82–83, 140–141
Liberty: Bilgrami on, 91–92; equality and, xii–xiii, 75–76, 83, 85, 94; of individual, 75, 83, 92; property and, 74–75, 77; self-governance and, 75, 83–85; self-interest and, 84
Life: alienated form of, 165–166, 182–183, 185, 196; animacy of, 202; Anthropocene and, 166; conditions of, 167–169, 181; as gift, 177–178; human, 167, 173, 183, 193, 221, 226, 229, 240, 248, 261; nonhuman forms of, 240; politics and, 15; unalienated, 76, 82–87, 165–166, 175–178, 182–183, 185. See also Earth, life on
Life sciences, 166, 173, 184, 186
Life systems, 182
Limits to Growth, The (report), 182
Lincoln, Abraham, 23
Lithosphere, 47
Lives of Animals, The (Coetzee), 257
Living Earth, 169–170, 184, 192–193, 196–198, 201–202

Living systems, 167–168, 170, 173, 186, 188, 244
Locke, John, xiii, 77–80, 82, 85–86, 90–91
Logical grammar, 86
Louv, Richard, 170
Lovell, Bryan, 41
Lovelock, James, 171, 199
Luther, Martin, 98
Lyrical Ballads (Wordsworth), 262

MacKaye, Benton, 133–134
Macy, Joanna, 202
Margulis, Lyn, 190
Margulis, Lynn, 167
Marshall, Robert, 134, 156n41, 156n43
Marx, Karl, 74–76, 78, 82, 228
Marxism, 237
Mass extinction, 37–38, 42, 164, 242
Material infrastructure, 144–146, 149–150
Mauss, Marcel, 177
Mazzocchi, Tony, 123–124
McColley, Diane, 109
McDonough, William, 200–201
McDowell, John, 70–71, 87n1
Meaning, 75–76, 98, 101
Medea Hypothesis, the, 170–173
Menely, Tobias, 102
Merleau-Ponty, Maurice, 201–202
Mesozoic Era, 26
Metaethics, 94
Metaphysics, of value, 225, 230–235, 237
Methodological norms, 49
Mexico, 39
Microplastic, 34
Midgley, Thomas, 20–21
Miller, D. A., 110
Mind, 72–73
Mineraloids, 34
Minerals, 34
Modernity, nature in, 68, 97
Modernization, 98, 164–167, 176, 190–191, 193
Morality, action and, 217–218, 222n13, 231–235
Moral perception, 74
Moral practice, 231–232
Moral relativism, 233–234
Moral sentiments, 70
More, Henry, 105–106
Mother Earth, 176–177, 192
Muir, John, 134

Murchison, Roderick, 31
Murder, 11
Muskie, Edmund, 138–139, 158n61, 158n65

Naess, Arne, 188
Nagasaki, x–xi, 2, 14
NASA. *See* National Atmospheric and
    Space Administration
National Atmospheric and Space
    Administration (NASA), 54
National Environmental Policy Act, 132
National parks, 148
Natural history: human history and, 63;
    physico-theology and, 99, 106
Naturalism, 248–251, 253–254, 260n24
Natural piety, xiv–xv, 14–15, 265, 273–275
Natural resources: as commodities, 192;
    conservationists on, 145; exploiting, 194;
    managing, 131; nature and, ix, xii–xiii,
    216
Natural Resources Defense Council
    (NRDC), 132, 140
Natural sciences, 66, 68, 249. *See also*
    Science
Natural selection, 4, 6, 15, 18, 275
Natural world, 13–14, 167–168
Nature: book of, 101, 110–111; conscience
    and, xiv–xv; debt owed to, ix;
    desacralization of, xii; design of,
    100–101, 105–107; disenchantment of,
    xii, 50, 97; gift of, xiv; humanity and, 19,
    23; human nature and, 261; humans
    and, 50–51, 55, 100 (*See also*
    Anthropocene); in modernity, 68, 97;
    natural piety and, 273–274; natural
    resources and, ix, xii–xiii, 216; natural
    sciences on, 68; in Newtonian science,
    50; normativity and, xv, 50–51;
    physico-theology on, 99–103, 105–107,
    109–110; relations to, ix–x; secular
    enchantment of, xii–xiii; state of, 77–78,
    90–91; universal laws of, 248; value and,
    xii, xiv; Wordsworth on, xiv–xv,
    263–264, 266
Nature Principle, The (Louv), 170
Negative Dialectics (Adorno), 252
"New Abolitionism, The" (Hayes), 119
Newtonian science, 50
Nichols, Kyle, xii, 63–67

Nietzsche, Friedrich, 13
Nihilism, 1–2, 13
Nitrogen, 127–128, 187–188
Nitrogen and phosphorus cycles, 27
Nixon, Richard, 132
Noise, in data, 48–49
NOMIS Foundation, x
Nonalienation, 91–94
Noncooperation, 81–82
Nondiscrimination: equality and, 207
Nonepisodic actions, 226–227
Nonhuman: agency, 244–255; in
    Anthropocene, 240–241; bodies, 251;
    causality and, 256; forms of life, 240;
    geology, 35; human and, 240–247, 252,
    254; in humanities and social sciences,
    242; vitality of, 247–248; world, value
    of, 250
Nonproductive activities, 226, 228, 236
Nonrenewable resources, 181–182, 194
Nonviolence, 172
Normative responsiveness, 252–256, 258
Normativity, 260n42; action and, 219–220;
    in infrastructure design, 149; nature
    and, xv, 50–51; objectivity and, 233–234;
    receptivity and, 258; science on, 50–51,
    67; values and, 74, 86, 231, 250
Norms, constitutive, 231–232
Northern Gateway Pipeline, 194
NRDC. *See* Natural Resources Defense
    Council
Nuclear weapons, x–xi, 2, 5–7, 17; in
    Anthropocene, 57; human extinction
    by, 22–23; ozone layer and, 20
"Nutting" (Wordsworth), 266, 269

Objectivity, 9, 54–55, 66, 98, 233–234
Occupational Tumors and Allied Diseases
    (Hueper), 135
Oil companies, 118–119, 124–125
Ong, Walter, 108
Ongoing action, 211–214
Ongoing activities, 213–214, 216, 218–220
On Violence (Arendt), 176
Open-source biotechnology, 150
Orbis spike, 54
Ordovician Period, 39–40
Ornithology (Willughby and Ray), 101
Ostrom, Elinor, 81, 90, 95n3

Our Ecological Footprint (Wackernagel and Rees), 169
Ozone layer, 20–21

Paleogene Period, 39
Paradigm, 75–76
Paris Climate Summit, 117
People, 100, 102–103, 110. *See also* Human beings
Perception, 93–94, 99
Perm, 31–32
Permian Period, 31–32
Personal freedom, 90
Personhood, 100, 107–108, 111
Persons, 102
Pesticides, 136–137
Phenomenology, 86
Philosophers, 69
Philosophical problems, 206–208, 216–217, 225
Philosophical solutions, 206–208
Philosophy, 205–206, 230–231
Phosphorus cycles, 27
Physico-theology, xiii, 99–103, 105–107, 109–111
Physico-Theology (Derham), 100, 109
Picciotto, Joanna, xii–xiii
Plants, 188
Plastic, 34–35
Pliocene Epoch, 28, 36
Poets, 262, 267–269, 274–275
Polanyi, Karl, 167–168, 177, 192–194, 200
Policing, 81–82
Political action, 215
Political economy, xiii–xiv
Political infrastructure, 142–143
Political rationality, 77, 79–80
Political will, 22
Politics: in ancient Greece, 3, 14–15; biology and, 15; conscience and, 149–150; ecology and, 144; ethics and, 249; geopolitics, 53; infrastructure and, 142–143, 149–152; natural selection and, 4, 15; science and, 22, 48; of scientists, 54–55
Pollin, Robert, xiii
Possessive individualism, 90
Posthumanism, 242–243
Poverty, 139, 157*n*57

Power, 19
Practical action, 92
Practical agency, 69–70, 72–73
Practical problems, 224
Practical solutions, 224–225
Practices: of dealienation, reengagement and, 197–201; moral, 231–232; of reconciliation, 187; sustainable, 201, 206, 220–221, 223*n*17, 224; unsustainable, 186, 223*n*17
Prediction, intention and, 70
Prelude, The (Wordsworth), 262–270, 276
Principles of Mechanics, The (Hertz), 89, 95*n*2
Prisoners' dilemma, 80–82, 90
Private and public ownership, in energy sector, 121–122
Private law, 145–147
Private property, 74–75, 78, 80, 90–91, 160*n*82
Privatization, 198–199
Problems: climate change, 206, 220–221, 224–225; frames for, 206–208; free-rider, 90, 92–93, 95*n*3; philosophical, 206–208, 216–217, 225; practical, 224; real-world, 206–208; of sustainability, 220–221
Production, food, 149, 151, 197–198, 226–227
Productive actions, 226, 228
Productive activities, 210–211, 236
Productive system, 237
Progress, 3, 15, 176
Property, 74–75, 77–78, 80, 192
Property law, 145–147
Property rights, 90
Proterozoic Era, 34
Public and Its Problems, The (Dewey), 252
Public lands, 148–149
Public ownership, 121–122, 145
Punishment, 81–82
Purdy, Jedediah, xiii

Qualitative dimension, of legal infrastructure, 148–149
Quaternary Period, 25–26, 36, 38

Rationality, 77, 79–80, 82, 87
Rawls, John, 111
Ray, John, 101, 104–106, 108

Realism, 68, 232–233
Real-world problems, 206–208
Reanimation, 201–202
Reason, instrumental, 97, 101
Receptivity, answerability and, 256–258
Reconciliation, 187
Reconnection, 197–201
Reddy, Sanjay, xii–xiii
Reengagement, reconciliation and, 196–201
Rees, William, 169
Regional planning, 134
Regulation, 81–82
Relativism, 68–69, 233–234, 238n5
Renewable energy, clean, 119–123, 125
Renewable resources, 181–182
Respect, attentiveness and, 253
Responsiveness, normative, 252–256, 258
Rist, France Gilbert, 177
Rocks: carboniferous, 31; fossils and, 30–31,
     39; in geological time, 38–39; Lovell on,
     41; stratigraphers, 46–47
Rovane, Carol, xiv
Royal Society, the, 97, 102
Ruddiman, Bill, 54
Ruisdael, Jacob van, 9

Sand County Almanac, The (Leopold), 175
Santiago Theory of Cognition, 176
Schell, Jonathan, x–xii, xiv, 163–167, 174–175
Schumacher, Fritz, 177
Science, xi, 17–18, 21, 23–24; of
     Anthropocene, 54–55; art and, 274–275;
     data collection in, 49, 55; Earth science,
     46, 166, 184, 186; experimental, 97–99;
     first-person approach to, 50–51; history
     of, 47; humanities and, 54, 58–59, 63–64;
     on humans, nature and, 50–51, 55;
     interdisciplinarity of, 56–58; life
     sciences, 166, 184, 186; methodological
     norms of, 49; natural, 66, 68, 249;
     Newtonian, 50; on normativity, 50–51,
     67; objectivity of, 54–55, 66, 98;
     plausibility in, 48; politics and, 22, 48;
     third-person approach to, 50; as
     value-free, 98. See also Geology;
     Geosciences
Scientific knowledge, 5, 17–19, 249
Scientific method, 51
Scientific research, 47

Scientists, 275; climate, 48–49;
     geoscientists, 46–47, 50, 52, 55–56, 64;
     humanity of, 51; politics of, 54–55
Scotland, 39–40
Sea level, 25–26, 36–37
Seasons, The (Thomson), 99, 102
Second nature, xiv–xv
Secular enchantment, xii–xiii
Self-governance, 75, 83–85
Self-interest, 84, 91–92
Sen, Amartya, 77
Shiva, Vandana, 177, 198–199
Short-timescale geosciences, 49–50, 59, 65
Sierra Club, 134, 136, 140
Silent Spring (Carson), 136–137, 175
Silurian Period, 39–40
Simpson, Michael, 188
Sixth great extinction, of life on Earth, 1–2
Smith, Adam, 110
Smith, William, 30–31, 43n2
Smoke, 41
Snow, C. P., 63–64
Social, the, 84–85
Social action, 86
Social belonging, 82–83
Social capital, 200
Social contract theory, 77–79
Socialism, 134–135
Social issues, ecological issues and, xiv
Social life, xiv
Social sciences, 54, 58–59, 242
Social system, 181–185, 187, 189; global,
     192–194; vicious, 191–194
Social traditions, 229–230
Solar energy, 120, 122, 125
Solutions, 206–208, 224–225
Spain, 121
Spinoza, Baruch, 248–249, 251, 253–254,
     259
Sraffa, Piero, 89
State, the, 129–130
State of nature, the, 77–78, 90–91
Steele, Richard, 107
Stoermer, Eugene, 38, 47–48, 51
Stone Age, 143
Stoppani, Antonio, 37
Strata, 33, 35–36
Stratigraphers, 46–47, 50, 55
Stratigraphic record, 46–47

Stratigraphy: of Anthropocene, xii, 16, 27, 32–33, 35, 38, 40–41, 52–53, 55, 57–59; in geological time, 39–40; GSSA, 53; GSSP, 52–53

Stratigraphy Commission of the Geological Society of London, 32

Strikes, 134–135

Subjectivity, 8–9, 168, 245

Subjects, 69

Sublimation, 262

Superfund, 132

Superfund, for fossil fuel workers, 123–124

Supreme Court, U.S., 139–140

Survey of the Wisdom of God in the Creation (Wesley), 99–100

Sustainability, 173, 182–183, 185, 194, 199; activities of, 205, 213–214; agency and, 225, 238; climate change and, 220–221; philosophy and, 205–206; practices of, 201, 206, 220–221, 223*n*17, 224; problems of, 220–221; unsustainable practices *versus*, 186, 223*n*17, 224; value and, 214–220, 227–228, 233; valuing, 214–217

Sustaining life on Earth, 163, 166–167, 170–171, 174–177, 186–189, 191–192

Sylva: A Discourse of Forest-Trees (Evelyn), 97

Symbiosis, interdependency and, 186–189

Sympathy, 70, 87, 88*n*5

Systems theory, 190, 194

Taylor, Charles, 109, 255

Technology, 274–275

Technosciences, 241

Technosphere, 27, 33, 127–129, 143–145, 150–151

Teles, Steven, 140, 159*n*71

Temporalities, xii

Tertiary Period, 35–36

Theory of Moral Sentiments (Smith), 110

Third-person approach, to science, 50

Third-person perspective, 70, 72

Thomson, James, 99, 102

Thought experiments, 111

Thyssen, Heini, ix–x

Timescale, human, 46–47, 50, 52, 58–59, 64–65, 67

Tipping points, 190–191

Tradition, 229–230, 232–233

Tragedy of the commons, 80–82, 85–87

"Tragedy of the Commons, The" (Hardin), 81

Transcendence, 3, 14–15

Triassic System, 31

Tully, James, xiv, 10

UN. *See* United Nations

Unalienated life, 76, 82–87, 165–166, 175–178, 182–183, 185

Unalienated worldview, 94

Unions, 135

United Nations (UN), 117

United States (U.S.), 121, 125, 132, 139–140

Universality, 233–234

Universal laws of nature, 248

Unsettling time, of humans, 240–244

Unsustainable practices, 186, 223*n*17, 224

Urban strata, 35

U.S. *See* United States

Value: action and, 216–219; activities and, 213–214; agency and, 69–73; Aristotle on, 70–71, 87*n*1; of beauty and goodness, 8–9; belief and, 71–73; Bilgrami on, 93, 250; desire and, 70–73; emotion and, 72–73; first-person perspective on, 70–71; human, 250–251; human extinction and, 8; McDowell on, 70–71, 87*n*1; meaning and, 98; metaethical perspective of, 94; metaphysics of, 225, 230–235, 237; natural sciences on, 68; nature and, xii, xiv; of nonhuman world, 250; normativity and, 74, 86, 231, 250; objectivity of, 9; philosophical views of, 230–231; realism about, 68; science and, 98; subjectivity in, 8–9; sustainability and, 214–220, 227–228, 233; of tradition, 230; world and, 70, 72–74, 84, 93

Value properties, 9, 12, 68, 71–74

Valuing, 8–10, 214–217

Vibrant Matter (Bennett), 244–245, 249

Vicious cycle, 194–196

Vicious system, 181–182, 190–194

Violence, 199

Vitality, 244–245, 247–248, 253

Volcanoes, 39, 42–43

Wackernagel, Mathias, 169
Ward, Peter, 170
Waxman-Markey legislation, 148
Weart, Spencer, 49
Weber, Max, xii
Wesley, John, 99–100, 106
Western imperialism, 181
Wilderness, 133–136
Wilderness Act of 1964, 131, 134
Wilderness movement, 134–137
Wilderness Society, 134, 136
Will, 261
Willughby, Francis, 101
Winstanley, 78–80, 82, 85–86, 90

Wittgenstein, Ludwig, 86, 89, 92
Wordsworth, William, xiv–xv, 261–270, 276
Worked Ground, strata of, 33
Workers: in fossil fuel industry, 118,
    123–124; health of, 135–136; striking, 134
Workers' Health Bureau, 135–136
Workplace health, 135–136
World, 85, 175–176; natural, 13–14, 167–168;
    nonhuman, value of, 250; value and, 70,
    72–74, 84, 93; Wordsworth on, 262–265
World-making, 129
Worster, Donald, 172

Zalasiewicz, Jan, xii, 52–53, 55–56